The Biofuel Delusion

The Fallacy of Large-Scale Agro-biofuel Production

Mario Giampietro and Kozo Mayumi

publishing for a sustainable future

London • Sterling, VA

First published by Earthscan in the UK and USA in 2009

Copyright © Mario Giampietro and Kozo Mayumi, 2009

All rights reserved

ISBN: 978-1-84407-681-9

Typeset by FiSH Books, Enfield, UK
Cover design by Emilio Vanni

For a full list of publications please contact:

Earthscan
Dunstan House
14a St Cross St
London, EC1N 8XA, UK
Tel: +44 (0)20 7841 1930
Fax: +44 (0)20 7242 1474
Email: earthinfo@earthscan.co.uk
Web: **www.earthscan.co.uk**

22883 Quicksilver Drive, Sterling, VA 20166-2012, USA

Earthscan publishes in association with the International Institute for Environment and Development

A catalogue record for this book is available from the British Library

Library of Congress Cataloging-in-Publication Data

Giampietro, M. (Mario)
The biofuel delusion: the fallacy of large-scale agro-biofuel production/Mario Giampietro and Kozo Mayumi. — 1st ed.
    p. cm.
  Includes bibliographical references and index.
  ISBN 978-1-84407-681-9 (hardback)
  1. Biomass energy–Economic aspects. 2. Biomass energy–Social aspects.
  I. Mayumi, Kozo, 1954–II. Title. III. Title: Fallacy of large-scale agro-biofuel production.
  HD9502.5.B542G53 2009
    333.95'39–dc22   2009010092

At Earthscan we strive to minimize our environmental impacts and carbon footprint through reducing waste, recycling and offsetting our CO_2 emissions, including those created through publication of this book. For more details of our environmental policy, see www.earthscan.co.uk.

This book was printed in the UK by TJ International, an ISO 14001 accredited company. The paper used is FSC certified and the inks are vegetable based.

Besides being invisible, your Highness, this cloth will be woven in colours and patterns created especially for you.

Hans Christian Andersen

To Howard T. Odum and Nicholas Georgescu-Roegen

Contents

List of boxes, figures and tables	vii
Forewords by Vaclav Smil and Jerome Ravetz	xi
Acknowledgements	xvi
List of abbreviations	xvii
1 Can We Solve the Agro-biofuel Riddle?	1
2 Learning from the Past	17
3 Not Everything that Burns is a Fuel	39
4 Pattern of Societal Metabolism across Levels: A Crash Course in Bio-economics	69
5 Assessment of the Quality of Alternative Energy Sources	105
6 Neglect of Available Wisdom	147
7 A Reality Check on the Feasibility and Desirability of Agro-biofuels	175
8 Agro-biofuel Production is Not Good for Rural Development	203
9 Living in Denial	227
10 Where Do We Go from Here?	253
Appendix 1: Basic Theoretical Concepts behind the Analysis of Societal Metabolism	261
Appendix 2: Examples of Grammars and Applications of Bio-economics	269
References	289
Glossary	307
Index	313

List of Boxes, Figures and Tables

Boxes

1.1	Examples of simplified solutions to the biofuel riddle	9
3.1	The chicken–egg relationship between the characteristics of society and its energy sector	47
3.2	The concept of grammar	49
4.1	Capitalization of the household sector of Spain	85
5.1	Elusive assessment of the energetic equivalent of human labour	114
6.1	Two epistemological issues that are key to the development of a multi-scale analysis of the exosomatic metabolic pattern of modern economies	154
6.2	The need to address the interface between biomass energy production and natural processes	161
7.1	Technical coefficients of ethanol production from corn in the US	180
7.2	Technical coefficients of Brazilian production of ethanol from sugar cane	183
8.1	The five steps of Cochrane's agricultural technology treadmill	213
8.2	Proposed certification scheme: major items	220
8.3	Example of a list of sub-items for each main item	221
8.4	Example of key guidance for each sub-item	222
8.5	The Concorde syndrome	224
9.1	Letter from US Senator Ken Salazar to the *Colorado Springs Gazette*	231
9.2	Passage from the policy statement of the Ecological Society of America	231
9.3	An example of granfalloon dynamics in scientific debate	233
9.4	Press release announcing a large donation by BP to research on biofuels	234
9.5	Cornucopians versus prophets of doom	239
9.6	Greenspan criticizing econometric models	241
9.7	Press reports on the benefits of companies producing agricultural inputs	248
9.8	An insider view of the process of decision-making about biofuels in the EU	249
9.9	European Parliament capitulates on biofuel deal	252

List of Boxes, Figures and Tables vii

Figures

2.1	Growth of the world's urban population	19
2.2	The input provided to the city by the countryside	19
2.3	Discontinuity in human development in relation to fossil energy use	22
2.4	Examples of metabolic patterns within ecosystems determining the formation of hierarchical structures	24
2.5	Set of expected relations over different compartments of a terrestrial ecosystem: relative size, rate of energy dissipation and density of energy dissipation	25
2.6	The dynamic budget of endosomatic energy in pre-industrial society	33
2.7	The flow of fossil energy from the cities to the countryside	36
3.1	Alternative energy sources: does anything go?	41
3.2	The boat that uses diesel fuel partly manufactured from human fat	42
3.3	Different energy forms in the energy metabolism of society	45
3.4	Expected pattern for a stable metabolism: a dissipative part sustained by a hypercycle	55
3.5	The dynamic budget of endosomatic energy analysed in terms of funds and flows	62
4.1	Percentage of labour force and GDP in agriculture versus GDP per capita	71
4.2	Relation between demographic structure and supply of work hours at the level of society	73
4.3	Monetary flows across levels of analysis; Spain, 1999	75
4.4	Exosomatic energy flows in Italian society, 1999	78
4.5	Endosomatic energy flows in the US	80
4.6	Exosomatic energy flows across levels of analysis; Spain, 1999	83
4.7	Example of impredicative loop analysis: mail delivery in society	90
4.8	Dendrogram of allocation of human activity over different compartments of society	93
4.9	Socio-economic metabolic pattern based on benchmark values	95
4.10	Characterization of the exosomatic metabolic pattern of the household sector in terms of tasks and energy carriers	97
4.11	An overview of the characteristics of the exosomatic metabolic pattern determining the bio-economic pressure	99
4.12	The correlation between the BEP indicator and standard indicators of development	99
4.13	The correlation between each of the three factors determining the value of the BEP indicator and standard indicators of development	102
5.1	Simple, mono-purpose grammar used to generate energy statistics: assessment of primary energy consumption in TOE of Spain, 1991	108
5.2	The map-flow of energy consumption in the US, 2007	110

5.3	The metaphor of heart transplant surgery	116
5.4	The two factors determining the quality of energy sources	125
5.5	Thought experiment: substituting society's primary energy source	130
5.6	Illustration of the high power level associated with the exploitation of fossil energy: Bagger 288	134
5.7	Illustration of the power level achieved in an oil palm plantation	135
5.8	Relevant qualities for characterizing the performance of a chain of energy conversions across hierarchical levels	136
5.9	Analytical scheme checking the possibility of using feeds of different qualities in meat production	138
5.10	Five distinct, contiguous hierarchical levels (n−2, ... , n+2) used in the narrative of the beef production system	142
5.11	Non-equivalent descriptive domains related to the beef production system	143
6.1	Network of flows and funds in post-industrial socio-economic systems	154
6.2	The grammar used by Farrell et al, 2006	159
6.3	Multi-purpose grammar useful for the energetic analysis of agro-biofuels	167
6.4	Meta-grammar useful for the energy analysis of fossil fuels	171
6.5	Meta-grammar useful for the energetic analysis of agro-biofuels	173
7.1	The density of typologies of supply and requirement of exosomatic energy	177
7.2	Overview of the benchmarks characterizing the requirement of society and the production process of ethanol	185
7.3	Illustration of the conceptual difference between peak oil and stock depletion using the metaphor of peak grain	194
7.4	Globalization and trade in an expanding pie and in a zero-sum game	198
7.5	After reaching peak oil: an increase in the size of human population entails reducing the pace of oil consumption per capita; an increase in the pace of energy consumption of a country entails reducing the pace of oil consumption in other countries	200
8.1	The metaphor of Flevoland	205
8.2	The density in space of monetary flows typical of different systems of agricultural production	215
9.1	Curve of success for researchers	237
9.2	Mechanisms of legitimization of the organization of state, after Funtowicz and Ravetz (1990)	242
A1.1	The concept of 'network niche' defined by the mutual information carried out by the M-R network on the identity of an internal element (also known as the 'sudoku effect')	264

A1.2	An overview of the exosomatic metabolic pattern	266
A2.1	Checking the dynamic budget for each of the five elements producing a critical supply	270
A2.2	Establishing a bridge between biophysical and economic analysis	271
A2.3	Looking at the interaction of the black box with the context	272
A2.4	The integrated dendrogram of fund and flow elements within the exosomatic metabolic pattern	273
A2.5	Using trigonometry to study forced relations over funds and flows	275
A2.6	Bridging levels in horizontal (same level)	276
A2.7	Bridging levels in vertical (on the PW side)	277
A2.8	A simple grammar establishing a relationship between demographic structure (structural types) and the profile of human activities (functional types)	278
A2.9	The grammar presented in Figure A2.8 applied to the study of Catalonia	279
A2.10	A grammar to map the direct and indirect consumption of household typologies	279
A2.11	The grammar presented in Figure A2.10 applied to Barcelona	280
A2.12	The dynamic budget of hours of paid work (requirement vs supply) between the household sector and the paid work sector	282
A2.13	Examples of different benchmark values found in different typologies of exosomatic metabolic patterns	283
A2.14	Comparing the exosomatic metabolic pattern of OECD countries and China	284
A2.15	Representing the exosomatic metabolic pattern of the world as determined by a set of metabolic patterns of country typologies	285
A2.16	The integrated analysis of Spain: historical series 1978–1998	286
A2.17	The integrated analysis of Ecuador: historical series 1970–1998	287

Tables

3.1	Different options for calculating the EROI in Figure 3.5	63
4.1	Allocation of human activity to paid work and household sectors for Italy and China in 1999	73
4.2	Reduction in size of fund and flow over levels of analysis	84
4.3	The correlation between BEP, exo/endo ratio, THA/HA_{PS} and ABM × MF over three different categories of indicators of development	103

Forewords

I

Modern high-energy civilization is overwhelmingly the product of burning fossil fuels, using their combustion directly for space and industrial heating and for transportation, and indirectly to generate most of the world's electricity. Renewable non-fossil energies have important but secondary roles. The combustion of biomass – the conversion that dominated the human use of energy during the millennia of pre-industrial evolution – now provides less than 4 per cent of the total primary energy supply in the world's affluent countries, while water power and wind contribute less than 3 per cent.

This pattern, as we are now incessantly reminded, cannot continue, and the global energy system must make a transit from its overwhelming dependence on fossil fuel to a new large-scale conversion of renewable energies. We are also told that two principal reasons make this grand energy transition not only desirable, but inevitable. The proponents of the argument of imminent peak global oil extraction tell us that we have either already reached the rate of maximum petroleum production, or that such an event will come very soon, and that it will be followed by continuous output decline. And, of course, the combustion of all fossil fuels, including coal and natural gas (whose resources are more abundant than those of liquid hydrocarbons), releases CO_2, the most important anthropogenic greenhouse gas, whose continuous emissions are bound to bring substantial global tropospheric warming with its worrisome and far-reaching consequences.

Given these concerns (and especially if we were to accept uncritically their simplified and misleading formulations), many green solutions assume an aura of providential inevitability. And while we are still far from being able to generate electricity on a massive scale by deploying photovoltaic cells or ocean waves or deep geothermal flows, we know how to turn plant components into liquid fuels, and we have been doing so over the past generation at steadily increasing rates. The easiest process – fermenting carbohydrates into ethanol – has already changed Brazil's supply of automotive fuel, and huge output increases of corn-based alcohol have been mandated by the US Congress. Turning plant lipids into biodiesel is not that productive, but Europe, with its large fleet of diesel-powered cars, still finds the idea very appealing. And then there is that holy grail of plants-into-fuels conversion: the promise of commercially viable cellulosic ethanol made from crop- and wood-waste by genetically engineered enzymes.

How appealing, how straightforward, how perpetually performing such ideas sound. The motives behind them may be understandable: who can argue with the desirability of moving away from fossil fuels, which are inherently final, polluting and emit greenhouse gases? But the promise of this green salvation owes more to fairy tales than to solutions based on hard energetic, engineering, agro-ecosystemic, agronomic and economic realities. In order to understand the real potential and the realistic limits of liquid biofuels – their advantages and their numerous drawbacks, their often surprisingly insignificant contribution to any reduction of greenhouse gas emissions, and their far-reaching impacts on agro-ecosystems, food prices, government budgets and consumer prices – it is necessary to take a closer look at most of these interconnected complexities, and do so in a systematic fashion.

This is exactly what this book by Giampietro and Mayumi does. As a result it is not light reading, and it demands a degree of scientific literacy and numeracy. But it repays the effort by imparting a solid appreciation of the fact that the currently widespread belief that liquid biofuels produced on large scale offer a promising replacement of fossil fuels is exactly what the book's title says: a delusion. Some delusions are relatively innocuous, but the biofuel delusion has already been costly in environmental and economic terms, and only informed societies can stop its further uncritical pursuit. In this sense, this book has a very sobering but also a very rewarding message.

Vaclav Smil, FRSC
Distinguished Professor
University of Manitoba
Canada

II

It is a sign of the times that this excellent study of agro-biofuels now belongs to history. As the world gets turned upside down, again and again with increasing speed, the story of this particular techno-fantasy gains significance as evidence of a syndrome, a systemic pathology that renders our present techno-civilization unsustainable even in the short run. We can understand it as the collapse of quality, resulting in the rule of fantasy and corruption, and the disintegration of the real technological systems whose design and operation are fatally compromised.

Mario Giampietro and Kozo Mayumi have very wisely turned to fiction as a source for their analytical terms. 'Granfalloon' and 'replicant' are two words that well describe the elements of our predicament. The former refers to a collective fantasy maintained by coercion; the latter to a special sort of unreality, a thing that looks real but has no identity or history. We might say that in technology, finance and politics alike, we have been governed by granfalloons, which have created a variety of replicants to maintain their rule. In all those areas, the lethal mixture of fantasy and corruption has ruled unchecked. In the US in particular, the Bush years imagined a total politico-military domination in all dimensions, supported by an ever-expanding bubble of financial values. The fantasy is obvious; the degree to which the whole venture was shot through by corruption must be guessed at. We have only isolated anecdotes, such as the stories of Bernie Madoff and Allen Stanford, and the various estimates of the billions that evaporated in Iraq.

The original supersonic airliner, Concorde, is mentioned as a comparative case. I had the honour to belong to groups opposing Concorde in the 1960s, and so I learned a lot about it. Compared to our contemporary mega-swindles, it was really rather innocent. It was born in confusion, and not without a certain idealism. But quite early on, the idea of a full fleet of supersonic transport was exposed as either a commercial disaster (if the US airlines didn't buy it for the transcontinental trade) or an environmental catastrophe (if they did). But policy-tetanus had already set in, and all interests (political, industrial and scientific) were locked in to its support, until the truth was finally exposed in the form of cancelled options from the US airlines. Concorde had to be given away to the national airlines, where it had an increasingly ignominious career as a novelty ride until a fatal crash in 2000 put an end to it. I was well prepared for this sort of technological tragedy, since I had already worked out that nuclear deterrence – the cornerstone of Anglo-American military and foreign policy – was not only absolutely evil, but also absolutely insane. (Later, I discovered that after the Cuban missile crisis of the early 1960s, the top planners had come to the same conclusion, and coined the acronym MAD: mutually assured destruction).

Another marker along the road to rule by self-destructive technologies was the US's Star Wars programme. This had been thoroughly exposed as a total impossibility, but the corrupted military-industrial-scientific complex took no heed. In a cosmic tragedy of misunderstandings, the Reykjavik talks between Reagan and Gorbachev were shipwrecked by the US's insistence on keeping this egregious junk-tech. Didn't the Russians know how bad it was? Or did they suspect that there must be another agenda, since no one could be so crazy as to place all their confidence in such obviously impossible systems? We don't yet know.

Civilian technologies have suffered from the same infection. Nuclear power does have reality testing at its operating core, but the problem of waste has turned out to be insoluble in practice, and is probably insoluble in principle as well. Undeterred, the proponents of nuclear power never give up, now donning the green mantle. Reality and fantasy are more intimately intertwined in information technology (IT) systems. In some respects they meet the most demanding standards of quality, as they perform arithmetical operations at unimaginable speeds and with unimaginable accuracy. Yet when the same core hardware is applied to complex socio-technological systems, the results are all too often disastrous. And again we have governments behaving like fanatics: as the goal recedes, they redouble their efforts. The ruling fantasy in the UK is of a super-panopticon, a database of all databases in which comprehensive information about every citizen is instantly available for surveillance or sale. All the evidence of botched systems, and of sensitive information being lost, is dismissed as irrelevant. The unnamed policy-makers behind the politicians continue as if the path to total control is smooth and simple.

All these collapses take time to work out. Like Concorde and Star Wars, some technologies can continue in a zombie state, still walking while effectively dead. But sometimes nemesis strikes suddenly and calamitously. This has happened with the IT of finance. It flourished in the context of the fantasy and corruption of a speculative market in hyper-sophisticated financial products. Computers and mathematics had three essential contributions to make to the expansion of the bubble. First, computers enabled the magnification of the market to a size hitherto unimaginable. Previously, each transaction had required its own bits of paper and people to carry them around. Now, files on computer disks could be multiplied indefinitely. Then, two famous theorems (Black-Scholes and Merton) justified the market in derivatives (bets on bets), in which trading had previously been banned as gambling. And finally, whenever someone created a financial product that was too complicated for any banker or trader to understand, a mathematician could be hired to produce a model for its value, quality or risk, yielding a magic number at which it could be traded. This edifice of fantasy has imploded, leaving behind trillions of dollars of toxic securities that may yet bring down the world financial system, and much else besides.

In this context, the story of agro-biofuels is entirely natural and understandable. Why shouldn't businessmen, scientists and bureaucrats all collude in pushing along this particular gravy train? In this case, as in others, reality is a minor inconvenience until such time as the whole game collapses. There remain just two questions: how did this situation come about, and what can be done about it? To answer the first, we need a theory that has yet to be developed; to answer the second, we need a strategy that is hard to imagine. The collapse of quality can be partly explained in terms of the cycle of growth and the decline of civilizations. Modern Europe hit its half-millennium just in time for the First World War, and (in spite of the temporary rise of the US and post-war prosperity) the last century has, in imperial terms, been going downhill at an accelerating pace. In our secular, democratic societies, Ibn Khaldun's cycle of moral degeneration operates not through dissolute princes but through corrupted men (and women) of power. But what is to be done? For that, new ideas, motivating and entangled with new social forces, will be necessary. For that, we must look, listen and engage.

Jerome Ravetz
Associate Fellow
The James Martin Institute for Science and Civilization
Saïd Business School, Oxford University

Acknowledgements

The authors would like to thank, first of all, Sandra Bukkens for editing the enormous amount of material generated for the preparation of this book. Her input was instrumental in cutting down the size of this book, which otherwise would have required several volumes. Thanks are also due to Ingrid Álvarez, who checked the English of the manuscript. Gonzalo Gamboa provided us with his preliminary analysis of the household metabolism of Barcelona, used in Appendix 2 (Figures A2.9, A2.11 and A2.12). Alev Sorman and Tarik Serrano – PhD students at Universitat Autònoma de Barcelona (UAB) – helped with the handling of data and figures.

In relation to comments received on earlier versions of the manuscript, we would like to thank Matteo Borzoni, Vasile Dogaru, Silvio Funtowicz, Joan Martinez-Alier, Giuseppe Munda, David Pimentel, Sergio Ulgiati and Jeroen van den Bergh for useful suggestions and advice that pushed us through three different drafts. However, since we did not manage to follow all of them, the final content of this book remains our own responsibility.

Mario Giampietro gratefully acknowledges the financial support provided by ICREA (Catalan Institute of Research and Advanced Studies), ICTA (Institute of Environmental Science and Technology at the UAB) and the EU (project DECOIN, no. 04428-FP6-2005-SSP-5A). Additional thanks for their help are due to Louis Lemkow, Agustin Lobo and Jesus Ramos-Martin at the UAB, and to Timothy Allen, Maurizio Di Felice, Tiziano Gomiero, Maurizio Paoletti, Andrea Saltelli, Joseph Tainter and Emilio Vanni. Finally, special thanks go to William Easterling for his invitation to spend a sabbatical year at Pennsylvania State University working on the development of the MuSIASEM methodology.

Kozo Mayumi gratefully acknowledges the financial support of Grant-In-Aid for Scientific Research (C) (18530178) and Grant-In-Aid for Scientific Research (B) (20330050), both provided by the Ministry of Education, Culture, Sports, Science and Technology in Japan. These indirectly contributed to creating the contents of the present book.

Kozo Mayumi would like to express sincere thanks to the following people for their spiritual and moral support, as well as their encouragement of the research contained in this book: Nicholas Georgescu-Roegen, Toshiharu Hasegawa, Hiromi Hayashi, Yasuo Oyama, Hiroki Tanikawa, Shunsuke Managi, Minoru Sasaki, Herman Daly, Joan Martinez-Alier, John Gowdy, Vincent Hull, Vasile Dogaru, Giuseppe Munda, Jesus Ramos-Martin, Mark Glucina, John Polimeni, Kate Farrell, Sylvie Ferrari, Heinz Schandl, Giancarlo Fiorito, Tori Mayumi, Katsuko Mayumi, Sumiyo Tashiro, Shigemi Igawa, Giichi Hirosawa, Akira Yamaguchi, Yoshie Okada and, last but not least, Yuki Okada.

List of Abbreviations

AG	agricultural sector
BEP	bio-economic pressure
BGH	bovine growth hormone
BTU	British thermal units
CEO	Corporate Europe Observatory
DDG	dried distiller grains
EC	energy carriers
EI	energy input
ELP	economic labour productivity
EMR	exosomatic metabolic rate
EROI	energy return on investment
ES	energy sector
ESIA	environmental and social impact assessment
EU	European Union
GDP	gross domestic product
GHG	greenhouse gas
GSEC	gross supply of energy carriers
GSI	Global Subsidies Initiatives
h	hour
HA	human activity
HEIA	high external input agriculture
HH	household
IT	information technology
J	joule (unit of energy)
LCA	life cycle assessment
LEIA	low external input agriculture
MuSIASEM	multi-scale integrated analysis of societal and ecosystem metabolism
NGO	non-governmental organization
NSEC	net supply energy carriers
OECD	Organisation for Economic Co-operation and Development
pcpy	per capita per year
PES	primary energy sources
PS	productive sector (sub-sector of the paid work sector)
PW	paid work sector
SG	service and government sector (sub-sector of the paid work sector)

TCE	tonne of coal equivalent
TET	total exosomatic throughput
THA	total human activity
TOE	tonne of oil equivalent
UN	United Nations
V	volt
W	watt

SI unit prefixes

n	nano (-10^{-9})
µ	micro (-10^{-6})
m	milli (-10^{-3})
c	centi (-10^{-2})
k	kilo (-10^{3})
M	mega (-10^{6})
G	giga (-10^{9})
T	tera (-10^{12})
P	peta (-10^{15})
E	exa (-10^{18})

Chapter 1

Can We Solve the Agro-biofuel Riddle?

The Various Pieces of the Riddle

The frantic development of biofuels

In the last two decades, the biofuel stage in Europe and the US has witnessed a whirlwind of political mandates, economic incentives and investments, as well as scientific research to promote the production and use of crop-based biofuels. The amount of direct and indirect private and public investment in incentives and subsidies is difficult to quantify.

According to the *Financial Times*, in the US:

> *Investors are sitting on billions of dollars [of] losses after buying into the corn-based ethanol industry that George W. Bush embraced as the answer to US energy woes ... Investor losses come as taxpayers have paid billions to support the ethanol industry. More than $11.2bn has been spent since 2005 on tax breaks for companies that blend ethanol into petrol. Billions more have been spent on direct state and federal subsidies for US ethanol production. 'We're looking at an industry that's cost $80bn to get to this point' said Bob Starkey, a fuels analyst at Jim Jordan &Associates, a research group in Houston.* (Allison and Kirchgaessner, 2008)

A similar view is given by a report from the World Bank:

> *Governments provide substantial support to biofuels so that they can compete with gasoline and conventional diesel. Such support includes consumption incentives (fuel tax reductions); production incentives (tax incentives, loan guarantees, and direct subsidy payments); and mandatory consumption requirements. More than 200 support measures, which cost around US$5.5 billion to US$7.3 billion a year in the United States, amount to US$0.38 to US$0.49 per litre of petroleum equivalent for ethanol.* (World Bank, 2008)

According to a study by Kutas, Lindberg and Steenblik (2007) – prepared for Global Subsidies Initiatives (GSI) as part of a series of reports addressing subsidies for biofuels in selected Organisation for Economic Co-operation and Development (OECD) countries – the direct subsidies within the European Union (EU) in the year 2006 amounted to €1.3 billion, but this is an underestimate since it does not include the support to investments in developing the biofuel industry. The direct subsidies in 2006 are equivalent to €0.74 per litre of biofuel (or about US$1 per litre).

The latest rush related to biofuel production is taking place in Asia, especially in Malaysian and Indonesian palm-oil production. According to an article by Patung (2007) in *Indonesia Matters*:

> *The government has offered huge incentives for companies to develop biofuels, primarily derived from palm oil, and many local firms that dominate in the logging, wood-processing and pulp industries have secured big contracts with such overseas companies as Chinese energy giant China National Offshore Oil Corp (CNOOC), which is one among 59 foreign and local energy investors who in January 2007 signed various biofuel-related renewable energy agreements worth US$12.2 billion.*

These investments and incentives have resulted in a boost in the pace of production of biofuels worldwide. 'Production of biofuels (ethanol and biodiesel) exceeded an estimated 53 billion litres in 2007, up 43 per cent from 2005' (REN21, 2007). It should be noted that, in 2006, 86 per cent of this production was ethanol, whose production was concentrated in the US (46 per cent of the ethanol was produced from corn) and Brazil (42 per cent of the ethanol was produced from sugar cane), while the remaining 14 per cent was biodiesel produced from crops, mainly in Europe (World Bank, 2008). 'In Asia, oil-palm plantations cover over 13 million ha, primarily in southeast Asia, where they have directly or indirectly replaced tropical rainforest' (Danielsen et al, 2008).

This frantic development of the agro-biofuel sector has been justified on the following grounds:

- The large-scale production of agro-biofuels can significantly improve energy independence and security, through the reduction of dependency on imported petroleum.
- The large-scale production of agro-biofuels can generate a significant reduction in greenhouse gas emissions.
- The large-scale production of agro-biofuels can help to improve rural development by supporting crop farm income.

The third motivation has been especially relevant in developed countries, where, independently from agro-biofuel production, a huge amount of subsidies is already poured into commodity support programmes in agriculture (about US$20 billion in the US and about US$80 billion in the EU).

Assuming these three grounds are valid, the governments of the US and the EU have focused on the implementation of this alleged win–win–win solution by setting specific targets, specific subsidies and specially arranged policy strategies. Attracted by the bonanza of a deluge of subsidies, investors jumped in. However, as the heroic efforts to quickly implement this world-saving strategy generated their first results, a lot of problems and unseen side-effects have popped out.

Negative effects on food supply

The most conspicuous of these unforeseen problems has undoubtedly been the worldwide food crisis. Paul Krugman (2008) described the crisis in the *New York Times*:

> *Over the past few years the prices of wheat, corn, rice and other basic foodstuffs have doubled or tripled, with much of the increase taking place just in the last few months. High food prices dismay even relatively well-off Americans – but they're truly devastating in poor countries, where food often accounts for more than half a family's spending. There have already been food riots around the world. Food-supplying countries, from Ukraine to Argentina, have been limiting exports in an attempt to protect domestic consumers, leading to angry protests from farmers – and making things even worse in countries that need to import food.*

This food crisis has several components, including the increasing demand for grains due to population growth and changes in dietary habits, a relative increase in meat consumption (Pingali, 2007) and the occurrence of unfortunate events (such as a couple of poor years of production). But there is a clear consensus that this crisis has been generated foremost by poor policies, and in particular by the frantic development of agro-biofuels.

World Bank President Robert Zoellick, in a letter written to Western leaders, said:

> *What we are witnessing is not a natural disaster – a silent tsunami or a perfect storm. It is a man-made catastrophe, and as such must be fixed by people.* (Spiegel Online International, 2008)

In fact, an internal report generated by the World Bank and leaked to *The*

Guardian was extremely clear in this regard:

> Biofuels have forced global food prices up by 75 per cent – far more than previously estimated – according to a confidential World Bank report obtained by The Guardian. *The damning unpublished assessment is based on the most detailed analysis of the crisis so far, carried out by an internationally-respected economist at the global financial body. The figure emphatically contradicts the US government's claims that plant-derived fuels contribute less than 3 per cent to food-price rises. It will add to pressure on governments in Washington and across Europe, which have turned to plant-derived fuels to reduce emissions of greenhouse gases and reduce their dependence on imported oil.* (*The Guardian*, 2008a)

(The World Bank's 'secret' study of April 2008 has now been revised and published, and can be accessed at www-wds.worldbank.org/external/default/WDSContentServer/IW3P/IB/2008/07/28/000020439_20080728103002/Rendered/PDF/WP4682.pdf.)

In spite of the controversy about the exact impact of crops for energy supply on food supply, this opinion has been expressed by many other important figures:

- The Secretary-General of the United Nations (UN), Ban Ki-moon, at a special meeting of the UN General Assembly, warned against investing too heavily in crops for biofuels at the expense of food production (BBC News, 2008a).
- The UN Secretary-General's Special Adviser on the Millennium Development Goals for reducing extreme poverty, Jeffrey Sachs, said: 'Europe's biofuels policy is having a "real impact" on food prices because wheat is being used on this continent to meet its energy demand, rather than to feed people' (Cronin, 2008).
- The previous UN Special Rapporteur for the Right to Food, Jean Ziegler, called biofuels a 'crime against humanity' (Spiegel Online International, 2008).
- The newly appointed UN Special Rapporteur for the Right to Food, Oliver de Schutter, has argued that the EU's policy is misguided: 'The production of rapeseed [and] palm oil destroys the forests in Indonesia. The use of one-quarter of corn in the United States is a scandal, in which taxpayers' money is used solely to serve the interests of a small lobby. I call for a freeze on all investment in this sector' (Cronin, 2008).
- Robert Bailey, policy adviser at Oxfam, says: 'While politicians concentrate on keeping industry lobbies happy, people in poor countries cannot afford enough to eat. Rising food prices have pushed 100 million people world-

wide below the poverty line, estimates the World Bank, and have sparked riots from Bangladesh to Egypt. Government ministers here have described higher food and fuel prices as "the first real economic crisis of globalization"' (Spiegel Online International, 2008).

The negative effect on CO_2 emissions

Also in relation to the second ground that biofuels are supposedly carbon neutral, there is increasing evidence that clearly contradicts this claim. Increased efforts to analyse the overall effects of large-scale agro-biofuel production point to several hitherto neglected side-effects.

A number of studies have focused on the initially neglected effects of the conversion of natural land covers, such as rain forest, grass land, peat land and savanna, into monocultures or plantations for biofuel feedstock production in Brazil, the US and Southeast Asia. According to a recent study by Fargione et al (2008) published in *Science*, this conversion can create a 'biofuel carbon debt' by releasing 17 to 420 times more CO_2 than the annual greenhouse gas (GHG) reductions that these biofuels would provide by displacing fossil fuels. Applying a similar method of analysis based on a worldwide agricultural model which estimates emissions from land-use change, Searchinger et al (2008) calculate that corn-based ethanol, instead of producing 20 per cent savings (as claimed by biofuel supporters), nearly doubles GHG emissions over 30 years and increases GHGs for 167 years. Biofuels from switchgrass, if grown on US lands, will increase emissions by 50 per cent.

Therefore, it is easy to predict that if a terrestrial ecosystem can no longer hold carbon (its structural biomass), this negative effect on the climate will also have a negative effect on biodiversity. A recent study, in fact, found that the conversion of forest to biofuel plantations not only will increase emissions for 75–93 years, but will also have a negative effect on biodiversity (Danielsen et al, 2008).

Analyses of the implications of land-use changes have also been performed for the UK. In a detailed report prepared by the Policy Exchange think tank (2008), a panel of scientists advised the UK government to immediately abandon targets and subsidies, since the best way to reduce CO_2 emission is to avoid further deforestation and protect existing natural covers as much as possible. Commenting on a request from Greenpeace and Oxfam to the British government to stop a policy demanding the inclusion of biofuel at pumps across the country, the UK's Chief Environmental Scientist Robert Watson called for a delay. He said: 'biofuel policy in the EU and the UK may have run ahead of the science' (BBC News, 2008b). As a matter of fact, the British Parliament has asked the EU Parliament to reconsider the attention being given to biofuels, which may be detracting from other, less expensive yet more effective ways to reduce transportation-sourced CO_2 emissions. The Commons environmental audit committee has come to this realization, and the MPs did ask the European Union to abandon its target of 10 per cent biofuel use by 2020 (*The Guardian*, 2008b). In addition, in February 2008 the Canadian Parliament's standing committee on agriculture and agri-food delivered a recommendation for the withdrawal of a bill imposing mandatory levels of biofuel in the fuels used for transportation (REAP Canada, 2008).

A commentary in the *New York Times* by Rosenthal (2008) neatly makes the point:

> *The destruction of natural ecosystems – whether rain forest in the tropics or grasslands in South America – not only releases greenhouse gases into the atmosphere when they are burned and ploughed, but also deprives the planet of natural sponges to absorb carbon emissions. Cropland also absorbs far less carbon than the rain forests or even scrubland that it replaces.*

To make life more difficult for biofuel proponents, a Nobel laureate in chemistry, Halmuth Michell, has recently cautioned the government of the Philippines against rushing into biofuel development because there is little energy to be gained from it. 'When you calculated how much of the sun's energy is stored in the plants, it's below 1 per cent. When you convert into biofuel, you add fertilizer, and then harvest the plants. There's not real energy gained in biofuel' (Burgonio, 2008). This last comment introduces the key aspect of the issue: are agro-biofuels a viable and desirable alternative to fossil energy? This aspect will be treated in detail in the following chapters.

Government reaction to this mounting evidence

Within the EU

After this series of events, the EU finally decided to change the Directive EC 2003/30, which had been issued in May 2003. It required Member States to place a minimum proportion of biofuels and other renewable fuels on their markets. The indicative reference value for these proportions in the old normative was 5.75 per cent of market share (in energy content) of all petrol and diesel for transport purposes by 31 December 2010.

A long discussion between EU Member States, the European Parliament and the European Commission over how to change the old directive generated an agreement reached in Brussels in December 2008. This increased the target for the proportion of renewable forms of energy in the EU's road transport fuel from 5.75 per cent to 10 per cent by 2020 (EUobserver, 2008)!

This decision prompted another wave of outrage among NGOs and civil society. Together with Friends of the Earth Europe, Lobbycontrol and Spinwatch, Corporate Europe Observatory (CEO) decided to confer the award for 'Worst EU Lobbying 2008' to the biofuel lobbyists in Brussels (www.worstlobby.eu/2008/home_en).

Within the US

The Energy Independence and Security Act of 2007 mandated the use of 36 billion gallons (136 billion litres) of biofuels by 2022, with significant requirements for cellulosic biofuel and biodiesel production. This is an increase of more than 500 per cent on current production levels.

No major changes in existing policies took place in 2008. As a matter of fact, the amount of money already committed to the ethanol cause in the future is quite impressive:

> *Total undiscounted subsidies to ethanol from 2006–2012 are estimated to fall within the range of $68 billion to $82 billion. Implementation of a higher Renewable Fuels Standard (e.g. 36 billion gallons per year by 2022) would increase total subsidies by tens of billions of dollars per year above these levels.* (Koplow and Steenblik, 2008, p96)

It should be mentioned that several groups are attacking the pro-ethanol policy in the US. A few titles found on the web say it all: 'Worse than fossil energy' by George Monbiot (2005); 'Smell of gigantic hoax in government ethanol promotion' (Hecht, 2007); 'Bio-foolery is causing "food shocks"' (Baker and Craig, 2007). It is unclear at present whether President Obama will maintain the existing subsidies for biofuels, but he has picked Steven Chu (a Nobel laureate in physics) as his Energy Secretary. Professor Chu is a well-known supporter of second-generation biofuels (from cellulosic biomass).

The Riddle to be Solved

Reading the various pieces of the riddle listed above cannot help but prompt a series of disturbing questions.

Are crop-based biofuels truly a viable and desirable alternative to fossil fuels?

It becomes difficult to suppress the uneasy feeling that nobody bothered to check the validity of the underlying and justifying grounds before joining the mad rush for the 'solution' to humankind's energy problem. As a matter of fact, in the rest of this book we will claim that none of the three grounds used to justify the frantic development of agro-biofuel is even close to being valid.

Is the energetic predicament confronting human societies something never faced before in human history? Can we learn something useful about agro-biofuels from our past experiences concerning energy and society?

The answer to the latter question is: 'Yes, we can' (illustrated in Chapters 2–7). There is a pretty good understanding of basic energetic principles applied to the functioning of human societies, and this experience clearly indicates that agro-biofuels are not a viable or desirable alternative to oil. Interestingly, as early as 1945 Samuel Brody, in the last chapter of his masterpiece on energy conversions in US agriculture, commented as follows on renewable energy sources:

> *it is said that we should use alcohol and vegetable oils after the petroleum energy has been exhausted. This reminds one of Marie Antoinette's advice to the Paris poor to eat cake when they had no bread.* (Brody, 1945, p968)

Since then, consistently, all those working in the field of energetics have ruled out the possibility of substituting fossil energy with biofuels. The actual large-scale production of agro-biofuels is only possible because of the large amount of oil being used in the production process. If the production process of agro-biofuels were to be self-sufficient, i.e. cover its own energy expenses with its own biofuel, then the net supply of agro-biofuel could not fuel even a negligible fraction of the transport sector of a developed economy.

If none of the assumptions used to justify the rush into agro-biofuels are true, and the well-established discipline of energetics considers the substitution of agro-biofuels for oil nonsensical, how can we explain the agro-biofuel madness?

Seeking the Solution to the Biofuel Riddle

On the type of solution we seek

This book has been written in an attempt to understand why and how society got carried away with the biofuel madness, and why this folly developed in the first place. In other words, ours is an attempt to solve the 'biofuel riddle'. Solving this riddle is anything but easy, since it requires putting together many pieces of a puzzle, pieces that belong to different typologies and dimensions of analysis. It requires a complex explanation, and complex explanations are not welcomed in a society like ours, addicted to reductionism. Nobody wants to waste time trying to understand issues; people look for plain facts, for the one and only 'truth', like the type of information presented on TV in the narrow span of time available between commercials.

Understanding requires time and reflection. One cannot really understand an issue for another person; at the most one can try to explain it. Real understanding of a complex issue must be achieved independently. It requires putting together various dimensions and analyses of different aspects in a coherent, complex picture. So if the reader is expecting us to simply say what is right and what is wrong with agro-biofuels, what should be done, and who is responsible for this mess, then he/she will be disappointed with this book.

This is not a book that will provide a series of straightforward explanations. Unfortunately, you cannot solve complex riddles with a simple key. In reality, it would be easy to give an explanation to the riddle by adopting simplifications. For example, two comments – by, respectively, Venezuelan President Hugo Chavez and 2008 Nobel Prize winner Paul Krugman (see Box 1.1) – provide pretty good simplified solutions to the biofuel riddle.

Box 1.1 *Examples of simplified solutions to the biofuel riddle*

Chavez's solution

Chavez said in a national address this week that increased production in Latin America would not help the region's poor or bring electricity to rural communities. Instead, crops like corn, meant for food production, will be diverted to create more biofuels so that 'illogical, absurd and stupid capitalism can continue its voracious growth'. (United Press International, 2007)

Krugman's solution

Where the effects of bad policy are clearest, however, is the rise of demon ethanol and other biofuels. The subsidized conversion of crops into fuel was supposed to promote energy independence and help limit global warming. But this promise was, as Time *magazine bluntly put it, a "scam". This is especially true of corn ethanol: even on optimistic estimates, producing a gallon of ethanol from corn uses most of the energy the gallon contains. But it turns out that even seemingly "good" biofuel policies, like Brazil's use of ethanol from sugar cane, accelerate the pace of climate change by promoting deforestation. And meanwhile, land used to grow biofuel feedstock is land not available to grow food, so subsidies to biofuels are a major factor in the food crisis. You might put it this way: people are starving in Africa so that American politicians can court votes in farm states. Oh, and in case you're wondering: all the remaining presidential contenders are terrible on this issue. One more thing: one reason the food crisis has gotten so severe, so fast, is that major players in the grain market grew complacent.* (Krugman, 2008)

Chavez's explanation is based on simplifications associated with socialist ideology about the market economy, while Krugman's explanation is based on simplifications performed by a smart economist analysing the functioning of the market economy. These two explanations can be considered 'very good' in terms of the narratives used. By adopting an easy narrative they end up providing a similar explanation: something went wrong and someone can be blamed for it. For Chavez, the culprit is illogical, absurd and stupid capitalism; according to Krugman, the culprit is the bad policies made by inadequate incumbents.

But how useful are these explanations? In relation to that given by Chavez, even China has realized that one must learn how to live with capitalism. We must tame it, not eradicate it. Since we cannot get rid of capitalism, it is important to explain which decision-making mechanisms within the capitalist system did not work properly with agro-biofuels, and why. Can we learn something from this failure, in order to do better in the future?

In relation to the explanation given by Krugman, the bad choices made by inadequate incumbents do not address the roots of the problem. Does the riddle of biofuels indicate a unique inadequacy among specific people writing policy about a very complex problem, or does it indicate a systemic inadequacy in the process used to generate decisions about very complex problems? If we want to know how to tame the capitalism with which we have to live, then we have to learn how to develop decision-making mechanisms capable of generating wise choices under pressure.

In fact, we claim that the systemic failure of quality control highlighted by the biofuel delusion is not due to bad people making bad decisions, but is rather the predictable outcome of a decision-making process that is called on to generate swift changes in an extremely complex socio-economic system in which several mode lockings are in place. The complexity of the system guarantees that many powerful forces operate against the proposed changes. Agro-biofuel is a perfect form of pseudo-change, suggested in order to preserve the status quo.

Can we learn something from the processes behind the promotion of agro-biofuel policies?

Yes. We believe that the agro-biofuel delusion represents a highly relevant case study in relation to future discussions of ambitious policies aimed at generating swift changes. In these future situations, scientific information will be mixed with social information in a super-charged political arena. It is extremely important to guarantee the quality of this mixing. Educated professors, world-class specialists, Nobel Prize winners: any of these could provide the 'right' explanation for a particular event, but they can handle only one issue and one perspective at a time. Nobody, no matter how smart and educated, can solve the complexity of the sustainability predicament by using only a single perspective or simple disciplinary knowledge.

In this book, we claim that the frantic implementation of policies on agro-biofuels has been possible because of a peculiar situation in which the establishment – governments, academic and financial institutions, and many non-governmental organizations – agreed that the naked emperor was wearing new invisible clothes. This situation has nothing to do with the 'bad' nature of capitalism; on the contrary, capitalism has proved pretty efficient at avoiding hoaxes. Nor were those involved necessarily bad people, or unperceptive. No, the situation was created because society as a whole did not want to see that the emperor was naked. Using available knowledge in the field of energetics, it would have been easy to perform a scientific analysis of the agro-biofuel proposal and detect imminent problems. However, the knowledge accumulated in the field of energetics was willingly ignored.

The explanation of the biofuel riddle requires an answer to yet another disturbing question: How is it possible that the academic establishment of developed countries has been buying the sloppy narrative about the advantages of agro-biofuels, to the degree that the scientific community has been generating a deluge of technical assessments backing up agro-biofuel policies?

The answers we provide in this book

The approach we use in this book is somewhat different from the approach generally adopted in conventional scientific journals engaging in the debate on agro-biofuel performance, or in reports or journal articles that aim to popularize these issues. In fact, conventional scientific debates on sustainability dilemmas tend to be carried out by two sides 'throwing numbers at each other over the fence' (an effective expression coined by Jeroen van der Sluijs). In these battles over numbers, each side worries only about checking the accuracy of its own calculations. This behaviour has led to a dialogue of the deaf, meaning that the numbers disputed by the opposing sides usually reflect the adoption of different conceptualizations or meanings. Numbers are often generated within different systems of accounting or, worse, reflect logically independent ways of characterizing different pieces of a bigger picture.

Any quantitative assessment of the sustainability issue must be necessarily developed within a particular narrative: a local definition focusing on just one aspect, referring to only one dimension and one scale of analysis (Giampietro et al, 2005; 2006). When dealing with different scientific analyses developed within different narratives, we find quantitative assessments which may give results that are non-equivalent, and therefore not reducible to each other. Therefore, when dealing with quantitative analysis applied to the issue of sustainability, the focus of the debate should be on the quality of the narratives chosen for making the calculations, rather than on numbers. That being said, we claim that the vast majority of the discussions found in the literature over the performance of agro-biofuels are based on numbers generated within irrelevant formal protocols, i.e. useless narratives.

We strongly believe that this obsessive focus on numerical details (syntax) rather than on the big picture (the semantics behind the numbers) has significantly contributed to weakening the scientific evidence produced and used to debate the feasibility and desirability of agro-biofuels. For this reason, we have chosen to focus first of all on the quality of the narrative – the robustness of the semantics – and the use of metaphors to appeal to the common sense of the reader. We felt it necessary to carry the reader on such a long journey because we are convinced that in order to fully understand the folly of crop-based biofuel production, we do not need more sophisticated mathematical models, more data or additional fancy calculations providing more accurate estimates. We believe that what is needed is a proper understanding of the issue, which allows us to see through the maze of numbers and make judgements using our common sense.

This book contains three interwoven threads, all of which are very relevant to the agro-biofuels riddle:

1 The idea of replacing fossil fuels with agro-biofuels is impractical. The fact that humans must find a substitute for fossil energy does not mean that this substitute should be, at all costs, crop-based biofuels. Indeed, we argue in Chapters 2 and 7 and part of Chapter 8 that, in respect to the task of powering a significant fraction of the metabolic pattern of modern economies, agro-biofuels are not a feasible (let alone desirable) substitute for fossil fuel.
2 The presentation of a general methodological approach to check the feasibility and desirability of alternative energy sources (not restricted to agro-biofuels). This method is based on concepts within the fields of energetics and bio-economics and is illustrated in Chapters 3, 4 and 5 and in the two appendices. Knowing from experience that energy is a notoriously elusive concept, as much as possible we use plain narratives and simple numerical examples when introducing technical concepts to make things more accessible to the reader. In particular, taking advantage of the similarity between economic analysis and energetic analysis – when defined within the framework of bio-economics – we attempt to illustrate basic concepts using economic examples, which are generally easier to follow for non-specialized readers. However, we could not avoid addressing the technical/scientific aspects of the analysis. Therefore, we warn the reader that these chapters deal with a theoretical analysis of how to frame the

discussion about the quality of alternative energy sources. This is certainly a relevant topic to the issue of agro-biofuel, but it widens the discussion to a more comprehensive analysis of the energetic predicament and the search for an alternative to oil. Readers who are not interested in this more comprehensive analysis can just skim the text of these chapters.
3   The discussion of the implications of the agro-biofuel riddle in relation to the use of science for governance. The riddle of agro-biofuels points to a systemic failure in the quality control of the science used for decision-making. Silvio Funtowicz and Jerome Ravetz (1990a, b) coined the label 'post-normal science' to refer to the predicament faced by science in this millennium. Chapters 8 (partly) and 9 deal with the sociological and procedural aspects of the use of science in the policy choice of the large-scale production of agro-biofuels.

There follows a brief outline of the book, chapter by chapter, giving the reader a better idea of how we will answer the questions raised above.

Chapter 2: Learning from the past
This chapter offers a historical view of the relation between socio-economic development and energy. It explores the big discontinuity in the basic energy metabolism of socio-economic systems that lies behind the transition from rural to urban society. The goal of this chapter is to illustrate that the technical progress associated with the Industrial Revolution was based on a dramatic reduction in the use of land, labour and ecological services in the productive sectors of the economy. This reduction was made possible by the increasing use of oil. The idea behind substituting agro-biofuels for oil is the opposite: to reduce the use of oil by increasing the use of land, labour and ecological services.

Chapter 3: Not everything that burns is a fuel
This chapter can be seen as a crash course in energetics. It provides a set of energetics-associated narratives that have been ignored by the proponents of agro-biofuels. These narratives are crucial for studying the feasibility and desirability of primary energy sources and energy carriers in modern societies. This chapter introduces the basic principles of energetics: the distinction between energy, power and useful work; the concept of metabolism; energy grammars; and the distinction between endosomatic and exosomatic metabolism. We arrive at the concept of net energy analysis, leading to the concept of energy return on investment (EROI).

Chapter 4: Patterns of societal metabolism across levels: a crash course in bio-economics
This chapter provides the basic principles of bio-economics, leading to the

concept of bio-economic pressure (BEP). In particular, it addresses the need to check internal and external biophysical constraints on the feasibility of metabolic patterns, and the need to perform analysis across the hierarchical levels of organizations. The metabolism of the parts must be compatible with the metabolism of the whole. Within this general framework, the chapter explains why the two concepts of EROI and BEP are essential in performing a quantitative assessment of the feasibility and desirability of energy sources in a developed society.

Chapter 5: Assessment of the quality of alternative energy sources

This is the most technical chapter of the book, addressing key theoretical issues. Energetics and bio-economics require two things: first, a definition of semantic concepts; and second, a formalization of these concepts into quantitative analysis. In relation to this task, we provide an overview of the treacherous quicksand of energy analysis and an example of the type of problems faced by scientists willing to check the feasibility and desirability of agro-biofuels as an alternative to fossil fuels. Believe it or not, 'energy' is a semantic concept that is very slippery when it comes to formalizations. Performing energy analysis requires deciding how to deal with the summing of 'apples and oranges'. For this reason, this chapter presents a set of basic conceptual distinctions that are required to perform a sound energy analysis. In particular, when it comes to the definition of quality for an energy source, it is essential to make a distinction between: primary energy sources and energy carriers; the output/input energy ratio of energy carriers generated and consumed in the operation of the energy sector; and the power level achieved in the energy sector.

Chapter 6: Neglect of available wisdom

This chapter starts with a quick overview of some of the most important authors, their contributions and the key concepts developed in this field. In fact, attempts to integrate economic analysis with biophysical analysis have a long history. These attempts aimed to improve our understanding of the functioning and evolution of human societies. This chapter provides a critical appraisal of the recent applications of energy analysis to the issue of agro-biofuels. The ongoing controversies on the 'right' quantitative assessment of output/input ratios are evaluated against the theoretical discussion provided in Chapter 5.

Chapter 7: A reality check on the feasibility and desirability of agro-biofuels

This chapter presents a quantitative analysis based on data sets derived from the two most impressive large-scale experiments yet established: ethanol production from corn in the US, and ethanol production from sugar cane in

Brazil. The quantitative assessment presented in Chapter 7 confirms that which is well known in the field of energetics: that is, that agro-biofuels are not even close to meeting the policy goals of energy security against the future consequences of peak oil and a reduction in GHG emissions. In the US, the low output/input ratio of energy carriers makes the solution unfeasible; in the case of Brazil, the low power level achieved in the process of ethanol production makes the solution undesirable. Towards the end, this chapter touches upon a related point: the implications of peak oil for the search for alternative energy sources.

Chapter 8: Agro-biofuel production is not good for rural development

This chapter checks the validity of the third policy goal justifying agro-biofuel mandates: the alleged positive effects on rural development and the support of crop farm income. In relation to this, we expose the current misunderstanding about the relationship between technical progress in agriculture and rural development. Technical progress in agriculture is currently based on high-input monocultures (the so-called 'paradigm of industrial agriculture'), which are required to boost the productivity of the production factors. The paradigm of industrial agriculture was developed to get rid of farmers; thus its historic goal, obviously, is incompatible with the goal of rural development. Our point is that the biofuel policy implemented by many developing countries will translate into the replacement of traditional land-use patterns – with their goals of sustaining and reproducing local communities – by large-scale monocultures for feedstock production. These plantations have the goal of boosting the production of commodities per hectare by eliminating local communities.

Chapter 9: Living in denial

This chapter addresses the last aspect of the riddle: If agro-biofuels are neither feasible nor desirable, why are we investing so much in them? This chapter points out the existence of a serious failure in the quality control of the scientific inputs used for decision-making. Indeed, in this chapter we argue that the above phenomenon can only be explained by a more systemic problem associated with three different types of lock-in taking place in society:

1   The ideological lock-in. When dealing with sustainability, nobody likes to acknowledge the obvious fact that sustainability has to do with change (ageing, death, turnover of incumbents in the power structure). Sustainability is often confused with the preservation of the status quo. This is at the basis of the ideological assumption of perpetual growth anchored in Western civilization.
2   The academic lock-in, generated by the formation of granfalloons (see glossary) within the bureaucracy of science. This leads to the generation of flawless formal analyses within invalid narratives.

3   The economic lock-in, caused by the efforts of private corporations to avoid losing economic resources invested in the biofuel sector. These forms of lock-in tend to generate a Concorde syndrome: the continued implementation of a given idea, even after it has been recognized as a failure.

Chapter 10: Where do we go from here?
In this chapter, we conclude that acknowledging the need to extricate ourselves from the folly of high-tech, crop-based biofuels does not mean that there is no future for bio-energy. On the contrary, a more critical and effective use of biomass for energy purposes is possible and desirable. But we should keep in mind the existence of severe socio-economic and ecological constraints. In regard to the use of science to govern sustainability issues, the agro-biofuel folly has unveiled a systemic challenge to the use of science in the field of technical innovation.

From energetic analyses of modern society, we can also learn something positive; the actual pattern of energy consumption is so extravagant that there is a large margin for readjusting to a lower level of energy expenditure that can still provide a more than decent material standard of living. The only requirement for this solution is wisdom. The world cannot be changed according to our wants; we ourselves have to adjust to the challenge of sustainability by using reflexivity.

Chapter 2

Learning from the Past

Going Forwards to the Past?

Agro-biofuels are being proposed as an innovative green solution to our emerging energy problems. It is unclear, however, whether the agro-biofuel proposal represents something new. For thousands of years, humankind relied on energy sources that were generated in a renewable way and that did not lead to GHG accumulation in the atmosphere. So perhaps we should consider first of all whether or not the agro-biofuel solution is a proposal to go forwards to the past. At first glance this might seem a trivial observation, but the delusion of agro-biofuels can only be explained by addressing this issue. As far as we know, the energy metabolism of pre-industrial society was mainly powered by crop-based energy inputs. Because of this, all pre-industrial societies shared a common set of biophysical constraints that limited their expansion and development. Hence, by studying the constraints that affected pre-industrial society we can better answer questions like: What were the limitations presented by crop-based energy inputs in the past? Will the proposed agro-biofuel solution be able to escape these constraints? Why were agro-biofuel solutions, being fully renewable and clean, abandoned in the past in favour of fossil fuels? Will it be possible to maintain the present characteristics of modern society, achieved with the use of fossil fuel, if we substitute agro-biofuels for fossil fuels on a large scale?

In order to answer these questions, we revisit in this chapter the transition from a pre-industrial rural society based on renewable energy inputs to a modern urban society based on fossil fuel. What changes in the basic pattern of energy metabolism enabled the move from pre-industrialism to a modern urban society? What type of complications can we expect when proposing the reverse, i.e. an energy transition from fossil fuels to the large-scale use of crop-based biofuels? (Note that by 'large scale' we mean a significant fraction of society's energy consumption, i.e. more than 10 per cent of the liquid fuels actually used for transportation.)

We start out with a general description of the discontinuity in the trajectory of human progress represented by the switch from predominantly rural agricultural society to modern urban life in the 19th and 20th centuries. In the following section we explore in detail the constraints that govern the metabolism of society when it relies exclusively on the natural conversions of energy flows (the fully renewable, zero-emission solution). This analysis draws on ecological theory. We then describe, from a historical perspective, the early attempts of pre-industrial societies to escape the ecological constraints associated with the pace of the conversion of solar energy into energy inputs compatible with societal metabolism. We show that by breaking the traditional link between energy input and colonized land, which is typical of the biomass energy solution, urban dwellers managed to achieve, at the same time, a large population and a higher material standard of living. This made it possible to have a global transition from predominantly rural to industrialized urban society. In this section we are obliged to introduce some theoretical concepts, namely fund-flow supply versus stock-flow supply, to represent the economic process of society in biophysical terms. The last section of this chapter concludes that the proposal to retreat from fossil fuel in favour of crop-based biofuel appears to overlook (or should we say chooses to ignore?) the changes in the basic patterns of societal energy metabolism that enabled the leap forwards in the last two centuries, as well as the key characteristics of the energy metabolism required to sustain a modern urban society.

From Pre-industrial Rural Settlements to Modern Urban Society: Discontinuity in the Trajectory of Human Development

According to the latest report, more than 50 per cent of the world's population was living in cities in 2007 (UNFPA, 2008) (Figure 2.1). This figure represents a dramatic break with the past; in the pre-industrial era the percentage of the population living in cities rarely exceeded 10 per cent. The unprecedented phenomenon of urbanization at the global level indicates a clear discontinuity in human development. But what is more relevant to our discussion is that this phenomenon is tightly linked to structural and functional changes that have taken place in society's energy metabolism.

It is well known that the structure of pre-industrial societies was strongly constrained by the ability of rural areas to produce food and fuel (wood) surpluses for the ruling classes in the cities (Cottrell, 1955; Tainter, 1988; Debeir et al, 1991) (Figure 2.2). The limited ability of rural activities to generate this surplus of resources represented a biophysical constraint to the expansion of cities.

This biophysical constraint, or bottleneck, was governed by two distinct urban–rural ratios, one relating to human time and the other to colonized land:

Learning from the Past   19

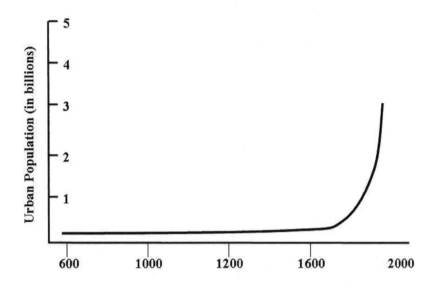

Figure 2.1 *Growth of the world's urban population*

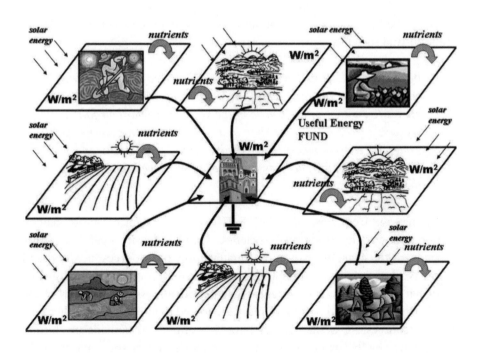

Figure 2.2 *The input provided to the city by the countryside*

1   the hours of activity of the urban rulers (lords with their courts and soldiers), who required inputs for their relative metabolism, in relation to the hours of activity of the farmers, whose activity provided the inputs for both the rulers and the ruled; and
2   the hectares of colonized land in the cities compared to the hectares of colonized land in the countryside.

Knowing, first, the profile of energy expenditures of a king and his court on the one hand and the farmers and their dependents on the other, and, second, the productivity of energy input of the working hours associated with food production, one can calculate a range of values for ratios concerning the rulers and the ruled; these ratios represent feasible configurations for society. That is, the size of the metabolism of the rulers depended on the surplus that could be generated by the activity and land use of the ruled. This implies a maximum rate of taxation beyond which farmers could not survive. Consequently, the relative proportion of the rulers within a society could not be expanded at will. A biophysical analysis of energy and material flows can assess with a certain degree of accuracy how many farmers are required on a given amount of land to sustain a king with his court.

As a matter of fact, the forced relations between farmers (generating a surplus), colonized land and lords (consuming the surplus) have long been known to be key factors in determining the ratio between rulers and ruled in pre-industrial societies.

> *Before the end of the first millennium AD ... the European kings and other big landlords were forced to move from one of their estates to another in the course of the year. They and their followers would remain for some time in a castle or manor, hunting the game and consuming the local wine, cereals, meat, fruits and so forth. When the surplus was consumed, they moved to another of their properties.*

In this passage from Boserup (1981, p97) we can recognize a more general pattern followed by traditional societies. Human density over a territory depends on the availability of an adequate supply of resources, i.e. food and energy (such as wood). As soon as the requirement of food and/or energy in a particular place and over a particular time period surpasses the density of the supply in that place and period, humans are forced to move on to another spot where resources are available; this paradigm applies to nomads or shifting cultivation. This behaviour keeps their overall density in space constant. In Boserup's passage above, nomadic behaviour is displayed by kings and major landlords, who are seen as actors exploiting the rural people, who operate on fixed land properties. In this example, the rulers are analogous to secondary carnivores in ecological systems.

As we will show in detail in this chapter, the movement from biomass exploitation to the use of fossil fuels implied a historic change in the nature of the relationship between humans and the ecological systems embedding them, and in the relationships between social classes. Indeed, the use of fossil fuels brought about a revolutionary emancipation from both land (Mayumi, 1991) and the social class system. There are two keys to this revolutionary change: first, fossil fuels (initially coal, later oil and gas) embody the activity of ecosystems (concentrated plant matter) of the past accumulated over millions of years; and, second, machines powered by fossil fuel provide society with the equivalent of the power of thousands of slaves.

With the switch to the use of fossil energy, the supply of energy to the cities was no longer dictated by the natural pace of the nutrient cycles of agroecosystems. After switching to fossil energy, humans were set free from the biophysical constraints represented by the limited surplus provided by the exploitation of natural energy flows occurring in ecosystems. The clear discontinuity entailed by the adoption of fossil energy is illustrated in Figure 2.3. It shows a sudden and simultaneous change in human population size, fossil energy consumption per capita, and colonized land per capita (the source previously supporting the expenditure) over the last 2000 years. The simultaneous increase in population size and per capita energy expenditure resulted in a spectacular increase in the overall energy expenditure of society. To use an analogy, it was like a blue-collar family with a humble lifestyle constrained by a meagre hourly wage adopting a Hollywood lifestyle supported by lottery winnings.

Learning From Ecology: Inherent Constraints to Renewable Solutions

Thus, pre-industrial rural society was based on a truly renewable energy supply, a solution that placed severe constraints on the development of society. These biophysical constraints – which dictated the relationship between the rulers and the ruled (humans versus humans) as well as spatial density (humans versus land) – have been described by anthropologists studying the structure and dynamics of pre-industrial societies (see also Chapter 6). However, the theoretical basis of the dynamics of these constraints can be found in the work of theoretical ecologists. Indeed, theoretical ecologists have provided a basic framework that explains the mechanisms determining biophysical constraints over the structural and functional organization of metabolic systems in general. Studying these constraints is extremely useful in understanding the limitations

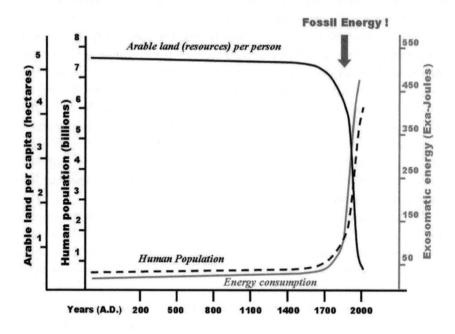

Figure 2.3 *Discontinuity in human development in relation to fossil energy use*

of truly natural and renewable solutions. For this reason, we think it necessary to take the reader on a little detour into the fields of theoretical ecology and energetics.

Theoretical ecology has studied the hierarchical organization of flows of energy and matter in natural ecosystems and in dissipative networks (e.g. Margalef, 1968; Odum and Odum, 1971; Odum, 1971; 1996; Ulanowicz, 1986; 1995). A brief overview of the theoretical concepts underlying this type of analysis is provided in Appendix 1. The most familiar example of a dissipative or metabolic network is undoubtedly an ecosystem, which we will use for our illustration. Considering an ecosystem as a dissipative network, we can first identify the various components. Plants fixing solar energy use solar energy, nutrients and water at a given rate and will produce a certain quantity of new plant biomass each year. This biomass is eaten by the animals operating within the ecosystem and hence represents their fuel. The herbivorous animals will eat a certain amount of plant biomass and drink a certain amount of water, and will produce a certain quantity of new herbivore biomass each year. The herbivore biomass in turn is eaten by the first-order carnivores operating within the ecosystem and, obviously, these dynamics repeat themselves until we arrive at the top carnivores. The mortality rate of all the various elements – plants, herbivores and carnivores – will provide a flow of dead biomass, together with excrement, which is used by detritus feeders to recycle the nutrients. Detritus feeders close the cycle, making the nutrients available again to the plants.

Having described the relationships between the different elements or compartments of the system across different levels and scales, one can calculate a set of expected relationships between the various compartments making up the internal structure and functioning (see Figure 2.4). These relationships refer to the relative size of elements and their levels of energy dissipation, energy flows, spatial densities and spatial sizes. It is important to note that this approach is only valid if the dissipative network is able to *maintain the identity* of its elements in time (see Appendix 1 for the special nature of dissipative networks). That is to say, the system must retain in time the original definition of the typology of inputs and outputs associated with the various conversions taking place in the network (for example, a plant is a plant because of its capacity to photosynthesize; a herbivore is a herbivore because it uses only plants as inputs), as well as the expected range of values for the output/input ratios (e.g. a tiger is expected to eat a certain amount of meat per unit of time and body mass).

A hierarchical structure is said to occur when different compartments of the system dissipate energy at different densities. The hierarchical structure of an ecosystem is similar to that of human society when analysing energy flows among its components. Indeed, all hierarchically-organized metabolic systems share a common pattern of energy utilization (Odum and Odum, 1971); the least abundant components (the 'elite', i.e. those with a smaller spatial density, such as top carnivores in ecosystems or rulers in pre-industrial societies) have the higher energy dissipation and the capacity to express more sophisticated structures and functions across a larger space–time domain, while the most abundant components (e.g. plant biomass in ecosystems or farmers in pre-industrial societies) have a lower energy dissipation rate and express structures and functions across a smaller space–time domain. The elite components are fed by flows of energy and matter made available by the components present in greater spatial density. Also, the pattern of feedback interaction is general. In exchange for the net gain in energy flow, the elite components act as an organizational centre for the others, providing information, control over a large space–time domain and the tuning of flows in the entire system, expertise, gathering and processing data, etc. In this way, the entire system benefits from a high level of organization while paying a relatively low energy cost for this service. Examples of this type of analysis, adapted from Odum and Odum (1971), are shown in Figure 2.4.

24 *The Biofuel Delusion*

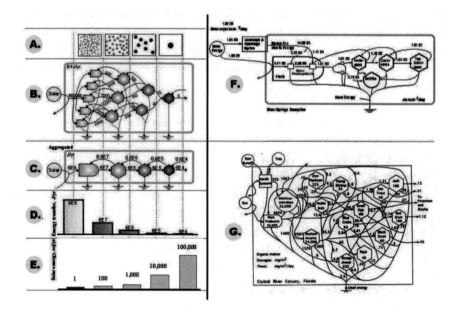

Figure 2.4 *Examples of metabolic patterns within ecosystems determining the formation of hierarchical structures*

To illustrate this general principle of biophysical constraints governing the structural and functional organization of metabolic systems, we provide a simple example in which we compare the density of tigers in their natural habitat and that of tigers found in a zoo. This example has been presented in more detail in Giampietro and Pimentel (1990) and is based on the seminal work of the Odum brothers.

The data provided in Figure 2.5 refer to a hypothetical closed forest ecosystem in which all the surplus produced by plants through photosynthesis (net primary productivity) is used to fuel the activity of animals (herbivores and carnivores) operating in the system, without the accumulation of additional plant biomass. Hence, we assume that this forest ecosystem is no longer growing in size (steady-state climax situation) and is operating at a density of standing biomass of 2.04kg/m². As illustrated in Figure 2.5, top carnivores require about 50,000m² of forest area per kilogram of body mass, which is equivalent to an average of eight tigers per 100km² or a tiger density of 0.00002kg/m². This large area is needed to support the activity of the plants, which is required in turn to fuel (directly and indirectly) the herbivores and carnivores that are eventually hunted by top carnivores.

The picture is quite different if we consider four tigers, of about 250kg each, confined in an artificial habitat of 10,000m² at a zoo. This situation trans-

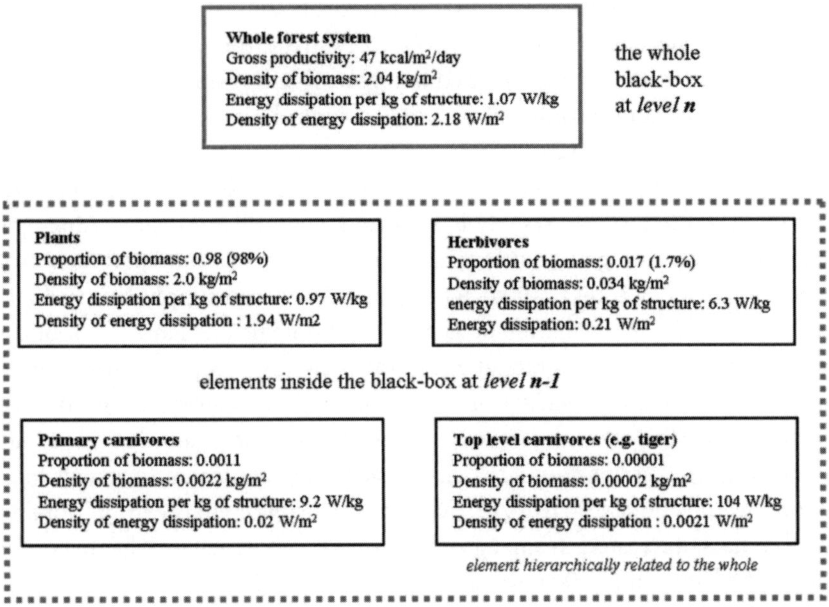

Whole forest system
Gross productivity: 47 kcal/m²/day
Density of biomass: 2.04 kg/m²
Energy dissipation per kg of structure: 1.07 W/kg
Density of energy dissipation: 2.18 W/m²

the whole black-box at *level n*

Plants
Proportion of biomass: 0.98 (98%)
Density of biomass: 2.0 kg/m²
Energy dissipation per kg of structure: 0.97 W/kg
Density of energy dissipation : 1.94 W/m2

Herbivores
Proportion of biomass: 0.017 (1.7%)
Density of biomass: 0.034 kg/m²
energy dissipation per kg of structure: 6.3 W/kg
Energy dissipation: 0.21 W/m²

elements inside the black-box at *level n-1*

Primary carnivores
Proportion of biomass: 0.0011
Density of biomass: 0.0022 kg/m²
Energy dissipation per kg of structure: 9.2 W/kg
Density of energy dissipation: 0.02 W/m²

Top level carnivores (e.g. tiger)
Proportion of biomass: 0.00001
Density of biomass: 0.00002 kg/m²
Energy dissipation per kg of structure: 104 W/kg
Density of energy dissipation : 0.0021 W/m²

element hierarchically related to the whole

Figure 2.5 *Set of expected relations over different compartments of a terrestrial ecosystem: relative size, rate of energy dissipation and density of energy dissipation*

Source: data from Odum, 1971, p75

lates into a tiger biomass density of 0.1kg/m². It is evident that this situation would be impossible if the food supply of the tigers were to be based solely on a natural chain of ecological transformations in that closed environment. In fact, the tiger density in the zoo is 5000 times higher than the natural density (0.00002kg/m²). Nevertheless, the tigers in the zoo can survive in a steady-state situation because the zoo is an *open system*. The zoo system does not actually produce the specific energy input (meat) needed to feed the tigers; it buys it from the market.

Given the robustness of the expected relationships between the relative sizes and metabolic rates of the various elements of an ecosystem (plants, herbivores, tigers), one can calculate the corresponding *embodied* ecological activity required to produce an input (Giampietro and Pimentel, 1990). For example, knowing the daily meat consumption of a tiger in the zoo and using the expected chain of conversions of solar energy – plant biomass – herbivore biomass – primary carnivore biomass, etc., we can estimate the overall forest activity required to maintain the tigers and map this onto a virtual requirement of space (the estimate of 5000 times the actual space). This approach, based on the study of natural chains of energy conversions within ecological systems (considered as metabolic networks), is based on the rationale proposed by H. T. Odum, and can be used as a tool to estimate the limits to the human exploitation of natural resources (Giampietro and Pimentel, 1990). This approach should not be confused with the ecological footprint, developed later on by Wackernagel and Rees (1996).

While a pre-industrial society represents a closed system (like our forest ecosystem), modern urban society is an open system (like the zoo). Analogous to the tiger in the zoo getting its energy from a stock of refrigerated meat from the wholesale retailer, modern societies get their energy supply from oilfields around the world. As a matter of fact, most typologies of land use found in modern urban societies operate at a density of energy dissipation that is much higher than the average density of energy flowing in the form of net biomass production within natural terrestrial ecosystems. For example, the typical density of energy dissipation in the city (residential, retailers, industry) is in the order of magnitude of $10-30 W/m^2$, while the density of energy input obtained from biomass harvested from agro-ecosystems can reach at the maximum about $0.05 W/m^2$ (Smil, 2003; see also Figure 7.1 in Chapter 7).

The difference in density between the expected energy flows consumed by an urban settlement and those produced by biological activity explains the low level of urbanization achieved in pre-industrial societies. In a closed pre-industrial society, one can have only limited urban settlements (the land-use typology with a high-density energy metabolism) because they require large rural areas (the land-use typology with a low-density energy input) to provide the required supply of inputs. In modern society, one can have large urban settlements because the energy input is obtained outside the system (ecosystem activity from the past, embodied in oil). Thus, powering a human society with a natural chain of energy conversions taking place within the system (the ultimate green solution) means concentrating the energy surplus, produced at the low density associated with biomass production, in a *limited* number of privileged spots.

Escaping the Natural Constraints to Energy Flows

In this section, we describe early attempts by complex societies of the past to overcome the biophysical constraints described above. We of course will have to introduce here some theoretical concepts related to the metabolism of human society. These are easy-to-grasp concepts proposed by Georgescu-Roegen (1971) to represent the economic process in biophysical terms. Given that energy is essentially a biophysical concept, there is no way to analyse its limitations and potentialities other than integrating the economic approach with a biophysical approach.

Early examples of the use of exosomatic energy

Human societies have two distinct forms of metabolism: endosomatic and exosomatic. Endosomatic metabolism refers to the food energy converted *inside* the human body to preserve and sustain the physiological activity of humans. Exosomatic metabolism refers to the energy converted *outside* the human body with the goal of amplifying the output of useful work associated with human activity (e.g. the use of tractors, the melting of metals, animal power). For example, when driving a tractor a farmer can deliver in one hour a flow of power that is a thousand times larger than the endosomatic power delivered in one hour of manual work.

This conceptual distinction between endosomatic and exosomatic metabolism was first introduced by Lotka (1956) and was later proposed as a working concept for Georgescu-Roegen's (1975) energetic analyses of bio-economics and sustainability. '[I]t has in a most real way bound men together into one body: so very real and material is the bond that society might aptly be described as one huge multiple Siamese twin' (Lotka, 1956, p369). The vivid image proposed by Lotka explicitly suggests that a hierarchical level of organization higher than the individual should be considered to describe the flow of energy in modern societies.

Both terms, endosomatic and exosomatic metabolism, refer to flows of energy and material inputs transformed under human control within the socio-economic process. The flow of endosomatic metabolism is fairly constant in time, especially when considered per capita, and is directly related to population size. Conversely, exosomatic metabolism is highly variable and depends on the amount of technical capital present in society. It therefore closely reflects the technical evolution of human society. In pre-industrial societies, exosomatic energy metabolism is mainly accounted for by fire, animal power, wind power and other occasional power sources such as waterfalls, river streams and geothermal events (Giampietro et al, 1997). For this reason, techniques improving the harnessing of fire, wind, waterfalls and animal power have been crucial in providing competitive advantages among competing pre-industrial civilizations (White, 1943; 1959; Cottrell, 1955; Cipolla, 1965; Debeir et al, 1991). Indeed, two crucial steps in the history of human civilization, the discovery of fire and the Industrial Revolution, can be directly associated with dramatic changes in the exosomatic metabolism of human societies.

Given that the endosomatic energy consumption of human society is subject to little change, the exosomatic/endosomatic energy consumption ratio of a society closely reflects changes in exosomatic energy metabolism, and hence is useful as an indicator of the processes of industrialization and capitalization in human society. For example, in a developed society the per capita endosomatic energy consumption lies in the range of 10–12 megajoules (MJ) per day (approximately 2400–3000kcal/day), whereas the per capita exosomatic energy consumption can be estimated at 500–900MJ/day (or 200–320 gigajoules (GJ)/year). Thus, the exo/endo energy ratio typical of developed societies falls within the range 50/1–75/1, while that of pre-industrial societies is typically only about 5/1 (including energy consumed for cooking, heating and illumination, as well as animal power and local sources of mechanical power, such as waterfalls or wind).

Stock-flows versus fund-flows

Definition of funds and flows

Georgescu-Roegen (1971; 1975) studied in detail the biophysical roots of the economic process and proposed a fund-flow model useful for representing, in biophysical terms, the metabolism of socio-economic systems. His terms, introduced here, will be used throughout this book in the representation of the metabolism of human society, whether rural or modern urban. Georgescu-Roegen (1971) is our source for the definitions of flows (below) and funds (opposite), although we have slightly reworded them.

> *Flows refer to elements disappearing and/or appearing over the duration of the representation; they enter but do not exit, or exit without having entered.*

Examples of flow elements are fossil energy, food or a new product consumed in the economic process. Within the fund-flow model, the energy inputs and the material flows used by society for its self-organization, the endosomatic food consumed by humans, and the exosomatic fuel used by the socio-economic process (such as raw materials and economic products) would all be classified as flow elements. Hence, flows include matter and energy that is already present, controlled matter and energy, and dissipated matter and energy. The pace of these flows is controlled by two types of factors: external and internal.

External factors refer to the accessibility of an adequate input flow from the environment or an alternative to the availability of a stock. In the case of a society, external factors would include natural resources and stocks of coal, natural gas or oil. As explained by Herman Daly (1994, p28):

> *The world is moving from an era in which man-made capital was the limiting factor into an era in which remaining natural capital is the limiting factor. The production of caught fish is currently limited by remaining fish populations, not by the number of fishing boats; timber production is limited by remaining forests, not by sawmills.*

Internal factors refer to the system's capability of processing the available flow during the relative conversion. In the case of a society, internal factors refer to human-made capital (available technology and know-how), and are related to the capability of a given society to process a larger flow of energy, material and resources, should these be available. Returning to Daly's example, internal factors would be represented by the number of fishing boats.

Given that flows are the elements that disappear within the duration of the analysis, we also need to define *those things that remain the same* within the duration covered by the scientific representation. These are the elements of the metabolism of the socio-economic system: the population, the buildings, the factories and the economic process. All these elements are described as *funds*.

> *Funds refer to agents that are responsible for energy transformations and are able to preserve their identity over the duration of the representation. They transform input flows into output flows within the time scale of the representation. Therefore, they enter and exit the process represented in the analysis while maintaining the same identity.*

Examples of fund elements are capital, people and colonized land (also called Ricardian land). Analogous to the representation of metabolic networks in theoretical ecology, fund elements must be able, at least during the duration of the analysis, to preserve their identity as converters. This is essential to guarantee the validity of the original definitions of 'inputs' and 'outputs'. The identity of the fund elements entails two important constraints.

First, fund elements can only be used at a specified rate and in relation to a specific definition of input and output. For example, human beings cannot eat more than a certain amount of food per day; they cannot eat gasoline instead of bread; nor can they generate more than a certain number of children in a lifetime. Similarly, a harvester cannot use meat as fuel, should it run out of gasoline; it cannot harvest more than a certain number of hectares of land per day; nor can it dig channels, should this be needed.

Second, fund elements must be periodically renewed, and this implies a given overhead on their performance. For example, individual human beings die, and to maintain the population they must be replaced. This implies the existence of a constant ratio of children to adults, and consequently an overhead on the flow of food, which cannot be entirely converted into adult labour. Moreover, human beings need rest and can get sick, which implies maintenance and hence additional overhead costs. Exactly the same applies to machinery: harvesters have a certain lifespan; they need maintenance and occasionally repair. This all affects the conversion of gasoline supply into hours of harvesting activity.

Thus, the variables used to represent funds and flows in a metabolic system reflect the set of attributes chosen by the analyst for defining *what the metabolic system is* and *what the system does*, respectively. The choice of flow elements defines the relation of the metabolic system with its context. Because flows disappear over the timespan covered by the representation, they have to be either consumed or generated by the investigated system, or made available by or absorbed by the context of the system. Funds refer to *the converters* of the system, elements that have to be preserved and reproduced by the metabolic process. In a metabolic network, funds and flows define each other to a certain extent, because the identity of a fund element entails an identity for the associated flows; the same is true in reverse.

Fund-flow and stock-flow energy

Given the definitions of funds and flows, we can appreciate the distinction between flows that originate from funds and those that originate from stocks. A flow originating from a fund does not entail a change in the characteristics of the system, while a flow originating from a stock does. We will illustrate this with examples.

We can milk a healthy cow – to be considered as a fund element – every day, and if we don't overdo it, the cow will remain healthy. If we consider an entire dairy farm producing milk, in which there are sufficient calves guaranteeing the replacement of cows and enough pasture for feeding the cows, then the flow of milk from this self-reproducing dairy farm represents a stable supply. Thus, as long as the fund (dairy farm) is able to repair and reproduce itself, the resulting flow can be considered a *renewable resource*. This rationale will be applied in Chapter 7 to check the renewability of cropping for biofuel.

A situation in which a flow originates from a stock is completely different. If we start with a stock – for example, an oilfield of 1000 units – and we consume for one hour a flow of 100 units per hour, then, after that hour, the stock from which we obtained the input will have changed its identity; the original stock of 1000 units will have been reduced to 900 units. The consumption of a stock-flow of energy entails a continuous change in the identity of the whole system. When considering both the stock providing the energy input and the system using the energy input, the representation of the whole system loses its validity in time. For this reason, we can call an input derived from a stock-flow a *non-renewable resource*.

Explaining the implications of the distinction between fund-flow and stock-flow in determining biophysical constraints on the option space of economic processes, Georgescu-Roegen (1971, pp226–7) writes:

> *The difference between the concept of stock and that of fund should be carefully marked, lest the hard facts of economic life be distorted*

at everyone's expense. If the count shows that a box contains twenty candies, we can make twenty youngsters happy now or tomorrow, or some today, and others tomorrow, and so on. But if an engineer tells us that one hotel room will probably last one thousand days more, we cannot make one thousand roomless tourists happy now. We can only make one happy today, a second tomorrow, and so on, until the room collapses. Take also the case of an electric bulb which lasts five hundred hours. We cannot use it to light five hundred rooms for an hour now. The use of a fund (i.e., its 'decumulation') requires a duration. Moreover, this duration is determined within very narrow limits by the physical structure of the fund. We can vary it only little, if at all. If one wishes to 'decumulate' a pair of shoes, there is only one way open to him: to walk until they become waste. In contrast with this, the decumulation of a stock may, conceivably, take place in one single instant, if we wish.

This passage points to an important trade-off when moving from stock-flow to fund-flow energy inputs. A large-scale move from fossil fuel (stock-flow) to agro-biofuels (fund-flow) is suggested to solve two problems generated by the excessive reliance on stock-flow energy: the accumulation of GHGs in the atmosphere, and the depletion of non-renewable fossil energy resources. However, by moving to a fund-flow energy solution we can revive different sustainability problems that were solved by using stock-flow fossil energy in the first place. For example, the energy produced by fund-flows would not be enough to serve the needs of large numbers of humans, if they are to enjoy the same pattern of exosomatic metabolism expressed by developed countries. That is to say, when considering the requirement of land as a fund element – in relation to demographic pressure – the supply of agro-biomass energy can be insufficient compared to the demand for exosomatic fund-flow energy (liquid fuel), and compete with the production of endosomatic fund-flow energy (food). The use of colonized land to produce exosomatic inputs for machines owned by the affluent could take priority over the use of colonized land to produce endosomatic food for those in need. Unfortunately, as discussed in Chapter 1, this hypothesis has already been confirmed by recent events.

For this reason, it is extremely important to develop a sound theoretical analysis and understanding of the basic energetic principles determining the feasibility and desirability of metabolic patterns in socio-economic systems in relation to both external and internal constraints. This is an essential prerequisite to the analysis of the feasibility and desirability of alternative energy sources.

Early examples of the utilization of stock-flows

Tainter (1988) studied in detail the growth and collapse of complex societies and has described beautifully how some pre-industrial societies managed to reach a certain critical level of complexity (for example, the formation of large empires) in spite of the innate weakness (the low density) of their primary energy sources (which were fully renewable, such as fund-type energy sources). However, these achievements were temporary and bound to collapse in the long run. The low density of energy fund-flows implied that the building of an empire had to be based on the initial exploitation of resource stocks, such as gold, slaves and other goods taken from neighbouring kingdoms conquered during the expansion phase. In fact, rather than relying on the small agricultural surplus of their own manors and estates, the growing empires resorted to boosting their internal metabolism by conquering other kingdoms. With these resources, the empires improved their armies so as to conquer additional kingdoms.

The tactic of drawing on readily available resources from neighbouring societies represents an early form of stock-flow exploitation. This tactic provided a much higher return than the alternative of expanding traditional fund-flow exploitation (boosting the surplus of agricultural production). Note that apart from its inherently low return, an additional obstacle to expanding agricultural production was presented by transportation costs. These costs were too high in pre-industrial times to allow the transportation of commodities over long distances by land (and in this context, 'long' refers to distances of less than 100km). Indeed, most large empires were dependent on the availability of suitable waterways to reduce the cost of transporting surpluses to the central administration (Cottrell, 1955; Debeir et al, 1991).

The low return of agricultural production also explains the massive exploitation of slaves in pre-industrial empires. It was common practice that young adults from the conquered kingdom were turned into slaves. Slave labour represented an attractive alternative to the fund local population because it required a much lower overhead for its reproduction than the farm labour of free people. Consequently, slaves could be considered a form of exosomatic energy for a given society, especially in early times when slaves were performing the function of animal power, albeit in a more sophisticated form.

Another standard solution adopted in virtually all pre-industrial societies was the stable and socially accepted differentiation of the expected rate of material consumption among social classes. In this way it was possible to maintain the difference between those generating a surplus (the ruled, who consumed less than the average) and those consuming a surplus (the rulers, who consumed more than the average).

The utilization of energy input (EI) within a pre-industrial socio-economic system is schematized in Figure 2.6a in order to illustrate the mechanism of control on the stability of the dynamic energy budget. For the sake of simplicity, we consider only endosomatic energy, that is, food production (for a more detailed analysis, see Giampietro et al, 1997). At any moment in time, the energy input available to society has to be distributed to meet:

1. the food energy needed to fuel the labour (working hours) that produces the food for society (W). This is a non-negotiable appropriation of the energy input (EI);
2. the food energy required by the ruled (in this example, farmers) to support their families and hence guarantee their own reproduction (FF); and
3. the food energy required by the rulers (K), which in complex pre-industrial societies refers to the consumption of the king and his government (royal court, army, administration and public works).

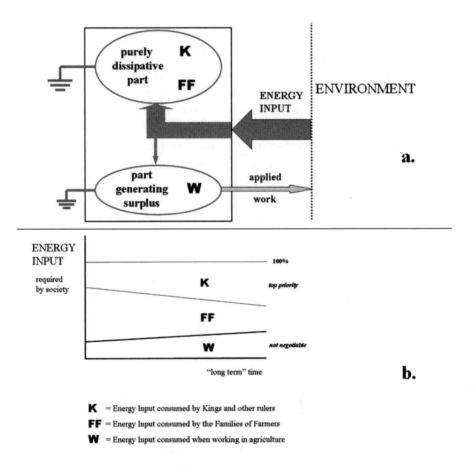

Figure 2.6 *The dynamic budget of endosomatic energy in pre-industrial society*

The mechanism leading to the collapse of pre-industrial societies in the long term is simple. In periods of ample surplus the consumption of rulers increased, thus expanding the K compartment. This was especially the case when the society expanded by conquering neighbouring kingdoms. In contrast, in periods of shrinking surplus the consumption of the FF compartment was compressed. Tainter (1988) has provided beautiful examples of this 'ratchet effect'. For instance, the Roman Empire experienced a continuous increase in the ratio of soldiers and administrative workers to farmers. When the external income from wars started to decline, this translated into a continuous increase in the tax pressure exerted on rural communities; farmers had to work for longer hours, and their family members (including children) were also forced to work.

The typical predatory behaviour of empires in the initial growth phase enhances the ratchet effect. In fact, young empires typically invest a sensible fraction of their energy throughput (EI) in armies and administration. In this initial phase, improving the efficiency of the army and the administration (expanding K) pays off better than improving the standard of living of farmers (expanding FF). Conquering kingdoms represents a form of resource stock exploitation and a means to boost the flow of EI well above that obtainable through resource fund exploitation (food provided by domestic agriculture). Moreover, conquered slaves provide labour with much lower overhead costs than farm labour (reducing FF). However, when the growing empire runs out of kingdoms to conquer, the burden of maintaining its total energy input (the sum of W, FF and K) comes to rest entirely on the shoulders of the farmers. At that moment, the gap between the required energy input and the actual return generated by its own primary sectors (the specialized part within the system expected to generate enough surplus) becomes painfully evident.

For a while, the gap can be filled by the increased taxation of farmers (squeezing the ruled), thus reducing FF consumption (Figure 2.6b). However, this solution makes the life of rural communities impossible in the long term, and their inhabitants cease to feel that they belong to the society. In spite of their vital role in keeping the empire alive, these rural communities are not given any attention or priority when the empire's resources are distributed. In these precarious conditions, even a small perturbation can generate the collapse of the entire system. For example, Tainter (1988) described how many rural communities in the Roman Empire welcomed the barbarian invasions that brought to an end the nightmare of continuously increasing taxation. Many farmers in the late Roman Empire simply left the countryside; rural areas became deserted, leading to a major deterioration in the empire's social and economic foundation. This analysis of the precarious dynamic equilibrium of the energetic metabolism of pre-industrial society shows the need to check continuously the relative sizes of the compartments of society in relation to the availability of the natural resources consumed by the social metabolism in fund-flow mode.

Another excellent example of the importance of avoiding the excessive expansion of the fund of human activity in pre-industrial society is represented by the essential role of pyramid building in stabilizing the Egyptian state (Mendelsshon, 1974). Cottrell (1955) calculated that the yearly endosomatic energy cost of pyramid building was approximately equal to the energetic surplus of Egyptian agriculture. Had this surplus of work been invested in an expansion of agricultural production (more food generated by the W sector, leading to more people working in the W sector, boosting the consumption in the overhead compartment FF), Egypt would have risked a Malthusian trap, with reduced returns and excessive environmental loading that could have resulted in instability in the short term (for example, famines and social unrest in less favourable years) and would have been fatal in the long term. We recall the collapse of the Mesopotamian civilization because of overpopulation that led to ecological degradation. This is why, in the past, complex pre-industrial societies tended to use their surplus of labour for non-essential activities, such as the construction of pyramids, great walls and cathedrals (a quantitative analysis of this point is provided in Giampietro et al, 1997).

Forward to the Past or Learning from the Past?

We have shown in this chapter that the dramatic discontinuity associated with the Industrial (and urban) Revolution can be explained by the abandonment of an integrated set of relations based on reciprocal constraints governing the interaction between human society and ecosystems. The term 'Industrial Revolution' refers to the point at which the flow of matter and energy consumed by the socio-economic process no longer respects the natural relationships among energy conversions typical of ecological fund-flows. The pace at which modern human society metabolizes energy is no longer compatible with the pace at which the environment can make available energy inputs and absorb the relative material wastes (such as CO_2). The natural pace of the nutrient cycles in agro-ecosystems, which regulated the supply of energy surplus from the countryside to the cities in the past, has been surpassed by the massive use of fossil energy stocks.

As a result of the Industrial Revolution, the energy required by modern society to generate useful work no longer flows from the countryside to the cities. On the contrary, the industrial sectors of developed societies, located near the cities, provides the flow of energy carriers (fuel for machinery and the production of fertilizers and pesticides) consumed in the rural areas (Figure 2.7).

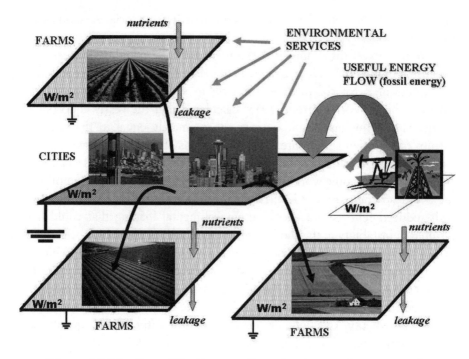

Figure 2.7 *The flow of fossil energy from the cities to the countryside*

A comparison of Figures 2.2 and 2.7 clearly shows that in pre-industrial society, both endosomatic and exosomatic metabolisms are dependent on the pace of surplus generation by funds (land, labour, animal power and technical infrastructures such as water mills or sailing ships). On the other hand, in post-industrial urban societies both types of metabolism are dependent on the pace of fossil energy flows (providing power and technical inputs) generated by stock depletion. The metabolism of modern socio-economic systems is based on flows with an extremely high energy intensity (per hour) and energy density (per hectare).

Comparing Figures 2.3 and 2.1, one can see that the flat part of the curve of human population in Figure 2.1 represents the limit of fund exploitation given the available technology. Note that the curve shows a slight increase in population due to minor improvements in fund-flow exploitation efficiency. The big discontinuity in Figure 2.1 is introduced by the Industrial Revolution and can be explained by the simultaneous removal of two types of constraints:

1   External constraints. By moving from fund-flow exploitation to stock depletion (of oil reserves), consumption growth in human society escaped the inherent limitation of renewable ecological flows.

2   Internal constraints. Through capitalization (building up a large amount of exosomatic devices per capita), humans were able to continuously and consistently increase the rate of energy consumption for more than a century.

To better understand the extraordinary achievements of human technical progress, two points need to be recognized. The human population has increased by 3 billion in the last four decades, more than in the previous 40,000 years; the per capita consumption of fossil energy has also been continuously increasing in this period. At the same time, there has been a reduction in natural resources per capita (ours being a finite planet), notably of colonized land per capita (Figure 2.3). Due to the special characteristics of fossil energy, notably the emancipation from land, it has been possible to increase dramatically both population size and per capita energy consumption in spite of the reduction in fund resources available per capita.

In the past, environmental problems related to energy metabolism typically surfaced on the supply side, such as the exhaustion of soil due to overexploitation or the lack of fuelwood because of deforestation. These problems typically led to the collapse of civilizations. For the first time in history, energy use has caused problems on the sink side, notably the accumulation of GHGs in the atmosphere. Obviously, filling the sinks is as serious a problem as shortage on the supply side.

Can we return to fund-flows and at the same time circumvent the biophysical constraints that made great civilizations of the past crumble?

Any discussion about the potential of fund-flow exploitation to partially replace fossil energy use in modern society should start from the individuation of the characteristics that make an energy source and/or an energy carrier *desirable* for a given society. History shows that human society has consistently switched to more convenient and versatile sources: from energy funds (biomass) to fossil energy inputs – first coal, later oil, and now natural gas. Experts are predicting that hydrogen from the steam-reforming of natural gas might soon be preferred over natural gas itself. From this perspective, the list of proposed potential bio-energy options – including animal manure, slaughter residues and other types of organic waste (see Hoogwijk et al, 2003) – may well be missing the point. Indeed, agreeing on the fact that humans must find a substitute for fossil energy does not mean that this substitute must be biofuel.

In this chapter we have focused on the problems related to the idea of returning from stock-flow to fund-flow exploitation as proposed by agro-biofuel proponents. We explained that the substitution of fossil energy stocks for biomass energy in the past led to emancipation from the land, and we underscored the enormous importance of this emancipation for the development of present human society. We explained the problems that can be expected if we reverse this process of evolution: the replacement of fossil energy with agro-biofuels aims to use land and the net primary productivity of terrestrial ecosystems in increasing quantities to cover our energy needs, and hence it is in direct competition with the use of these fund resources for our food needs. The amount of land required to cover 3 per cent of the total energy consumption of developed countries is assessed in Chapter 7.

Chapter 3

Not Everything that Burns is a Fuel

Introduction

In this chapter we point out that in order to make a fruitful contribution to the discussion of alternative energy sources as a substitute for oil, it is highly recommended that one should have at least a basic knowledge of energetics. Energy is a physical entity whose transformations are regulated by thermodynamic laws; it cannot be analysed by economics alone. This chapter may be seen as a crash course in energetics, the field that provides systemic analyses of energy transformations associated with the maintenance and reproduction of metabolic systems such as human society. Indeed, we argue that it is essential to use a comprehensive, organic approach that addresses the full complexity of the network of energy transformations that are behind the successful maintenance and reproduction of human societies. When looking for alternative energy sources, it is essential that the proposed energy sources can express the characteristics required by the metabolic pattern of society. We further show that we must acknowledge the existence of the general principles of energetics that dictate the features of metabolic patterns, and hence the characteristics of the socio-economic systems that are based on the expression of these metabolic patterns. These general principles can be used to define the quality of an energy source, such as agro-biofuel, in relation to its proposed use in stabilizing the activities of a given socio-economic system.

For these reasons, we strongly believe that a basic knowledge of energy conversions and metabolic patterns is a prerequisite for anyone willing to get involved in discussing and/or developing energy policies, especially given that humankind is running out of high-quality energy sources. Moreover, even if new primary energy sources should (magically) become available, several additional sustainability problems (such as the loss of biodiversity, the shortage of water, the shortage of critical minerals and the deterioration of ecological services due to demographic pressure) still justify the need to seek a better understanding of the interaction between the metabolic pattern of modern societies and ecosystem metabolism.

As we mentioned in Chapter 2, several scientists have developed comprehensive theoretical frameworks by applying the general principles of energetics to the analysis of metabolic patterns of society. The most sophisticated theoretical foundation for developing these approaches has undoubtedly been provided by Georgescu-Roegen's bio-economics, which makes it possible to apply basic principles of energetics to the study of the feasibility and desirability of metabolic patterns associated with socio-economic systems (Mayumi and Gowdy, 1999; Mayumi, 2001). Georgescu-Roegen's concept of bio-economics makes it possible to integrate the representation of monetary flows with matter and energy flows within scientific analyses. In a way, it is similar to the structural economics approach proposed by Leontief (1941), which represents various economics sectors as interacting with each other in terms of their relative inputs and outputs. The most important differences are:

- the epistemological stand: Georgescu-Roegen calls for the necessity of remaining semantically open at the moment of choosing the representation; and
- the key focus on the need to reproduce fund elements, in particular human activity.

As described in Chapter 6, Georgescu-Roegen's approach can also be associated with the well-established tradition of energy analysis of human societies.

Thus, in this chapter we discuss, after a short preamble, the basic concepts of energetics, defined as the systemic analysis of energy transformations associated with the maintenance and reproduction of metabolic systems. We then discuss the implications of these concepts for the evaluation of the quality of a proposed fuel. Finally, in the last section, we elaborate in detail on the concepts of net energy and energy return on the investment (EROI).

When Looking for Alternatives to Oil, Does Anything Go?

A quick review of the rapidly growing literature on biofuels and many initiatives looking for alternative energy sources shows an amazing variety of options proposed by experts and visionaries. All these options are being proposed as feasible and/or highly desirable solutions to replace our use of oil. As illustrated in Figure 3.1, proposed solutions include running stoves on corn and producing biodiesel from salmon oil.

To ridicule this trend towards bizarre solutions, the first author decided to invent a new energy source for a conference presentation: biodiesel from human fat obtained as a by-product of liposuction (Figure 3.1). The idea was to present it as a joke to the audience. However, shortly after the presentation we found out that someone in New Zealand had already been working on its implementation; see Figure 3.2!

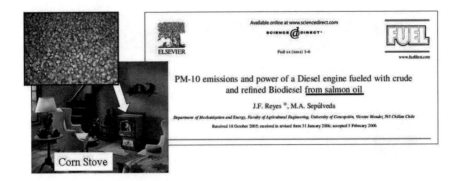

And what about refined biodiesel from human fat obtained by liposuction?

Figure 3.1 *Alternative energy sources: does anything go?*

Photo of the corn stove courtesy of Harman Stove Company, www.harmanstoves.com/gallery5.asp. Photos of obesity, Wikimedia commons, http://commons.wikimedia.org/wiki/File:Central_Obesity_008.jpg; http://commons.wikimedia.org/wiki/File:Central_Obesity_011.jpg.

This rapid proliferation of potential alternative energy sources generates information noise that seriously hampers the evaluation of priorities in research funding in this field. It is important for policy-makers and public alike to realize that not everything that can be burned can be considered a potential energy source in a developed society. This statement obviously raises the question of how we can distinguish between useful and non-useful potential energy sources. To this purpose, in the field of energetics we employ the concept of quality of primary energy sources.

The next question is: Do we have a methodological approach to define the key characteristics of a high-quality or a low-quality energy source? This approach should also allow us to exclude things that can be burned (such as $100 notes) to provide energy (heat), but that should not be considered an energy source at all.

In this chapter we build on the idea that in order to define the quality of a potential energy source, it is necessary to first establish a relation between the characteristics of the energy input *expected* by the society metabolizing energy,

"There's an interesting business model: link a biodiesel plant with the cosmetic surgeons," says Mr. Bethune. "In Auckland we produce about 330 pounds of fat per week from liposuction, which would make about 40 gallons of fuel. If it is going to be chucked out, why not?"

"Mr. Bethune is the founder of Earthrace, a project to promote the use of biofuel as a renewable energy source. The round-the-world record attempt is part of Earthrace's campaign to highlight concerns about diminishing supplies of fossil fuels and the environmental advantages of biodiesel."

Peter Bethune

"A large liposuction operation involves removing 10 pounds of fat, which would drive a car about 50 miles once converted."

Figure 3.2 *The boat that uses diesel fuel partly manufactured from human fat*

Photos and text courtesy of CalorieLab; http://calorielab.com/news/2005/11/11/biofuel-promoter-to-power-boat-using-human-fat/

and the characteristics of the process generating the supply of the potential energy input. In studying the compatibility between what is required and what is supplied as energy input, the concept of metabolism is extremely useful. In fact, as we have seen in Chapter 2, the concept of metabolism conveys the idea of an expected set of relations between the energy input and the various fund elements using it. For example, herbivores are different from carnivores because of a systemic difference in the definition of their energy input. Perhaps this may seem obvious to the reader, but it is seemingly ignored by a large number of self-proclaimed energy experts.

In order to ascertain whether or not a certain source of energy would be an appropriate input for a system, one must first carefully observe the characteristics of the energy system. We cannot feed gasoline to humans, or power a refrigerator with pizzas. When refuelling a vehicle in a modern gas station, the driver must first select the appropriate type of fuel: gas or diesel, regular or premium octane rating. In the same way, US electric plugs with a voltage of 110 volts (V) deliver an energy input incompatible with EU appliances, which require 230V. All these examples clearly show that it is impossible to judge the feasibility and desirability of a potential energy source by studying *only* its characteristics in terms of a generic determination of the amount of joules (or kilocalories) it can deliver.

Common sense tells us that it is important to check the compatibility of the potential energy source and the system that is supposed to use it. It is therefore hard to understand why many focus only on the characteristics of the production process of the energy input, e.g. the overall output/input ratio of the production process of ethanol from crops. They seem to make a naive assumption: if it can be burned, and if it can be produced with an output/input energy ratio higher than 1, then it is a feasible and desirable energy source for both developing and developed societies.

As a matter of fact, this assumption has been proved wrong by the evolution of the energy sources used by humankind. Wood was almost entirely replaced by coal at the beginning of the Industrial Revolution, even though both can be burned and both have an output/input energy ratio higher than 1 (and much higher than the one achieved with biofuels). In the same way, coal was largely replaced by oil in the middle of 20th century, and in the last few decades, wherever possible, oil has been replaced by natural gas. This evolution strongly suggests that there are other characteristics that define the quality of an energy source, besides its ability to burn and an output/input energy ratio greater than 1. However, an organic discussion of what makes an energy source high or low quality is never provided by those investing their time in calculating the second decimal of the output/input energy ratio of corn-ethanol production.

For this reason, we strongly believe that it is essential to explore some basic concepts of energetics before getting involved in any discussion of the feasibility and desirability of a large-scale move toward agro-biofuel. This step is a prerequisite if we want to have a more informed discussion about our future energy sources.

Basic Concepts of Energetics

In this section, we present six points that we believe are needed to clarify the discussion of the quality of energy sources.

Point 1: Human technology does not and cannot produce energy

According to the First Law of Thermodynamics, neither matter nor energy can be created or destroyed. Therefore, it is impossible to have an output/input energy ratio greater than 1 when considering a closed set of energy transformations. To make things more difficult, according to the Second Law of Thermodynamics, any conversion of an energy input into useful work implies an unavoidable loss. This simple observation points at the existence of two important misunderstandings.

First, when dealing with the biophysical analysis of the conversion of energy *sources* into energy *carriers*, human technology can produce energy carriers but *it cannot produce primary energy sources*. Therefore, when performing energetic analysis, whenever we are dealing with an output/input energy ratio greater than one, we are dealing with an assessment referring to flows of energy *carriers*; the output is the gross flow of energy carriers and the input is the internal consumption of energy carriers (see Point 6). However, these two flows (input and output) of energy carriers can only be sustained in time because of a larger consumption of another flow of primary energy sources. When considering the overall conversion of primary energy sources into a net supply of energy carriers, the result is *always* an *energetic loss*.

The second misunderstanding concerns the economic analysis of the conversion of energy sources into energy carriers. Energy is unlike any other good or service exchanged in the economic process. Even if we find different prices for different energy forms, it is important to always remember that primary energy sources cannot be produced, no matter how much capital and know-how are invested in this task.

Point 2: Distinguishing between primary energy sources, energy carriers and end-uses in the analysis of societal metabolism

The distinction between energy sources, energy carriers and end-uses within the network of energy conversions taking place in society is illustrated in Figure 3.3. As shown in the figure, the energy sector is the specialized compartment of society in charge of the conversion of primary energy sources into energy carriers. The energy carriers are then required by the various compartments of societies (primary sector, industrial sector, service sector, household sector and the energy sector itself) to perform useful activities: end-uses. Examples of primary energy sources include underground fossil energy reserves (coal, gas and oil), wind, moving water, the sun and the process of biomass-generating photosynthesis. Energy carriers include oil in a furnace, gasoline in a pump, electricity and hydrogen in the tank of a car. Only the energy sector can use primary energy sources. The energy carriers produced by the energy sector are employed by society to perform various functions. In fact, all sectors of society convert energy carriers into end-uses in relation to their specific tasks. Indeed, end-uses imply a flow of applied power to perform a useful task, such as moving goods, melting iron or regulating the temperature of a room. A more detailed discussion of these concepts is given in Chapter 5.

When converting a primary energy source into an energy carrier, there is an unavoidable loss of energy. As a consequence, within the set of conversions associated with the energy metabolism of society, 1MJ of an energy carrier has a higher quality (i.e. is more valuable) than 1MJ of the primary energy source

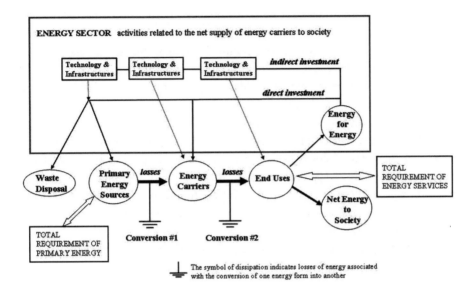

Figure 3.3 *Different energy forms in the energy metabolism of society*

used to produce it. For example, international energy statistics state that 1MJ of electricity (an energy carrier) obtained through a hydroelectric plant is equivalent to 3MJ of tonnes of oil equivalent (TOE, a primary energy source). Put another way, international energy statistics adopt a specific form of energy – TOE, a primary energy source – as the reference point for accounting. In fact, in order to compare different forms of energy it is necessary to establish, first of all, a standard energy form of reference, and then a criterion of equivalence (see also Chapter 5). For this reason, and as explained in the next chapter, one must be careful when calculating output/input energy ratios of a given process, especially when dealing with the production of agro-biofuels. Very often, assessments of the energetics of agro-biofuel systems are based on the summing of apples and oranges; primary energy sources are wrongly equated with a variety of energy carriers of different qualities.

Implications of the first two points

The only thing that human ingenuity and technology can do in order to make available to society a net supply of energy carriers is to *transform* a supply of primary energy. For example, to deliver the required flow of gasoline (an energy carrier) to motorists for transportation (the end-use), the energy sector requires an adequate amount of accessible crude oil below ground (the primary energy source) to start with. To deliver the required amount of electricity (an energy carrier) to households for air conditioning (the end-use), the energy sector requires an adequate amount of falling water (the primary energy source if this electricity is to be generated with hydroelectric power).

Therefore, in order to increase the net supply of energy carriers in society, the energy sector must have access to a larger supply of primary energy sources. Alternatively, by improving the efficiency of the transformation process, a society can obtain the same set of end-uses while requiring fewer energy carriers. Technology by itself cannot create energy. The implications of this observation are particularly relevant to the issue of sustainability, because it is simply not true that if the price of oil continues to rise, or oil reserves are seriously depleted, human ingenuity will necessarily provide a substitute energy source capable of prevailing over any type of biophysical shortage. Indeed, the energy security of a society depends on its ability to match two relevant flows of energy:

1. the flow of energy *required by the whole society*, which is determined by its socio-economic identity: its population structure, the material standard of living, and the diversity of activities performed inside and outside the economic sectors; and
2. the flow of energy *supplied by the energy sector* of that society, which is determined by its biophysical and technological identity: the mix of accessible primary energy sources, available technology and know-how, and the mix of specific energy carriers required by society for its various end-uses.

Human ingenuity and prices can prevent collisions between the socio-economic system and its biophysical constraints by adjusting mismatches between these two flows. These two flows can be matched by adjusting the identities of either of the two systems involved: the whole society and/or the energy sector (see also Box 3.1). The material standard of living in the society (the flow of energy required by society) may have to be adjusted because of biophysical constraints; in this case, the identity of the energy sector affects that of the entire society, and society has to adapt to the constraint. Technical innovations (human ingenuity in using available natural resources) might ease some of the existing restrictions in the energy sector imposed by biophysical constraints; in this case, the identity of the society affects that of the energy sector. Note that the second solution requires the availability of an adequate energy source supporting the change. This latter solution is exemplified by the Industrial Revolution, which was enabled by the development of appropriate technologies to exploit newly discovered fossil energy stocks (a primary energy source) (see Chapter 2).

> **Box 3.1** *The chicken–egg relationship between the identity of society and its energy sector*
>
> Using technical jargon, we can say that the relationship between the identity of the socio-economic system and that of the energy sector is defined over an *impredicative loop*. We deal with an impredicative loop 'when a property is possessed by an object whose definition depends on that property' (Kleene, 1952). In this case, the term means that the identity of the whole (society) affects the identity of the parts (energy sector), and vice versa, in a pattern of causality recalling the chicken–egg riddle. This circular causality challenges the conventional assumptions of reductionism typical of economic analyses (Giampietro, 2003; Giampietro et al, 2006).
>
> In spite of this difficulty, this concept can be handled by addressing in semantic terms the expected characteristics of the class, which can be described using meta-models of the expected patterns of metabolism in society and their interactions with their context. This is discussed in more detail in Chapter 4 (and particularly Figure 4.7).

According to Point 1 above, the energy sector, like the other sectors of society, consumes energy carriers in the course of guaranteeing energy security; it transforms energy, but does not produce it. Therefore, investing in the energy sector in an effort to make more energy available to the rest of society entails consuming more energy in the energy sector itself. Depending on the availability of primary energy sources and the overall loss of the relative conversion, this may reduce the net flow of energy carriers available for consumption by the other sectors of society. Therefore, in case of a shortage of high-quality primary energy sources, society should carefully decide *how* to invest its available flow of energy carriers, because it is not necessarily true that investing more in technology and energy consumption in the energy sector will generate a larger net supply of energy carriers to society. When the primary energy sources to exploit are no longer abundant and/or of good quality, investing in the energy sector does not necessarily pay off. For example, since the year 2000, oil companies working in the US have doubled the number of wells drilled per year, with a glaring lack of results (COL, 2008). Doubling the investments did not generate a net increase in energy carriers because there are no new fields to exploit. Note that this phenomenon also applies to other sectors dealing with finite natural resources, such as forestry or fishing. For example, after an over-exploitation of the fish stocks – as happened off the Atlantic coast of North America in the 1990s – it is the size of fish stock that matters, and not the size of the fishing fleet. As observed by Daly (1994), when exploiting a finite amount of natural resources, additional capitalization does not necessarily result in more output.

It is well known that the usefulness of any investment depends on its return. When dealing with the energy sector, the overall return on investment depends on the availability of primary energy sources and on the specific set of activities performed within the energy sector. The overall return on investment will reflect the overall performance – the net increase in energy carriers supplied to society – achieved because of the investment. However, if the increased energy consumption in the energy sector is itself in direct competition with the energy consumption of other sectors of society, additional investments in low-quality energy sources may very well result in a decrease in the actual material standard of living. For this reason, it is of utmost importance to explore the existing relations between the material standard of living reached in modern societies, and the quality of primary energy sources exploited by developed societies. This relationship has been explored in general terms in Chapter 2, and will be explored in theoretical terms and with real data in Chapter 4.

Point 3: Quantification of the semantic concept 'energy'

Any comparison of two different energy forms will depend on the reasoning chosen by the analyst. For example, it is not true that 1MJ is always equal to 1MJ. To use technical jargon, quantifications of different forms of energy are never substantive. This means that any number found in energy analysis reflects the choice of a given scale and a given converter used to represent change. The quantification of energy requires choices of accounting reflecting the semantic perception of energy transformation. This entails that we need to use different methods of accounting depending on the type of change we want to describe. For example, the quantitative method used to assess the energy input of a virus (based on the lexicon of biochemistry, in which fuel is represented by energy-rich molecules) is not reducible to the quantitative method useful to assess the energy input for a society (TOE, based on the lexicon of technical energy conversions). Therefore, whenever we decide to use qualitative indices referring to comparable energy forms, such as the overall output/input energy ratio of a biofuel system, we are analysing the overall effect of the interaction of different energy forms. For example, primary energy sources such as solar energy and rain are transformed into energy carriers such as crops, via the consumption of fossil energy carriers used for making the fertilizers. Finally, there are other energetic inputs very difficult to quantify, such as the input of human labour. Because of this heterogeneity of the energy forms interacting in the process, the final number will depend on a series of pre-analytical choices made by the analyst when defining the underlying grammar (see Box 3.2).

Thus, in order to integrate the various forms of energy in a common system of accounting the analyst has to define the following:

- An energy graph (a lexicon). What is the finite set of energy conversions considered in the chosen analysis? For example, when studying the energetic balance of agro-biofuels, we do not consider gravitational energy, the energy metabolized by a virus or the energy of percolating water in the ground.
- The viability domain for the graph. What are the acceptable conditions in which the chosen set of conversions can operate? For example, when studying the energy balance of agro-biofuels we have to know the range of admissible values of certain conversions – that is, the technical coefficients – such as the pace at which solar energy is fixed in biomass, the speed of fermentation and the maximum power deliverable by machines.
- The production rules. What are the best technical coefficients and inferential systems (models) to use to represent changes of flows when moving across numerical assessments over the graph?

Box 3.2 *The concept of grammar*

The conceptual tool of *random grammars* as a key feature of the possibility of autopoiesis (see glossary) in complex adaptive systems has been proposed within the field of complex systems theory by Kauffman (1993). In general, we can say that the concept of grammar can be associated with the pre-analytical declaration of a set of expected relations between the following:

- A set of semantic categories (the meaning attached to the value of the variables by those who will use the numbers). The meaning of these semantic categories has to be understood by all those using the grammar.
- A corresponding set of formal categories; proxy variables used in formal representation, to which we can assign quantitative values.

Using the language of software development, we can say that a grammar requires the definition of two things:

1. A lexicon: the finite set of categories that will be used to represent the possible energy transformations (the identities assigned to fund and flow elements). These categories can be divided into tokens and names. Tokens derive values directly from measurement (data inputs), whereas names depend on both the values of the tokens and the set of production rules established in the grammar.
2. The set of production rules determining the value assigned to names when data are entered in the category of tokens. The application of the concept of grammar to energy analysis is discussed more in detail in Chapter 5.

In conclusion, the quantification of the semantic concept of energy requires the preliminary definition of a grammar, based on a pre-analytical choice of a lexicon and production rules to represent the integrated set of energy conversions.

Point 4: Energy flows of endosomatic and exosomatic metabolism cannot be summed

We briefly introduced the concepts of endosomatic and exosomatic metabolism in Chapter 2. Endosomatic metabolism refers to the physiological conversion of food within the human body into human activity, while exosomatic metabolism refers to the technical conversions of different types of energy inputs into useful power. Such conversions take place outside the human body.

The criteria used for the definition of various energy forms characterizing exosomatic and endosomatic metabolisms refer to two distinct typologies of metabolic pattern; that is to say, two distinct autocatalytic loops (see Appendix 1). This implies that joules accounted within these two metabolisms cannot be summed; they belong to different lexicons, and therefore they require the adoption of different grammars. The case is that, in order to sum two different energy forms, it is necessary to use a criterion of equivalence (more on this point in Chapter 5) within a given grammar.

Endosomatic metabolism refers to the physiological processes taking place within the human body. These processes are well defined; nutrients from food (the primary energy source) are catabolized (broken into smaller components) and transformed into molecules (energy carriers) fuelling biochemical reactions that generate activity (the end-use) in the human body. The human activity thus generated is (partly) used to make food accessible for consumption, bringing new nutrients into the cycle and thus creating an autocatalytic loop.

Exosomatic metabolism is based on a chain of energy conversions taking place outside the human body. Also, in this case, the identity of the energy carriers (such as gasoline, steam, electricity) reflects the identity of the exosomatic converters of energy (cars, boilers, refrigerators) using these carriers. Therefore, the exosomatic metabolism refers to another type of autocatalytic loop; the activity of exosomatic devices generates accessible energy carriers for exosomatic devices. Then, the conversion of exosomatic energy by exosomatic devices generates the applied power required to maintain and reproduce the activity of exosomatic devices and to make available the required fuels.

In Chapter 5 we provide a more detailed discussion of the epistemological challenges of energy analysis associated with this point. Additional concepts about dissipative networks and autocatalysis are discussed in Appendix 1.

Here, we only want to emphasize that the quantitative representation of two distinct autocatalytic loops of energy forms requires the use of two distinct systems of accounting:

1. food → nutrients → physiological conversions → useful work → more nutrients; and
2. primary energy sources → exosomatic energy carriers → specific technical converters → useful work → more exosomatic energy carriers.

In conclusion, food energy and fossil energy, even if measured using the same unit (MJ or kcal), cannot and should not be summed when assessing the desirability and feasibility of metabolic patterns. They belong to different networks of autocatalytic transformations; that is, they refer to different narratives about self-organization and evolution.

Point 5: Energy and power are distinct concepts

'Energy' indicates (in general terms) a given amount of energy, with no reference to time, and it is measured in joules. In energy analysis it can refer to a given amount of a primary energy source, or to a given amount of an energy carrier. 'Power', on the other hand, indicates the given pace of an energy conversion in time – the rate at which useful work is performed – and its unit of measure is watts (joules per second). Power is intrinsically linked to the characteristics of the energy converter (generating power) and the useful work performed with such a power.

Unfortunately, in many applications of energy analysis the distinction between energy and power often becomes blurred because of the way data on energy flows are presented. In fact, when dealing with the analysis of the metabolism of human beings (endosomatic metabolism) or socio-economic systems (exosomatic metabolism), one gets easily confused because data on energy inputs are usually expressed on a time basis. For example, when dealing with food energy we'll typically find the consumption of food energy per year; when dealing with commercial energy we'll find the consumption of TOE per year. These data are expressed as rates of energy over time, for example GJ/year, and therefore have the same dimension (energy per unit of time) as power levels. Such confusion should be carefully avoided.

Therefore, it is critical that a meaningful energy analysis has both types of information: the consumption of energy input *and* the power level at which the energy conversion is expected to take place. For example, one gallon (3.87 litres) of gasoline – a given amount of energy – can be used to fuel a small motorbike for more than 200km, a minivan for about 50km, and a Formula 1 car for no more than a few kilometres. If we do not specify the purpose of the vehicle – which will be linked to different requirements of power level – at the outset of our analysis, it is impossible to perform any meaningful energetic assessment. Indeed, it is essential to first define the power level required by the given task.

It is important to keep in mind that missing the crucial difference between energy and power can lead to macroscopic errors in judgement. This is the capital sin behind the biofuel delusion. How to decide whether one needs information about power level or about energy input?

One has to focus on power levels when checking the capacity of a given converter to convert an energy input into work at a given rate (the internal constraint). For example, the power level determines the capability of a given engine to generate enough power to perform a given task. In more general terms, an assessment of power level is required to check an internal constraint of the system. That is, after ascertaining or assuming that the required energy input is available, a certain level of power is required to define whether or not a given system is able to perform a given function over a given period of time (for example, a truck required to transport a given load uphill, or a person capable of running over a given distance in a given time, or the ability to pull a given weight). An assessment of power level refers to the expected characteristics of the converter, or – using the terminology proposed by Georgescu-Roegen (1971) – an assessment of power level is required to define the expected characteristics of fund elements.

One has to focus on energy input when checking the availability of a required flow of energy resulting from the interaction of the metabolic system with its context (an external constraint). For example, the availability of gas stations along the road from New York to San Francisco will determine whether a truck carrying 20 tonnes of material will be able to travel from one city to the other in less than a week. That is, after assuming the capability of delivering enough power for the task to be performed, it is necessary to check the existence of favourable boundary conditions – in this case, the continuous availability of gas along the selected road – to define whether or not a given system is able to perform a given function over the given period of time. In respect to the analytical approach proposed by Georgescu-Roegen, an assessment focused on energy input is required to define the expected characteristics of flow elements (a compatibility with boundary conditions).

In conclusion, information about power levels is crucial to detect the viability of a metabolic process in relation to internal constraints (the characteristics of the metabolic system → funds), whereas information about the amount of energy flowing is crucial to detect the viability of a metabolic process in relation to external constraints (the characteristics of the interaction of the metabolic system with its context → flows). This generalization can be used to look for systemic characteristics: a converter (fund characteristic) capable of expressing a high power level tends to require a concentrated flow of energy input (flow characteristic). For example, it is difficult to imagine a jumbo airliner powered by straw; the required storage capacity for a trans-oceanic flight would be huge.

Point 6: Distinction between gross and net energy supply of energy carriers

Agro-biofuel production requires the availability of sun, water, soil and biota (primary energy sources), as well as liquid fuels (energy carriers). Therefore, the total production of energy carriers in the agro-biofuel system – the amount of biofuel produced in the process – should be considered the *gross energy supply*. In order to calculate the *net energy supply* to society, one has to subtract from the gross supply the internal consumption of energy carriers within the process itself.

It should be noted that the original discussion over gross and net energy analysis in the 1970s has generated some confusion in relation to this issue (Georgescu-Roegen, 1979). In fact, in the original formulation the distinction between gross and net energy was associated with the energy loss referring to the conversion of primary energy sources (gross) into energy carriers (net). This being the case, the output/input ratio must always be smaller than 1.

As explained in Chapter 5, this confusion can be avoided if the calculation of the overall energetic losses addresses the existence of two steps in the conversion: first, primary energy sources are converted into energy carriers; and second, the gross flow of energy carriers is reduced to a smaller net flow of energy carriers, because of the internal loop of energy carriers used to generate energy carriers.

Defining internal and external constraints on potential energy sources

The above six points convey the general idea that when dealing with the complex set of energy conversions taking place within a socio-economic system, one cannot expect to perform a meaningful and useful energy analysis by adopting a simplistic representation based on a simple ratio of energy outputs and energy inputs. Even more naive is the idea that by looking only at the calorific value of salmon oil or human fat, one could define them as potential alternative energy sources. An integrated analysis of the feasibility and desirability of alternative energy sources should address the whole set of energy conversions that make up the metabolism of the entire society in question.

Clearly, such an integrated analysis is no simple task. It can only be performed by adopting a multi-scale analysis of the dynamic budget of a society. In fact, the stabilization of the metabolic pattern of a society implies the existence of different energy flows transformed at different paces in different compartments, operating at different hierarchical levels. Such an analysis should consider both what is required by society, according to the set of functions expected to be performed by the various compartments making up the society; and what can be supplied by the energy sector (which is only a small part of society), according to its specific characteristics: power level (internal constraints) and the availability of energy inputs (external constraints).

In practice, in order to judge the desirability and feasibility of a potential fuel for a socio-economic process, we have to perform two distinct checks that examine these two different types of constraints: internal and external biophysical constraints.

The Dynamic Energy Budget of Metabolic Systems and the Energy Return on the Investment (EROI)

The dynamic energy budget of metabolic systems

The very survival of metabolic systems entails their ability, first, to gather energy sources from their context; and, second, to process the flow of energy inputs in order to maintain and reproduce themselves (Appendix 1). For example, human societies can maintain their structures and functions only by continuously converting a flow of energy inputs taken from the environment (food, wood, fossil energy) into useful work (White, 1943; 1959; Cottrell, 1955; Odum, 1971; 1996; Pimentel and Pimentel, 1979). In primitive societies, the cycle of energy dissipation is mainly related to a loop of endosomatic energy, whereas industrialized societies are mainly stabilized by loops of exosomatic energy.

As illustrated in Figure 3.4, a metabolic system must use the energy input it consumes for carrying out simultaneously two different tasks (Conrad, 1983; Ulanowicz, 1986):

1 gathering the energy input consumed by the system for its current metabolism, using only a fraction of the energy consumed in the hypercyclic part; and
2 performing additional activities needed for its survival in the long term, such as reproduction, self-repair and development of adaptability in the purely dissipative part.

Hence, the flow of energy gathered from the environment – that is, the energy input available to a society – cannot be destined entirely for discretional (dissipative) activities. A certain fraction of this input – that used by the hypercyclic part – *must* be spent in the very process required for the gathering and processing of this energy input. This fraction is the overhead cost, represented by the metabolism of the energy sector. This overhead is the rationale behind the concept of net energy analysis.

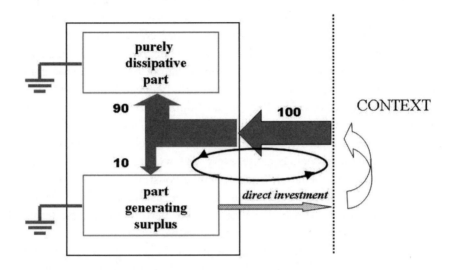

Figure 3.4 *Expected pattern for a stable metabolism: a dissipative part sustained by a hypercycle*

The idea of dynamic energy budget can be related to basic economic analyses that are probably more familiar to the reader. For example, let's imagine that the scheme of Figure 3.4 refers to a family of four spending a certain amount of money: say, 100 units per month. We further presume that within this household, only one person is working. This person is the part of the family generating the surplus: the hypercyclic part. The other three people represent the purely dissipative part. To check the viability of the family budget we have to examine two characteristics of the system:

1. the performance of the part generating the surplus – the person working – which depends on two factors: the salary (100 units/month) and the money spent in relation to work (travel to the office, lunch at the office: 10 units/month); and
2. the overall consumption of the whole family (100 units/month), which is determined by both parts of the family: the surplus generator (our worker when at work, consuming 10 units/month) and the purely dissipative part (the other three family members plus the worker when at home, consuming 90 units/month).

The dynamic economic budget of this household immediately points at the existence of internal and external constraints. That is, the dynamic budget entails a forced relationship between the relative size of the two compartments and their specific rate of consumption. For example, if the salary is low, it will be impossible for our worker to support a larger family. Depending on the salary, the household can decide either to reduce consumption, to have the worker work extra time, or to send additional family members into paid work activities. The main message is that the family cannot freely decide on combinations of family size, expenditure and the ratio between hours spent in paid work and non-paid activities. Rather, this combination must be compatible with a set of internal constraints (there are limits to the hours a family can work) and external constraints (salary levels and job opportunities). This message may seem trivial, because every day we are confronted with the constraints related to the family budget. Similar constraints apply to the energy budget of socio-economic systems, as illustrated in this book, but strangely enough this trivial and obvious analysis is never performed in these terms to examine the viability and desirability of alternative energy sources.

Let's now move to a biophysical reading (actually a thermodynamic reading) of this same dynamic budget. As soon as one describes the functioning of ecosystems or socio-economic systems in this way, one discovers that this very pattern of interaction is found for the entire class of metabolic systems. In fact, both ecosystems and socio-economic systems – as well as individual organisms – have parts that generate a surplus of energy carriers for consumption by the whole system, and purely dissipative parts.

In ecological jargon, the part generating the surplus is called the 'hypercyclic part' (Ulanowicz, 1986). The term 'hypercycle' indicates a positive autocatalytic loop generating a net surplus of energy that is needed to drive and keep the system out of thermodynamic equilibrium. In our example of the family budget, the hypercycle is represented by the chain of conversions (indicated by a closed loop in the figure) formed by the person getting a salary and spending only 10 economic units per month. This money makes it possible to perform a working activity that produces 100 economic units, which enter the household. This hypercycle therefore entails that for each economic unit invested in supporting the working person, the system gets a return of 10 economic units. When representing the hypercycle in energetic terms, we can say that a metabolic system must be able to express a set of activities aimed at gathering an energy input from the context (remember, the energy input is not produced, it must already be available somewhere in the context). The cost of these activities, providing energy carriers to the rest of the metabolic system, must be much lower than their return.

The hypercyclic part is required to keep the dissipative system in a situation of non-equilibrium (Eigen, 1971). That is, a positive autocatalytic loop is required to preserve a set of metabolic structures and functions that can only be stabilized by a continuous flow of inputs taken from the context, and a continuous flow of wastes dumped into the context. Note, however, that the hypercyclic part is not just a source of positive return for the whole system, but also a major liability. Network analysis shows that a hypercycle in a network may very well be a source of trouble, like a microphone feeding back to the amplifier to which it is connected (Ulanowicz, 1986). This is true for all metabolic systems. Any metabolic system that keeps growing in size just by amplifying the same basic process of dissipation will become unsustainable. Consider the story of Zhu Yuan-Chang's chessboard: if you put one grain of rice on the first square, two on the second, four on the third, and keep doubling the number on each square, an astronomical number of grains will be required before the 64th square is reached. This metaphor says it all. Ulanowicz (1986) nicely makes this point: when hypercycles operate without a coupled process of control (and damping), they do not survive for long; they 'just blow up'.

So, when looking for a general metabolic pattern, we can say that the stability of metabolic systems, in the long run, depends on the ability to couple wisely a hypercyclic part to a purely dissipative part. Therefore, within this metabolic complex, the purely dissipative part is not just a burden on the hypercycle, but rather is a crucial element, essential for the long-term sustainability of the system. When dealing with living systems, the dissipative part is in charge of the adaptability of the metabolic system in the long run: reproduction, education and social relations. Adaptability and flexibility are crucial qualities for the sustainability of adaptive systems (Conrad, 1983; Ulanowicz, 1986; Kampis, 1991; Holling, 1995). They both depend on the ability to preserve diversity in the long run.

As a matter of fact, when studying the metabolic pattern of post-industrial developed societies, the purely dissipative part is represented by us: humans! The hypercycle is generated by exosomatic devices using fossil energy for their self-production in order to gather more energy for making more exosomatic devices. Within this framework, the performance of the exosomatic hypercycle is related to its ability to generate enough surpluses of both endosomatic and exosomatic energy carriers to sustain a large number of human beings at a high standard of living (this is at the basis of the concept of bio-economic pressure discussed in Chapter 4).

Net energy analysis and primary energy sources

Within this general analytical framework, we can use the concept of net energy analysis to characterize the quality of an energy input. An energy input has a high quality when it implies a small overhead for its own gathering and processing. We will describe how to quantify this concept in Chapter 5. Going back to the example of the economic budget of our hypothetical household, a high-quality job would translate into a high salary and low expenses for the working individual.

Georgescu-Roegen (1975, p354), in his discussion of the quality of energy sources, proposes a distinction between *available* and *accessible* energy: 'There certainly are oil-shales from which we could extract one tonne of oil only by using more than one tonne of oil. The oil in such a shale would still represent available, but not accessible, energy.' Other examples of available but inaccessible energy are the energy contained in human fat from liposuction, or the energy obtained from burning a $100 note; the generation of the net supply would require much more energy than is obtained.

Thus, the concept of energy availability refers to the energy content of a given amount of energy carrier. It reflects an assessment that deals only with the characteristics of the energy carrier. The concept of energy accessibility refers to the net energy gain that can be obtained when relying on a given amount of an energy carrier obtained by exploiting an energy source. This latter assessment deals with the overall pattern of generation and use of energy carriers in the interaction of the metabolic system with its context. The concept of accessibility requires a contextualization of the characteristics of the energy source with the characteristics of the user in the exploitation process.

Another example proposed by Georgescu-Roegen to convey this concept is that of pearls dispersed in the sea. These pearls may represent, in theory, a huge economic value. However, the practical value of these pearls depends on the cost of their extraction and their gathering in a reliable supply. Similarly, there may be thousands of dollars in coins in the sofas and car seats of US families, yet no businessman would consider starting up an economic activity based on the extraction of this potential resource.

These examples automatically bring us to the many assessments in the literature of the huge potential of biomass energy as an alternative to oil. Like the pearls dispersed in the ocean, there is a huge amount of biomass dispersed over this planet. The problem is that the extraction of this biomass and its conversion into suitable energy carriers requires the investment of human labour, capital and energy, something that is usually overlooked in these assessments, as are the environmental costs related to the process. It is not the total amount of pearls, biomass or coins that matters, but the ability to generate a net supply of the required resource at the required pace, with a high return on the investment.

The energy return on the investment (EROI)

The application of the concept of net energy analysis to the characterization of energy sources leads to the use of the concept of energy return on the investment. We will start out again with an example of return on investment drawn from classic economics, before moving to a biophysical application using energy. Our experience is that elementary concepts are easily understood by anybody when formulated in economic terms, but remain obscure when formulated directly in energetic terms.

Let's imagine that we are asked to evaluate an economic investment that will yield a gain of $10,000. It is obvious to anybody with some experience in financial affairs that this piece of information alone is not enough to perform a sound evaluation of the proposed investment. This may be a good or a bad deal, depending on the size of the required investment and the timespan involved. For example, it is a good deal if it requires a $20,000 investment and pays back in a year, but it is a pretty bad deal if it requires a $1,000,000 investment and pays back in 30 years. Thus, intuitively, we look for three crucial factors when evaluating economic investments: the output/input ratio (the amount of money gained in relation to the initial investment required); the amount of money that we have to invest; and the time taken to pay out.

While the economic concept of the return on investment is extremely clear to anybody, when it comes to the evaluation of investments referring to energy transformations (such as the potentiality of agro-biofuels as alternatives to oil), the concept of EROI is often ignored by those performing the energy analysis. For example, the comprehensive review of controversial assessments found in the literature on the performance of biofuels by Farrell et al (2006) totally ignored the issue of EROI (Cleveland et al, 2006; Hagens et al, 2006; Kaufmann, 2006; Patzek, 2006).

On the other hand, it should be noted that when applied to energy analysis, the EROI index is difficult to define in exact terms. The concept of EROI is a semantic one, which refers to a meta-model of analysis inspired by the concept of economic return on investment. In financial analysis, this concept addresses at least two relevant issues beside the output/input ratio: the feasibility of the investment (the investor must be able to handle the required investment); and the payback time of the investment (the investment is justified only if pays back within an acceptable period). For this reason, it is essential when applying this concept to energetic analysis to consider the time element; one has to address the crucial importance of critical power thresholds.

However, in general, applications of the EROI concept to energy analysis do not address explicitly the time dimension. The EROI is just defined as the ratio between the energy delivered to society by an energy system, and the quantity of energy used directly and indirectly in the delivery process over a given period of time. This index has been introduced and used in quantitative analysis by several scientists (Cleveland et al, 1984; 2000; Hall et al, 1986; Gever et al, 1991; Cleveland, 1992; Hagens and Mulder, 2008). When ignoring the issue of power levels, the implementation of this concept in quantitative analysis remains problematic. In fact, in order to perform a quantitative assessment of EROI we must be able to characterize the implications of, first, the distinction between primary energy sources, energy carriers and end-uses (see Point 2 above) and, second, the distinction between the gross and net energy supply of energy carriers to society (Point 6 above). That is, the losses determining an overall EROI can belong to different categories:

- losses due to the conversion of a form of energy into a different form of energy; for example, MJ of primary energy source (solar energy) converted into MJ of energy carriers (biomass); and
- losses due to an overhead referring to the same energy form; for example, MJ of energy carriers required to produce MJ of energy carriers (an output/input energy ratio of a fuel produced and used in the process of production).

An overview of the set of relations over different energy forms behind the definition of EROI is given in Figure 3.3. The total energy consumption of a society depends on its aggregate requirement of useful work or final energy services, the box represented on the right of the graph. In turn, this aggregate requirement of useful work has to be split, according to the overhead associated with the EROI, between the following:

- Energy-for-energy: used for the internal investment within the energy sector and needed to deliver the required energy carriers. This refers to the energy consumption (or metabolism) of the energy sector, a part of the whole society, in the internal loop energy-for-energy.
- Net energy to society: used for the production and consumption of non-energy goods and services. This refers to the energy consumption (or metabolism) of the rest of society. This can be assessed by looking at the consumption of energy carriers in the various sectors, but to compare it with the total consumption of society it has to be expressed in terms of primary energy source equivalent (for example, J of TOE).

From this figure, we can appreciate the crucial role that EROI plays in determining the relationship between the requirement of end-uses determining the material standard of living associated with a given metabolic pattern, and the overall requirement of primary energy sources determining the pressure on the environment and the requirement of natural resources associated with a given metabolic pattern.

In particular, Figure 3.3 illustrates two points determining a systemic conceptual problem faced by those attempting to operationalize the concept of EROI using a single output/input ratio (Giampietro and Mayumi, 2004; Giampietro, 2008). First, when expressing the EROI as a simple output/input ratio, we are forced to ignore the time dimension. This complicates the question of how to deal with the various overheads associated with the internal loop of energy-for-energy that operates across hierarchical levels. In fact, as discussed in Chapter 5, depending on the power level required in the network the internal loop can change in a non-linear way. An output/input energy ratio is out of scale. A spider can get a very high energetic return by fabricating and operating a web, but this does not imply that we can power the US by adopting the primary energy source (flies) so useful to spiders. Second, the analysis of this internal loop presents a major epistemological problem, because the quantification of the EROI can be done in non-equivalent ways depending on the boundaries chosen to define this internal loop.

Dealing with living systems, one can easily get caught in an analytical loop in which everything depends on everything else. For example, several options may be considered to calculate the required energy input for generating the supply of energy carriers:

- Option A: Include only the energy spent on the operation of the exosomatic devices used in the energy sector.
- Option B: Option A plus the energy used to fabricate and maintain the exosomatic devices used in the energy sector. This requires considering the turnover of the exosomatic devices used in the energy sector.
- Option C: Option A plus Option B, plus the energy used to reproduce the energy sector as a whole. This requires including the infrastructure and the demand of services of the energy sector.
- Option D: Option A plus Option B plus Option C, plus the energy used to reproduce the humans working in the energy sector. This requires including the leisure time of the workers and education and assistance to their families.

Expanding the list of inputs for inclusion leads to the well-known problem of the infinite recursive loop, to which we dedicate the next section.

The systemic problem encountered when trying to calculate the EROI

To illustrate the relevance of the issue of net energy analysis and EROI to the stability of the metabolic pattern of societies, we recall the example discussed earlier in Chapter 2 of the collapse of pre-industrial empires. According to the analysis of Tainter (1988), a large empire ceases to be sustainable when it runs out of small kingdoms to exploit. At that point, the flow of energy consumed by the rulers, which was previously obtained through the exploitation of the resources of conquered kingdoms, has to be provided by the surplus generated by the internal compartment represented by farmers. In other words, the stock-flow exploitation is over, and the empire has to switch to fund-flow exploitation (farmers' surplus). Consequently, the ratio of rulers to ruled (the farmers paying taxes) becomes too high, and the farmers are over-taxed (see Figure 2.6). Over-taxing means that the fraction of the energy input destined for purely dissipative activities (supporting the rulers) becomes too high relative to that destined for hypercyclic activities (investments in reproducing the farmers). In the long run, over-taxing leads to the disappearance of farmers (the fund element) and consequently the primary energy supply. This will eventually cause the collapse of the whole system (the empire).

We can further illustrate this problem by calculating the EROI of the energy sector of a hypothetical complex pre-industrial society operating only on endosomatic energy. Hence, this hypothetical society uses food as the only energy input, and to simplify matters further we assume the only food produced is cereals. The energy sector of this society thus coincides with the sector in charge of producing the cereal consumed during the year. A diagram representing this society is provided in Figure 3.5.

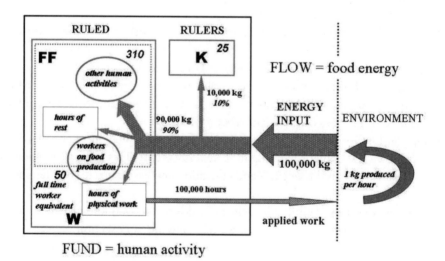

Figure 3.5 *The dynamic budget of endosomatic energy analysed in terms of funds and flows*

We further assume that there are 385 people in our society: 25 rulers (who do not produce any useful work in relation to energy inputs) and 360 ruled (who are involved in various activities, including the production of food). The average consumption of grain per capita in this society is 260kg per year, equivalent to 100,000kg of grain per year for the whole society. However, this consumption is unequally divided between the two compartments: the rulers consume about 400kg per capita per year, including the indirect consumption of cereals for animal products, whereas the ruled, on a strict vegetarian diet, consume only 250kg per capita per year.

Assuming a labour productivity (typical of pre-industrial agricultural societies) of about 1kg of grain per hour of labour, we estimate that the ruled compartment has to invest 100,000 working hours per year in food production, which is equal to the activity of 50 full-time workers, assuming a workload of 2000 hours per year.

We adopt the same analytical framework of pre-industrial societies illustrated earlier in Figure 2.6. That is, the society is divided in three sectors: the K sector (25 rulers), the W sector (50 workers generating the surplus) and the FF sector (farming families, comprising 350 people, who can be thought of as the overhead of the W sector). Given this breakdown, we can calculate the EROI of the energy sector of this society.

Table 3.1 *Different options for calculating the EROI in Figure 3.5*

	Whole society (level n)	Compartments (level n – 1)	
		Rulers	Ruled
Fund Human activity	385 people	25 people	360 people
Flow Food energy	100,000kg/year	10,000kg/year	90,000kg/year
Endosomatic metabolic rate	260kg pcpy	400kg pcpy	250kg pcpy

Conversions used for the calculation:

- food energy for adults: 10MJ/day = 2400kcal/day = 0.7kg grain/day = 250kg grain/year (1kg grain = 3400kcal);
- hours of work in food production: 100,000 per year (coming from a productivity of labour of 1kg of grain/hour);
- food energy allocation in workers: 60 per cent in basal metabolism, 40 per cent in extra energy for work;

Looking at Table 3.1, it is easy to assess the energetic output of the W sector:

output = food energy produced in an hour of work in food production = 14MJ = 3400kcal

Unfortunately, it is less easy to assess the energetic input to the W sector:

input: how much energy is consumed to generate an hour of work in food production?

64   *The Biofuel Delusion*

In fact there are four possible options for such an assessment:

A   0.5MJ (considering only the energy for work, but not the basal metabolism of worker): EROI = 28/1
B   1.25MJ (considering the whole metabolism of the worker): EROI = 9/1
C   12.6MJ (considering the whole metabolism of the ruled compartment): EROI = 1.1/1
D   14.0MJ (considering the whole metabolism of rulers and ruled) achieving dynamic equilibrium: EROI = 1/1.

In this example, in order to calculate the value of EROI, we have to calculate the food energy input required to generate each hour of grain-producing work, and the food energy output generated by each hour of work. Actually, we know the value of the output: 1kg of grain is equivalent to 3400kcal of food energy, or 14MJ of endosomatic energy. Then we can define the set of conversions between primary energy sources and energy carriers (yield per hectare can be related to the conversion of solar energy into food energy), and the conversion of energy carriers into end-uses (the consumption of food energy per hour of labour and the labour productivity of farming). However, after having specified the network of conversions over the autocatalytic loop, we have to decide how to define the overhead affecting the net supply of energy within this network. Table 3.1 illustrates the set of possible choices for assessing the energy embodied in an hour of work in this society.

Choice A
We define the energy consumed in the energy sector as the food energy required for the working hours in food production, excluding the basal metabolism of the worker. In analogy, in a developed country based on exosomatic energy this would be an assessment considering only the energy consumed by the machines operating in the energy sector (like farmers working in agriculture), without considering the energy spent on building and maintaining the machines. With this assumption, in our simple example, the energy spent in crop production is 0.50MJ/kg of crop produced, and therefore the EROI is 28/1.

Choice B
We define the energy consumed in the energy sector as the food energy required for the whole metabolism of the workers producing the food, including consumption when out of work. In a developed country based on exosomatic energy, this would be a life-cycle assessment of the machines used in the energy sector, which includes the energy spent in building and maintaining the machines. Given this assumption, the energy spent on crop production is 1.25MJ/kg of crop produced, and therefore the EROI is 9/1.

Choice C
We define the energy consumed in the energy sector as the food energy required for the metabolism of the whole compartment of the ruled. We therefore consider as overheads the energy used by the family (dependents) of the workers in reproduction (children) and maintenance (activities needed to stabilize the worker, such as fetching water, cooking, cleaning, etc.). In a developed country based on exosomatic energy, this would be an assessment including embodied goods and services produced in other economic sectors that make possible a correct operation of the energy sector. With this assumption, the embodied energy spent in crop production is 12.6MJ/kg of crop produced, and hence the EROI is 1.1/1.

Choice D
We define the energy consumed in the energy sector as the food energy required for the metabolism of the whole society. With this choice, we account as embodied energy in the work of the W sector the food consumed by both the FF and the K sector. In this case, the whole amount of energy consumed by society is considered as being used, directly and indirectly, for the operation of the energy sector. With this choice of accounting, we are assuming that the activity of control and direction provided by the rulers (the K sector providing the protection of soldiers) is necessary to keep the integrity of the social system (and hence the energy sector) and to drive the evolution of society when considering a larger timescale.

As illustrated by this simplified example, in order to escape an infinite recursive loop – the tautology associated with Choice D – the calculation of a given number used to characterize the EROI must ignore some of the overheads associated with at least one (or more) fund element defined at a certain level of analysis. It should be noted, however, that Choice D is a very defendable assumption; every form of energy consumed by society is indeed directly or indirectly consumed in useful activities necessary to the operation of the energy sector. This level of analysis provides relevant information about the system: the overall power level (metabolic rate) achieved in the dynamic budget.

We want to make clear that we have not set out this example to dispute the usefulness of the concept of EROI in energy analysis. On the contrary, we are exploring the nature of the problem in order to provide in Chapter 5 a method for the systemic implementation of this concept. The point we want to make here is that the concept of EROI, like the concept of energy, is a semantic one, which can be implemented in quantitative terms only within an integrated analysis of the complexity of the metabolic pattern of socio-economic systems. This preliminary integrated analysis is required to enable an informed choice about how to calculate the EROI in a useful way.

In more general terms, we can say that when dealing with metabolic systems it is impossible to rely on a simple number for assessing the quality of the energy sources. So if we characterize the EROI just in the form of an output/input ratio, we are back to throwing numbers over the fence. Moreover, this number might not yield particularly useful results. As illustrated in Chapters 4 and 5, a quantitative characterization of the metabolic pattern of a society requires a more elaborate representation across multiple levels. Therefore, the particular choice leading to a quantitative definition of the EROI of energy sources has to be tailored to the specific characteristics of the dynamic budget of society. When dealing with complex metabolic systems, one size does *not* fit all. We have, first of all, to choose a useful typology (such as the exosomatic dynamic budget of a developed country, or the endosomatic dynamic budget of a pre-industrial rural community) that reflects a specific problem of how to balance a dynamic budget between a given requirement of energy carriers and a given availability of energy sources. Only at that point does it become possible to look for a useful implementation of the concept of EROI.

Final remarks about the quality of fossil energy

In the community of energy analysts, there is absolute consensus that the major discontinuity in all major trends of human development (population, energy consumption per capita, technological progress) associated with the Industrial Revolution (see Figure 2.3) was generated by the extremely high quality of fossil energy as a primary energy source, compared with biomass (discussed in Chapter 5). The tremendous advantage of fossil energy over alternative energy sources is easy to explain: when considering the energetic costs of the production of energy carriers, oil does not have to be produced; it is already there. This means that to avoid another major discontinuity in existing trends of economic growth (this time in the wrong direction), it is crucial to replace fossil energy with an alternative energy source providing similar performance in terms of useful work delivered to the economy per unit of primary energy consumed. If not, be prepared for a significant readjustment in the current metabolic pattern of modern societies.

What is missing in the conventional analysis of EROI is the analysis of the implications of the power level of the metabolic pattern, within which the EROI is calculated. The same EROI can be achieved by investing 1MJ/hour and getting back 20MJ/hour, or by investing 1000MJ/hour and getting back 20,000MJ/hour. A large difference in scale of the pattern of energetic metabolism considered does imply a completely different set of external and internal constraints to be considered. This point will be discussed in Chapter 5. In developed economies operating at an elevated power level, the high EROI means that the conversion of oil into an adequate supply of energy carriers (such as gasoline) and their distribution absorbs only a negligible fraction of the total energy consumption of a society, the total hours of paid work of a society, and the total amount of colonized land of a society.

This small overhead associated with the energy sector makes it possible for a large fraction of the total energy consumption to go to covering the needs of society, with very little of it absorbed by the internal energy-for-energy loop. Moreover, due to the high spatial density of the supply of energy flows in oilfields and coal mines, the amount of land required to obtain a large supply of fossil energy carriers is negligible.

Chapter 4

Pattern of Societal Metabolism across Levels: A Crash Course in Bio-economics

Societal Metabolism and Bio-economics

According to the basic rationale of bio-economics, changes in the structure and function of socio-economic systems can be studied using the metaphor of societal metabolism. The metabolism of human society is a concept going back to the mid-19th century (Martinez-Alier, 1987; Fischer-Kowalsky, 1997). It has been applied in the fields of industrial ecology (Ayres and Simonis, 1994; Duchin, 1998) and to conceptualize matter and energy flow analysis (Adriaanse et al, 1997; Fischer-Kowalski, 1998; Matthews et al, 2000).

The concept of societal metabolism suggests that the various characteristics of the different sectors (or compartments) of a socio-economic system must relate to each other as if they were organs in a human body. Indeed, as illustrated in the rest of this chapter, it is possible to establish a mechanism of accounting within which the relative size and performance of the metabolism of energy and material flows of the various sectors – the organs – must be congruent with the overall size and metabolism of the socio-economic system as a whole. By performing an integrated analysis across levels (society, sectors, sub-sectors) and dimensions (energy, monetary and material flows) of these constraints, we can obtain a richer and more useful picture of the option space for adjustments in socio-economic systems than can possibly be provided by classic economic analyses.

We have developed a methodological approach – multi-scale integrated analysis of societal and ecosystem metabolism (MuSIASEM), originally presented as MSIASM – that can be used to implement the basic concepts of bio-economics into quantitative analysis (Giampietro, 1997b; 2000; 2001; 2008; Giampietro et al, 1997; 2001; 2006a; 2006b; Giampietro and Mayumi, 2000a; 2000b; Giampietro and Ramos-Martin, 2005; Ramos-Martin and Giampietro, 2005; Ramos-Martin et al, 2007). In particular, the MuSIASEM approach can be used to perform a congruence check across levels and dimensions of analysis over proposed changes in the characteristic of a given part (an alternative energy system) or the characteristics of the whole (a metabolic pattern with lower consumption of energy).

Two key concepts are central to our approach and to bio-economics in general. The first refers to a key characteristic of the metabolic pattern of developed societies: it requires an extremely high labour productivity in the productive sector. The second concept is a methodological one. It is about how to represent patterns of societal metabolism across levels using the fund-flow model proposed by Georgescu-Roegen.

The key role of labour productivity in modern societies

To illustrate the crucial importance of labour productivity in modern economies, we start the discussion with a mental experiment. Imagine that in order to reduce the unemployment in rural areas of a developed country, a politician suggested abandoning mechanized agriculture and returning to pre-industrial techniques of tilling and harvesting by hand. According to the politician, implementing this strategy would generate millions of job opportunities overnight.

Hopefully, the suggestion would be immediately dismissed by opponents as a stupid idea. It is well known that during the Industrial Revolution, the mechanization of agriculture made it possible to free a large fraction of the rural workforce. This, in turn, enabled the investment of human labour into other economic sectors able to generate monetary flows at a higher pace than in the agricultural sector. That is, if the workforce of society is entirely engaged in producing its own food – at a low level of productivity of labour – society will never be able to diversify the set of products and services produced and consumed; it will never become a developed economy. This explains why no developed country has more than 5 per cent of its workforce in agriculture, and why the richest countries have less than 2 per cent of their workforce in agriculture (Giampietro, 1997a) (Figure 4.1).

Figure 4.1 shows that economic development entails a systemic and dramatic reduction of the percentage of farmers (those generating endosomatic energy input for society) in the workforce. In the same way, as is argued later on, the mechanism generating internal biophysical constraints on the relative sizes of different economic sectors in a developed society, also entails a systemic and dramatic reduction of the fraction of the workforce employed in the energy sector (those generating exosomatic energy inputs for society). In order to understand this mechanism, it is necessary to focus on quantitative and qualitative socio-economic changes associated with economic development. In particular, we focus on the consequences of changes associated with economic development using the ratio between hours of work in the paid work sector and total hours of human activity.

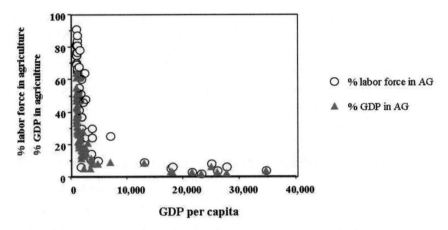

Figure 4.1 *Percentage of labour force and GDP in agriculture versus GDP per capita*

Values in 1991 US$.
Giampietro, 1997a.

Returning to the example of the economic family budget presented in Chapter 3 (see Figure 3.4), balancing the dynamic budget of that household implies a forced relation between:

- the ratio between the hours invested in paid work (generating a monetary income) and the total remaining hours spent by the household in activities implying expenditure;
- the average level of consumption of the family; and
- the hourly wage of the money-earning household part.

With the numbers provided earlier for this example (Figure 3.4), we can calculate:

- total household activity = 4 people × 24 hours/day × 30 days/month = 2880 hours;
- expenditures/hour = 100 economic units/2880 = 0.034 unit/hour;
- hours in paid work = 1 worker × 150 hours/month = 150 hours; and
- hourly wage = 0.66 unit/hour.

We can now formulate the dynamic budget as follows:

total household activity × expenditures/hour = hours in paid work × hourly wage
(2880 hour/month) × (0.034 unit/hour) = (150 hours/month) × (0.66 unit/hour)

72 The Biofuel Delusion

On the left of the equation, we have the total requirement of money flow referring to the hierarchical level n, the whole household, while on the right we have the *supply* of money flow referring to the hierarchical level n−1, the 'specialized' compartment generating the supply.

Can we perform the same type of analysis if we move up the scale from the household to an entire socio-economic system? The answer is 'yes', and we show below how to do so.

Multi-level system of accounting based on hours of human activity

In order to calculate the ratio of hours in paid work to total human activity over a period of a year for an entire society, we must abandon the conventional accounting approach adopted in available socio-economic statistics. These statistics are based on a default unit; the 'individual citizen' is the hierarchical level of analysis. For example, the total size of society is always characterized by the number of individuals (population); employment is measured in terms of the number of people employed; and GDP per capita is measured as monetary flow per individual per year. The same applies to other indicators of development, such as the number of doctors per person, or the number of pupils per teacher.

It is the case that the per capita system of accounting is incompatible with a multi-level analysis of the economic dynamic budget of a country according to the general scheme of Figure 3.4. Indeed, per capita accounting tends to miss important differences between countries. For example, the very same assessment of 'per 1000 people' (equivalent to a per capita assessment) can imply quite different supplies of work hours per year into the economy depending on the demographic and social structure of society. This difference is illustrated in Figure 4.2 and Table 4.1.

As illustrated in Figure 4.2 and Table 4.1, in 1999, Italy supplied 680,000 hours of work to the economy per 1000 people, while China supplied more than double that: 1,650,000 hours of work per 1000 people. In China, 1 out of every 5 hours of human activity was allocated to paid work, while in Italy this was only 1 out of every 13 hours (Table 4.1).

In fact, more than 60 per cent of the Italian population is not economically active, including children, students and the retired elderly. The human activity associated with this part of the population is therefore not used in the production of goods and services, but is allocated to consumption. Furthermore, the active population – the 40 per cent of the population included in the workforce – works less than 20 per cent of its available time (yearly workload per person of 1700 hours).

A Crash Course in Bio-economics 73

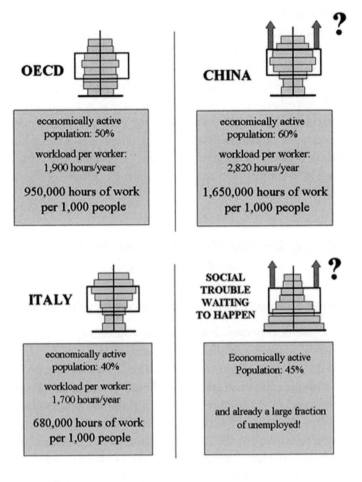

Figure 4.2 *Relation between demographic structure and supply of work hours at the level of society*

Adapted from Giampietro, 2009.

Table 4.1 *Allocation of human activity (HA in hours) to paid work (PW) and household (HH) sectors for Italy and China in 1999 (per 1000 people per year)*

	Italy	China
Total human activity (THA) in hours/year	8,760,000	8,760,000
Paid work sector (HA_{PW}) in hours/year	680,000	1,650,000
Household sector (HA_{HH}) in hours/year	8,080,000	7,110,000
Ratio of paid work/total human activity (HA_{PW}/THA)	1/13	1/5

THA=1000 people × 8760 hours/year, and THA= HA_{PW} + HA_{HH}

This comparison shows the importance of adopting a multi-level system of accounting based on hours of human activity. Describing the size of societal compartments in hours makes it possible to define, at the same time, the size of the whole society (level n), the size of a sector (level n−1), the size of subparts of sectors (level n−i), the size of households and even the profile of time allocation of individuals over the day. This is especially important if the analysis across levels is also used to characterize the various fund elements (such as humans) in terms of monetary flows or energy flows per hour of human activity. In contrast, a classic economic per capita assessment reflects only average values for a whole country and can miss important differences between internal parts of the system.

Reading Societal Metabolism across Hierarchical Levels of Analysis

Matching requirement with supply: monetary flows in society

As with the dynamic economic budget of the household, it is possible to study the feasibility of a given pattern of the metabolism of society in terms of a dynamic budget of monetary or energy flows per hour of human activity. We start again with an economic assessment based on monetary flows, which is usually easier to grasp. A biophysical reading of the dynamic budget of society, based on energy flows, is provided further on in this chapter.

As illustrated earlier for the household budget, we can write a relation of congruence between what is consumed per hour of human activity at the level (n) of the whole society (GDP), and what is produced per hour of human activity by the paid work sector (level n−1, sectorial GDP) in terms of monetary flows (Figure 4.3). To express this relation in numerical terms, we present an example, illustrated in Figure 4.3, based on the dynamic budget of Spain in the year 1999.

We started out by multiplying the population of Spain in 1999 (39 million) by 8760 (the hours in a year) to obtain the total human activity (THA) in that year: 344Gh (gigahour: 10^9 hours). We then calculate the consumption of society; dividing the GDP (US$15,800 per capita per year) by the total human activity, we obtain the GDP per hour (US$/h1.8) (Figure 4.3). We then turn to the supply side. On the basis of socio-economic variables such as demographic structure, workload per year per worker, education level, unemployment level and age brackets for entering and leaving the workforce, we calculate the hours allocated to the paid work sector (HA_{PW} = 23Gh). This figure implies that 1 out of every 15 hours was allocated to paid work in 1999 (HA_{PW}/THA = 1/15). Dividing the total GDP of Spain by the

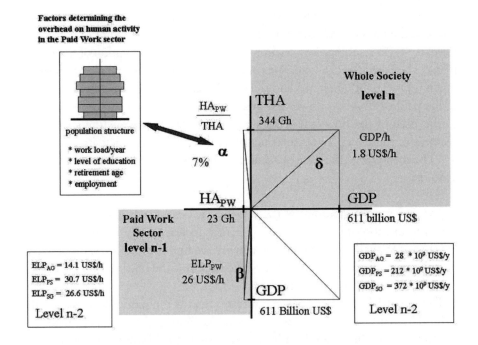

Figure 4.3 *Monetary flows across levels of analysis; Spain, 1999*

hours of work allocated to the paid work sector, we finally can define an economic labour productivity of the paid work sector as $GDP/HA_{PW} = ELP_{PW} = US\$26/h$.

All this information can be organized in a set of relations of congruence as illustrated in Figure 4.3. The forced internal relation of congruence between the length of the segments and the size of the angles shows the existence of constraints on the feasibility of the dynamic equilibrium between what is spent by society (on the left of the equation below), and what is generated by the paid work sector (on the right of the equation below):

    total human activity × GDP/hour = hours in paid work × economic labour productivity
        (344Gh)   ($1.8/h)   (23Gh)   ($26/hour)

The system of representation of the dynamic budget of monetary flows per hour within a society adopted in Figure 4.3 is based on the fund-flow model suggested by Georgescu-Roegen and discussed in Chapter 2. In fact, the upper vertical axis and left horizontal axis refer to the *fund* human activity; the horizontal axis to the right and the lower vertical axis refer to monetary *flow*. The upper right quadrant refers to the pace of the flow (flow/fund) calculated at the hierarchical level of the whole society (level n), while the lower left quadrant refers to the pace of the flow (flow/fund) calculated for the sector of paid work (level n−1). Put another way, Figure 4.3 provides a representation of the forced relation between the characteristics of the metabolism of the whole economy of Spain (upper right quadrant) and the metabolism of one (very) relevant part of it, the paid work sector (lower left quadrant). Moving from one representation of the pace of the flow (the whole) to a lower level element (the specialized sector), we must account for a reduction in the fund human activity.

The meaning of the four quadrants of Figure 4.3 can thus be summarized as follows:

- The upper-right quadrant of the figure provides a fund-flow representation of the metabolism of the whole referring to level n. By dividing the total GDP of Spain by the total human activity, we find the ratio flow/fund, which is the pace of metabolism of the entire society – the black box – in relation to the monetary flow (US$1.8/h).
- The upper-left quadrant of the figure represents the reduction in size of the fund human activity when moving from level n to level n−1, a part inside the black box. In this example, human activity in the paid work sector is a part of the whole, and much smaller than the whole society. The rate of reduction (HA_{PW}/THA) is 1/15 and is proportional to the angle α (a discussion of how to use trigonometry within this type of figure is given in Appendix 2).
- The lower-left quadrant of the figure provides a fund-flow representation of the metabolism of the paid work sector (flow/fund ratio) referring to the element, the part inside the black box, defined at level n−1. The rate of this metabolism is characterized by the economic labour productivity of the paid work sector ($GDP/HA_{PW} = ELP_{PW} = US\$26/h$).
- The lower-right quadrant of the figure provides a representation of the reduction in size of the monetary flow when moving from level n to level n−1. In the chosen example, the entire Spanish GDP is assumed to be produced by the paid work sector, and therefore there actually is no reduction (angle of 45°) when moving from the whole society at level n to the paid work sector at level n−1.

This choice of the system of accounting entails a tautological definition of the various terms; we will discuss later that this implies adopting an impredicative loop in the representation of the characteristics of parts and the whole. However, a closer look at Figure 4.3 shows that the value of the angle β at level n−1 (lower-left quadrant) depends on the characteristics of lower level compartments (defined at level n−2) the ELP_i of the compartments making up the PW sector. This provides a way out from the tautology, since we can express the characteristics of the different parts as referring to the characteristics of other parts, in a network of expected relations, across levels. When expressing the same relation between the characteristics of the PW sector and its sub-sectors in terms of metabolic rate (flow/fund over the considered elements), the economic labour productivity of the paid work sector (ELP_{PW} = US\$26/h) depends on:

- the ELP_i of the sub-sectors making up the PW sector (the three sectors defined at level n−2 considered here are: AG = agriculture, PS = productive sector and SG = service and government); and
- the profile of distribution of the workforce over these three sectors.

This implies that the overall value of ELP_{PW} is affected not only by changes in individual ELP_i, but also by changes in work-time allocation across different sub-sectors. In fact, $GDP_{PW} = \Sigma GDP_i$ (sum of the sectoral GDP of the industrial, agricultural and service sectors).

This in turn implies that if demographic changes will determine a lower fraction of human activity in the paid work sector, the economy will have to increase proportionally the ELP_{PW} in order to keep the same level of GDP. In the same way, if ageing implies that the household sector requires more services, it is obvious that the job profile in the paid work sector has to change towards an increased number of jobs in the service sector. The feasibility of this move will depend on the ability to guarantee the same flow of consumed products, either by producing more per hour in the productive sector or by importing. An example of this analysis in biophysical terms follows. The requirement and supply of work hours across compartments of society have to be adjusted in a coordinated way across a series of economic and biophysical constraints. An overview of how to perform this integrated analysis across levels and dimensions – economic and energetic – is provided in Appendix 2.

Matching requirement with supply: energy flows in society

Exosomatic energy flows

Can we do the same type of analysis for energy flows as for monetary flows? The answer is 'yes', and we present here two examples: one referring to exosomatic energy flows and one to endosomatic energy flows.

Our first example deals with the exosomatic metabolic pattern of Italy (Figure 4.4). At the hierarchical level of society (level n), the Italian economy can be seen as a black box (right-hand side of Figure 4.4). In 1999, the population of 57.7 million Italians represented a total of 503.7Gh of total human activity (THA). In that same year, the Italian population consumed 7EJ (exajoules; exa = 10^{18}) of commercial energy. We call this flow of exosomatic energy the total exosomatic throughput (TET). We can then calculate the average exosomatic metabolic rate of the entire society (EMR$_{AS}$) as the exosomatic energy consumption of society per unit of human activity, which equals 14MJ/h (mega = 10^6).

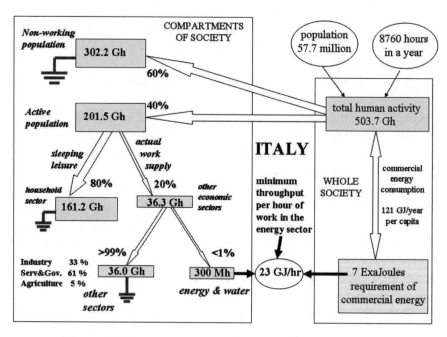

Figure 4.4 *Exosomatic energy flows in Italian society, 1999*

We now open the black box and examine the different compartments making up the Italian economy (left-hand side of Figure 4.4). We use a multi-level matrix of compartments measured in hours of human activity. In doing so, we discover that out of the 504Gh of total human activity available to Italian society in 1999, only 38Gh (or less than 8 per cent) were invested in the production of goods and services (paid work), while 467Gh were allocated to the consumption of goods and services (household sector) (see also Figure 4.3). Thus, about 12 hours of human activity were invested in consumption for each hour invested in production (467Gh/38Gh).

We now examine the profile of the distribution of work time over the various sub-sectors of the paid work sector, including the energy sector in charge of the mandatory task of supplying exosomatic energy carriers for society. Out of the 38Gh (8 per cent of total human activity) allocated to the paid work sector, 60 per cent was allocated to the service and government sector. The industrial and agricultural sector absorbed another 39 per cent, leaving the energy sector less than 1 per cent of the time allocated to the paid work sector (the already tiny 8 per cent of the total). The extreme fragmentation of work activity over a diversity of tasks within the paid work sector is a well-known characteristic of complex modern societies (Tainter, 1988).

In conclusion, in Italy in 1999, only a fraction equal to 0.0006 (less than one-thousandth) of total human activity was used for supplying the energy carriers guaranteeing the 7EJ of primary energy consumed in that year. Thus, dividing the total energy consumption of Italy by the hours of work allocated to the energy sector, we find that the energy sector was able to deliver no less than 23,000MJ of exosomatic energy per hour of work.

In the US, the productivity of the energy sector is even higher. With an exosomatic energy consumption of 333GJ per capita per year (in 2005) and a fraction of just 0.007 of the workforce (about 1 million workers) in the energy sector (50 per cent of the population being economically active, and a workload of 2000h/yr), the resulting productivity of the energy sector is about 47,000MJ of exosomatic energy supplied to society per hour of labour!

Endosomatic energy flows

The same analysis that is performed in Figure 4.4 in relation to exosomatic energy flows, can also be performed in relation to endosomatic energy flows (Figure 4.5). We illustrate this with an analysis of food production (endosomatic energy) in the US.

For the sake of simplicity, we focus on the production and consumption of cereals. The per capita consumption of cereal used for human alimentation in the US is about 1000kg/yr. This includes direct and indirect consumption. Indirect consumption includes feed for animal products, the production of alcoholic beverages (beer and spirits) and other uses in the food industry. Considering that the economically active population in the US is about 50 per cent and the workload about 20 per cent of workers' time, we find a ratio HA_{PW}/THA of about 1/10 (level n−1/level n). Thus, out of 876,000 hours of the total human activity of 100 US citizens (level n), only 88,000 working hours are available for the paid work sector (level n−1).

The various economic sub-sectors (level n−2) all compete for this supply of working hours. It is the case that more than 98 per cent of the US workforce is employed in sectors other than agriculture; hence we find a ratio HA_{AG}/HA_{PW} of about 1/50 (reduction from level n−1 to n−2).

In conclusion, in the case of the agricultural sector – the specialized sector in charge of generating the flow of endosomatic energy carriers required by society – we find that only a tiny fraction of total human activity (less than 0.002) is invested there. This implies the need of achieving high labour productivity in the agricultural sector. The minimum threshold calculated in this analysis – 67kg of cereal produced per hour of labour (Figure 4.5) – is much

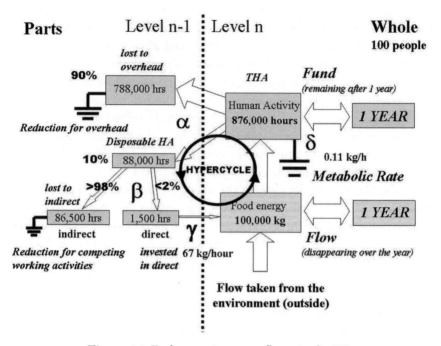

Figure 4.5 *Endosomatic energy flows in the US*

lower than the actual labour productivity of cereals found in the US as a technical coefficient, which is in the order of several hundreds of kg per hour (Giampietro, 2002). This is because US farmers do not produce only grains; they also produce other crops requiring higher labour inputs. Still, this threshold of labour productivity in grain production is more than 50 times higher than the level typical of pre-industrial societies.

The Concept of Bio-economic Pressure

The term 'bio-economic pressure' was proposed by Franck-Dominique Vivien (personal communication), building on Georgescu-Roegen's (1971) ideas. Bio-economic pressure refers to the biophysical (side-)effects of economic growth, where economic growth or development is defined as the continuous increase in monetary flows within the economy with the goal of augmenting the quality of life of society's members. Bio-economic pressure is defined as an increase in the intensity of the throughput of matter and energy per hour of labour (human activity) in the productive sub-sectors of the economy (including the primary sectors of agriculture, forestry, fisheries, and energy and mining, as well as building and manufacturing) in response to economic growth. Bio-economic pressure forces the productive sector to increase the supply of energy and material products to the rest of society (household and service sectors) to meet growing demand, while at the same time the work hours allocated to the productive sector are shrinking because of that same economic development.

We briefly comment here on the important relationship between economic growth, demographic structure and bio-economic pressure. Returning to Figures 4.2 and 4.3, we see that in Europe, less than 8 per cent of total human activity goes into the paid work sector. On the other hand, in China, almost 20 per cent of human activity is allocated to the paid work sector. This marked difference is due to differences in population age structure and socio-economic characteristics related to economic development. In general, economic development is associated with longer life expectancy, prolonged education and early retirement, all of which imply a smaller fraction of total human activity allocated to paid work.

On top of this, a greater share of the human activity allocated to paid work is directed to the service and government sectors in response to economic development (such as jobs in education, leisure, insurance and finance). That is, in biophysical terms, more human time is dedicated to consumption than to production. Thus, in terms of overheads (see Figures 4.4 and 4.5), the reduction in work hours allocated to the productive sector in response to economic growth and development is manifested as follows.

- Within the whole society (level n), the ratio of hours of non-working activity to hours of working activity goes up. Development means a higher value for HA_{HH}/HA_{PW}. Similarly, the ratio of hours of working activity to total hours of human activity (HA_{PW}/THA) goes down.
- Within the paid work sector (level n–1), the ratio of hours of work in service and government to hours of work in the productive sector goes up. Development means a higher value for HA_{SG}/HA_{PS}.

As concerns energy consumption, an increase in bio-economic pressure is reflected in an increased value of the overall ratio TET/HA_{PS} and the allocation of an increased share of total energy consumption to the service and government sector as compared to the productive sectors (Giampietro and Mayumi, 2000a; 2000b). As concerns the former, the increasing total exosomatic energy consumption of society – referring to the entire autocatalytic loop of exosomatic energy – has to be generated with the limited and declining amount of work hours allocated to the energy sector, which is part of the productive sector. This translates into the need to reach an extremely high supply of energy per hour of labour in the energy sector typical of developed countries, as illustrated earlier for Italy (Figure 4.4)

82   *The Biofuel Delusion*

In relation to the latter phenomenon, we see that in developed economies a significant fraction of energy is used to support human activity in the household sector (final consumption). Developed countries are locked in to a high standard of living, meaning they are forced to buy an increasing amount of oil to sustain the large requirement of energy of their final consumption sector – a large societal overhead – in addition to the oil used to generate added value in the productive sector. On the flip side, developing countries such as China tend to buy additional barrels of oil only if they can be used to produce added value. China, in its phase of rapid development, mainly purchases oil to sell cheap products to developed countries, but in the future may very well develop its own internal demand. This comparative advantage, exemplified by China, is based on the low dependency ratio of the population and the small share of energy invested in final consumption (a small societal overhead on the economic process). This comparative advantage, however, will last only until the current wave of adults becomes elderly (see Figure 4.2) and/or the desire of citizens for a better standard of living forces an increase in the societal overhead, causing the bio-economic pressure to build up.

In the rest of this section, we explore the causes and implications of the increase in bio-economic pressure associated with economic development, using the concepts and approaches introduced earlier in this chapter. We start out by addressing the decoupling of investments of human activity from investments of exosomatic energy, the process that lies at the roots of bio-economic pressure.

Decoupling of investments of human activity and exosomatic energy

The same type of assessment illustrated in Figure 4.3 for monetary flows – a four-angle figure looking for congruence over fund and flow elements – is now applied to exosomatic energy flows (Figure 4.6a). Note that here we do *not* seek to assess the congruence between the supply provided by a part relative to the requirement of the whole; in this figure we focus purely on the characterization of the consumption pattern across levels (parts and whole). Data presented below and in Figure 4.6a refer again to Spain in the year 1999.

The upper vertical axis in Figure 4.6a refers to the *fund* total human activity (THA = population × 8760 = 344Gh, in 1999); the right horizontal axis refers to the *flow* total exosomatic throughput (TET = 4200PJ (petajoules; peta = 10^{15})). Both values in the upper right quadrant refer to the hierarchical level of the whole society (level n). Dividing TET by THA, we find the flow/fund ratio; the average exosomatic metabolic rate of Spain (EMR_{AS} = 12.3MJ/h).

Next, we focus on the productive sector (level n−2, a sub-sector of the paid work sector), which is responsible for the generation of the autocatalytic loop

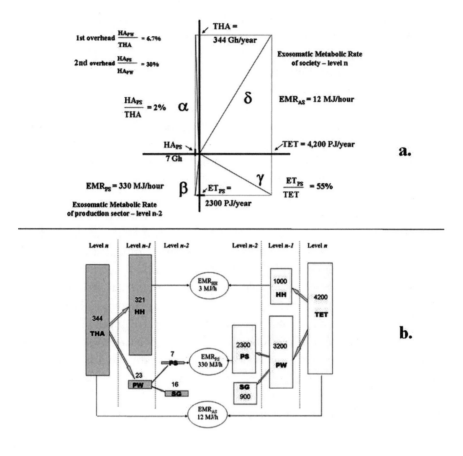

Figure 4.6 *Exosomatic energy flows across levels of analysis; Spain, 1999*

of exosomatic energy (lower-left quadrant in Figure 4.6a). We go again briefly through the description of the four quadrant/angle representation, this time focusing not on the particular chosen variables, but on the functional role that the various parts of this integrated representation provide for the assessment.

- The upper-right quadrant of Figure 4.6a provides a fund-flow representation of the metabolism of the whole black box (Spain) referring to level n.
- The lower-left quadrant of the figure provides a fund-flow representation of the metabolism of its productive sector (the metabolic rate of a part within the box), referring to level n−2.
- The upper-left quadrant of the figure provides a representation of the reduction in size of the fund human activity when moving from level n to level n−2. This reduction depends on two overheads: the ratio of human activity

allocated to the household sector over that allocated to paid work (HA_{HH}/HA_{PW}), and the ratio of human activity allocated to the production sector over that allocated to the overall paid work sector (HA_{PS}/HA_{PW}).
- The lower-right quadrant of the figure represents the reduction in size of the flow of exosomatic energy consumption when moving from level n to level n−2. This again has to be calculated on the basis of two overheads: the ratio of exosomatic energy consumed in the household sector over that consumed in the paid work sector (ET_{HH}/ET_{PW}), and the ratio of exosomatic energy consumed in paid work sub-sectors other than the productive sector over that consumed in the productive sector (ET_{PS}/ET_{PW}).

To better visualize the various overheads, we provide a different view of the same metabolic pattern in Figure 4.6b. Moving across levels, we find different patterns of reductions determining the various compartment sizes for the *fund* human activity (on the left in Figure 4.6b) compared to the *flow* of exosomatic energy (on the right); see also Table 4.2 below. In particular, at level n−2, the PS sub-sector gets only a tiny fraction of total human activity (THA), whereas it gets the largest chunk of total exosomatic energy throughput (TET). Indeed, the asymmetric pattern of reduction in relation to relative sizes of funds and flows across different levels entails that the exosomatic metabolic rate of the productive sub-sector at level n−2 (EMR_{PS} = 330MJ/h) is almost 27 times higher than the average exosomatic metabolic rate of the whole society at level n (EMR_{AS} = 12.3MJ/h). This implies dramatic differences in the resulting power level for the different compartments (Table 4.2).

Table 4.2 Reduction in size of fund and flow over levels of analysis

	Fund size HA (Gh)	Flow size ET (PJ)	EMR or power level (MJ/h)
Whole society (level n)	344	4200	12
PW sector (level n−1)	23	3200	138
PS sub-sector (level n−2)	7	2300	330

PW = paid work sector; PS = production sub-sector; HA = human activity; ET = exosomatic energy throughput; EMR = exosomatic metabolic rate.

We recall here the distinction made earlier between power level and energy input. The overall average consumption rate of exosomatic energy of society (EMR_{AS}) is related to society's ability to produce and consume goods and services at a given pace. This is the reason why EMR_{AS} correlates well with GDP (an empirical analysis of this point is given in the last section of this chapter; see Figure 4.12 and Table 4.3). The value of the exosomatic metabolic rate tells us that:

- society has an adequate power level (technology and know-how) to effectively control the metabolism of this amount of energy per hour – feasible against internal constraints; and
- society has access, in one way or another, to an adequate input (supply) of exosomatic energy sources (TET) – feasible against external constraints.

Thus, at level n, the pattern of metabolism reflects the ability to balance a dynamic budget of exosomatic energy maintained by a network of investments of monetary, energetic and material flows controlled by a set of funds (compartments) of human activity, interacting at the lower levels.

However, if we want to understand what makes it possible to balance this dynamic budget, or if we want to look for the option space for possible changes or substitutions of one flow with another, we have to open the black box and look at the characteristics of and relations between lower-level compartments. This is especially the case when we are interested in the critical thresholds of power level in specialized compartments of the system in charge of the stabilization of individual flows; see the example of the critical threshold of labour productivity for the energy sector (Figure 4.4) and the agricultural sector (Figure 4.5). By opening the black box, it is possible to study the internal biophysical constraints on the feasibility and desirability of a given pattern of metabolism.

Biophysical definition of capital

At this point, it is useful to introduce the biophysical definition of technical capital proposed by Hall et al (1986) following the original ideas of Cottrell (1955) and other pioneers of energetics (see Chapter 6). The rationale behind the definition is that the effectiveness of human labour can be enhanced by exosomatic energy converted by exosomatic devices under the control of the worker. Thus, the exosomatic metabolic rate of a given compartment of society can be used as a proxy of the biophysical capital of that compartment. That is, we consider the value of EMR_i (flow/fund) of a given compartment as a characteristic benchmark of the fund elements associated with that compartment, since this value reflects the power level available to the fund human activity. More specifically, the EMR_i found in the paid work sector (level $n-1$) and in its sub-sectors (level $n-2$, $n-3$, etc.) are suitable proxy variables for investments in capitalization (machinery and tools) in these (sub-)sectors.

Cleveland et al (1984) observed a link between energy consumption per hour of work (EMR_i) and economic labour productivity (ELP_i; the monetary flow in US$ generated per hour of human activity in the compartment) within the paid work sector. This relationship is explained by the fact that an increase in EMR_i implies the use of more machinery and tools, beside more direct consumption of energy inputs, which in turn entails an increase in both economic and biophysical cost to be paid by the fund. In fact, to produce and use machinery and tools, more energy and resources have to be used. Obviously, one would go for this larger economic investment only if one gets an adequate economic return.

The assumption of a strong relationship between the aggregate requirement of machinery, tools and direct energy consumption and the resulting generation of monetary flows has certain limitations. When studying evolutionary changes, it does not account for improvements in energy efficiency. However, experimental data for the US (Cleveland et al, 1984; Hall et al, 1986), Ecuador (Falconi-Benitez, 2001), Spain (Ramos-Martin, 2001) and Catalonia (Ramos-Martin, 2009) show that a relationship between the two exists at a given point in time and in historical series: changes in EMR_i tend to map onto changes in ELP_i (examples are presented in Appendix 2).

Accepting that the ability to use more exosomatic energy to boost the efficacy of one hour of human activity is associated with the concept of biophysical capital, we must point to an important difference between the bio-economic analysis of societal metabolism and conventional economic analysis. Conventional economic analysis considers the concept of capital *only* in relation to the step of *production* of either products or services. On the contrary, bio-economics addresses the entire process of production *and consumption* associated with society's metabolic pattern.

Thus, in biophysical terms, the capitalization of an economy refers to an expected characteristic of the economic development of society as a whole. For example, if society consumes 150–300GJ of exosomatic energy per capita per year, which means an EMR_{AS} of 17–35MJ/h, we can say that it belongs to the typology of a developed country. On the contrary, if a society consumes less than 50GJ/year of exosomatic energy per capita, which means an EMR_{AS} <5MJ/h, then we can say that it belongs to the typology of a developing country. Due to the pattern of relations across parts within the whole, we can apply the same reasoning also to the organs of these different types of societies. That is, a bio-economic analysis can look for benchmarks (expected levels of power or capitalization) of the lower level compartments of socio-economic systems – typologies of metabolic pattern of whole countries can be associated with expected benchmarks of the metabolism of their individual organs. For example, the paid work sector in developed countries has an EMR_{PW} in the range of 100–300MJ/h, whereas developing countries typically have an EMR_{PW} smaller than 50MJ/h. The same use of benchmarking can also be applied to the household sector (see Box 4.1).

This crucial aspect of bio-economic analyses was first introduced by Zipf

(1941) in his description of nations as 'bio-social forms of organization'. Zipf proposed a basic principle of socio-economic development: if an economy wants to be able to produce more, it has to invest in consuming more. This principle implies that socio-economic development must be based on achieving an internal balance between parallel investments of both human activity and exosomatic energy over the two compartments of production *and consumption* of goods and services. Thus, it is important to recognize that the final sector of consumption, the household sector, must also be capitalized in order to be able to increase the overall rate of production and consumption of goods and services at the level of the whole society (see Box 4.1).

Box 4.1 *Capitalization of the household sector of Spain*

Using a comparative analysis based on benchmarks, Ramos-Martin (2001) found an anomaly in the trajectory of the evolution of the energy metabolism of Spain in the 1990s. In contrast to other European economies, the historical trajectory of development in Spain did not exhibit a decrease in the energy intensity of its economy after reaching a given performance in terms of macro-economic indicators. When checking the various benchmarks describing the metabolic rate of the compartments of the Spanish economy, Ramos-Martin found a very low level of capitalization of the Spanish household sector; the EMR_{HH} in Spain was only about 1MJ/h, much lower than the benchmark typical of other European countries (4–6MJ/h). This difference in capitalization of the household sector pointed to the existence of what he called a 'Franco-effect' in the trajectory of evolution of the Spanish economy. Three decades of compression of the sector of final consumption in favour of investments in the paid work sector guaranteed the adequate capitalization of the various sub-sectors of the productive economy, but left the Spanish household sector undercapitalized. Since the household sector makes up 90 per cent of total human activity, the subsequent increase in the metabolism of the household sector, which eventually took place in Spain from the 1980s onwards, markedly increased the total energy consumption of the country. An overview of the data of this study is given in Appendix 2.

Therefore, any sound theory of economic growth should explicitly address the changes taking place within the household sector, because it is a key component of the autocatalytic loop leading to the qualitative and quantitative expansion of the economic process. With remarkable anticipation, Zipf introduced the idea that self-organizing systems express a trend toward critical organization (an expected pattern) when considering the relative size of compartments across hierarchical levels. Combining the wisdom of Georgescu-Roegen and Zipf, we can say that an economy *does not produce goods and services, but rather it produces the processes required for producing and consuming goods and services* (see Appendix 1). Thus, in order to raise the average rate of exosomatic energy consumption of the whole society (EMR_{AS}), the sectors of production (PW) *and* consumption (HH) have to increase in a coordinated way their endosomatic and exosomatic activity. The increased capitalization of the paid work sector (EMR_{PW}) will reflect a greater number of exosomatic devices used during work hours to supply goods and services to the household sector and for internal consumption. The increased capitalization of the household sector (EMR_{HH}) will reflect a greater number of exosomatic devices used outside work activity, such as private cars, air-conditioning, refrigerators, home appliances, etc. As noted earlier, since human activity allocated to the household sector amounts to 90 per cent of total activity, small changes in EMR_{HH} can have important consequences for the overall energy intensity of society (Ramos-Martin, 2001; 2009; Ramos-Martin et al, 2007).

Capitalism in relation to stock-flow energy exploitation

The decoupling of investments of useful energy from investments of human activity has been the key to capitalism as we know it (Figure 4.6b). In fact, by simply injecting capital it is possible to maintain a pattern of metabolism, typical of modern society, in which only a tiny fraction of human activity is invested in those specialized compartments generating surplus. Contrast this situation with the ratio of rulers over ruled (compartment generating surplus) in pre-industrial society (Figures 2.6a and 3.5). In modern society, exosomatic devices (biophysical capital) have taken the place of the ruled in pre-industrial society. Power generation with engines has implied an unprecedented power level *controlled* by humans, but no longer generated manually, and has dramatically reduced the overhead costs of power generation. By continuously accumulating capital, developed societies managed to lift the internal biophysical constraint (limited power level) which previously prevented the decoupling of the supply of useful energy from investments of human activity. Modern society no longer needs human slaves or a social organization based on distinct classes (rulers versus ruled).

However, it is important to recognize that these dramatic changes were possible only because of an adequate supply of suitable energy inputs to sustain the exosomatic hypercycle driven by machines. The continuous increase in capitalization to ease the internal constraint never arrived to crash against the existence of an external constraint. The flow of energy used by the growing capital was a stock-flow. Thus, the strategy of continuously lifting internal constraints by increasing the level of capitalization worked because the external constraints (on the energy input) were removed by the adoption of a stock-flow supply. This is an important point, since if there is no possibility of taking advantage of accumulated capital, like the Roman Empire having a powerful army but no more small kingdoms to conquer, there is no incentive for building up or expanding capital. This observation brings us back to the discussion in Chapter 2. The abundant availability and accessibility of fossil energy removed the historical external constraint – that is, the impossibility of expanding fund-flow energy inputs (from colonized land) as much as necessary – and permitted humans to control quantities of energy unthinkable in pre-industrial times. The move from fund-flow (land-tied energy inputs) to stock-flow (fossil energy) opened up the era of exploitation of concentrated flows of energy carriers. The only bottleneck faced to get a larger metabolic flow was that of achieving a larger power level. When dealing with stock-flow exploitation, the more capital is accumulated, the more fossil energy can be used.

It was the extraordinary strength of the autocatalytic loop of exosomatic energy that made it possible to successfully implement the ideological imperatives of 'maximization of profit' and 'perpetual growth'. The acceptance of these ideologies translated into a powerful and simple strategy: 'survival of the first' (Hopf, 1988). Societies that were faster in accumulating capital have won the battle for control over more energy. In fact, they could generate more useful work, thanks to their higher power level. As a consequence, they were able to use more resources than others. There is nothing intrinsically wrong with this strategy. As a matter of fact it represents exactly the series of events expected in the evolution of living systems as described in the field of theoretical ecology as the 'maximum power principle' (Odum and Pinkerton, 1955; Lotka, 1922; 1956). The emergence of capitalism in human history was a natural event and, just like other large-scale natural events, it had positive and negative side-effects.

A Crucial Concept in the Viability of Dynamic Budgets: Impredicative Loop Analysis

Thus far, in this chapter, we employed the four-angle figure to demonstrate the forced relationship between relative sizes of funds and flows described over a dynamic equilibrium and across hierarchical levels (parts and whole). Now, our aim is to show that this type of representation provides a general template for meta-analysis, versatile in its possible applications. The representation is a quantitative scheme based on the forced congruence over numbers defined in a system of accounting, but it remains semantically open. The congruence can be achieved by changing the technical coefficients determining the numbers, but also by changing the categories used to describe what the system is and what the system does. This systemic property allows us to use this tool to explore the option space of the different solutions that can be adopted for obtaining congruence over the pattern. Obviously, these solutions are not equally desirable in relation to different criteria of performance. We will illustrate these with examples.

In our first example in Figure 4.7, we have at level n of society the *fund* human activity and a *flow* of letters written and received over the course of one year by the people living in society. We assume this society has 1000 people, hence THA=8.76Mh (over one year). We further assume that the flow is of 24,000 letters per year (per 1000 people), resulting in a 'letter metabolic rate' of 0.003 letters/h (or two letters per person per month).

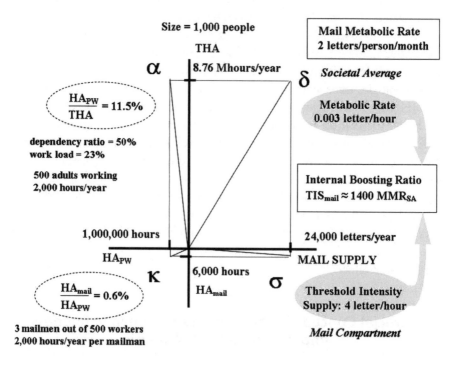

Figure 4.7 *Example of impredicative loop analysis: mail delivery in society*

As illustrated in Figure 4.7, starting from total human activity (THA at level n), we again have to go through a series of overheads to find the size of the specialized part, mail service, at level n−3. The first overhead (HA_{PW}/THA) leaves only 1 million hours for the paid work sector (per 1000 people). The second overhead (HA_{mail}/HA_{PW}) entails that only 0.6 per cent of paid work hours are allocated to mail service, corresponding to 6000 hours of human activity (or only about 0.07 per cent of total human activity).

At this point, one can appreciate the versatility of this figure. The forced congruence over the lower right quadrant can be interpreted as a *threshold value* which is required to obtain congruence between the requirement of the whole (the characterization of society's metabolism in the upper-right quadrant) and the supply by the mail compartment (the technical performance of a specialized sector in charge of delivering the flow under consideration). Given the characteristics provided so far, the mail sector must correctly deliver 24,000 letters in 6000 work hours. Therefore, the threshold for delivery is a minimum of 4 letters per hour of work, as a yearly average.

The term 'impredicative loop analysis' derives from the existence of *reciprocal constraints*, which are reflected in the four-angle representation (again, the use of trigonometry to perform this analysis is presented in Appendix 2, Figure A2.15).

- In the upper-right quadrant of the figure (the metabolism of the whole society at level n), we have the angle δ, which is proportional to an arbitrarily chosen, desirable characteristic of society. The value of this angle is related to the pace of the metabolized flow under consideration, an attribute of performance for the whole.
- In the upper-left quadrant, we have the angle α, which is proportional to the reduction in fund size from level n to n−1: that is, the first overhead. In this example, this is the ratio HA_{PW}/THA, defining the share of total human activity allocated to paid work versus final consumption.
- In the lower-left quadrant, we have the angle κ, which is proportional to the reduction in fund size from level n−1 to n−3: that is, the second overhead (in this example HA_{mail}/HA_{PW}). The value of this angle is related to the relationship between service sector and productive sector at level n−2, and the relationship between the service under consideration (postal service) and other sub-sectors of the service sector at level n−3.
- In the lower-right quadrant we can represent either an expected value for scenario analysis or a technical coefficient (actual value) characterizing a given situation. Thus, the value of the angle σ can be related to either the expected or required value for the performance (letter delivery) of the sub-sector under consideration ($\sigma_{expected}$), or the actual performance of that sub-sector ($\sigma_{achieved}$).

If we use this closed set of reciprocal relations over the elements of the graph, we can perform an impredicative loop analysis to discuss future scenarios. That is, we can look for a configuration of dynamic equilibrium over the four angles that is feasible according to characteristics of the supply. In this scenario, society has to adapt to existing constraints determined by the characteristics of the sub-sector. Alternatively, we can start by defining an expected pattern of metabolism for society which is considered as desirable, and then look for the characteristics that would be required in the compartment under consideration to guarantee such a pattern. In this case, the characteristics of the sub-sector must be adapted to those of the society.

The important feature of this type of analysis is that the whole set of characteristics used to find congruence over the dynamic budget can be adjusted according to either actual or desired values. For example, we can analyse proposed changes in:

- the attribute of performance at the level of the whole (for example, in Figure 4.7, this means writing more or fewer letters at the level of society);
- population characteristics, such as the age structure of population through proposed regulations for immigration or changes in retirement age (to alter the ratio HA_{PW}/THA);
- distribution of work hours over the various economic sub-sectors, such as through government incentives (subsidies) to stimulate expansion of selected sectors; hire more postmen or pay more for overtime work; and
- technical coefficients of the specialized compartment under consideration through implementation of new technologies, mechanization, training, etc.

Thus, the impredicative loop analysis does not provide any deterministic prediction of what will happen – any combination of changes can generate a viable dynamic budget – but is useful for an informed deliberation over proposed solutions. It can eliminate proposals that are not biophysically feasible, establish a link between integrated patterns of change, and indicate what is gained and lost – using indicators of desirability – in relation to different tasks defined at different levels and compartments.

Typical Pattern of the Exosomatic Metabolism of a Developed Society

Combining the various concepts discussed so far in this chapter, we claim that it is possible to define a typology of exosomatic metabolic patterns in a developed society. This can be used to discuss the quality of alternative energy sources (see also Chapter 5).

Dendrogram within the multi-level matrix of human activity

All developed societies exhibit a standard pattern of distribution of total human activity over lower-level compartments (Figure 4.8). As illustrated in Figure 4.8, the sectors responsible for the production of endosomatic (agricultural sector) and exosomatic (energy sector) energy input use only a negligible fraction of the available paid work hours. This is a crucial point that must be considered when looking for alternative energy sources, and more in general when proposing changes in the agricultural and/or energy sector (see also Figure 4.1).

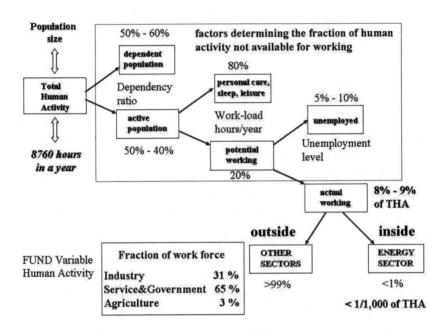

Figure 4.8 *Dendrogram of the allocation of human activity over different compartments of society*

The structure of socio-economic systems reflects a given pattern of metabolism, which requires a given profile of investments of human activity, biophysical capital and flows of energy and matter across lower-level compartments of the system. This structure is directly related to the set of functions expressed by society. If we want to introduce in this pattern of metabolism a system of generation of energy carriers (an alternative energy source) that would require a dramatic readjustment of this complex structure, it would be wise to find out whether these changes will represent an improvement or not. Any proposed change should carefully take into consideration the internal constraint related to labour productivity within the energy sector; the supply of energy carriers must be guaranteed while using only a negligible fraction of the paid work hours.

The overall pattern of exosomatic metabolism associated with this dendrogram

The characterization of the typical pattern of exosomatic metabolism of a developed country is provided in Figure 4.9. In spite of the use of quantitative assessment for the metabolism for the whole (level n) and the various organs (lower level), Figure 4.9 does not describe any country in particular. The characterization is based on a set of benchmark values that may be expected for a developed society. They refer to the profile of distribution of the *fund* of human activity, and the *flow* of exosomatic energy consumption over lower level compartments (see Figure 4.6b). The simultaneous definition of these two profiles generates the emerging profile of exosomatic metabolic rate per hour of human activity over the chosen set of compartments (a more detailed explanation in Appendix 2, Figure A2.4).

The standard benchmark values used in this representation have been found in empirical studies applying this approach (see also Chapter 7 and Appendix 2). But what Figure 4.9 is really about is not so much the numerical values, but the grammar useful to characterize in quantitative terms an expected set of relations between the profiles of distribution of a chosen set of funds of human activity and a chosen set of flows of exosomatic energy.

Chosen set of funds of human activity

In this grammar, the fund human activity is made available (reproduced) by the household compartment. Its total size (THA) refers to the whole society (level n). It is then split over various elements depending on the choice of the analyst. We have chosen a first distinction between household sector and paid work sector (level n−1). In this grammar, we adopt a further distinction (at level n−2) within the paid work sector between PS*, ES and SG, where PS* is the productive sector minus the energy sector, and ES is the energy sector (singled out on purpose because it is the sector under consideration).

The constraint of closure over the fund human activity at each level requires that $HA_{PW} = HA_{PS*} + HA_{ES} + HA_{SG}$ (at level n−1) and $HA_{PW} + HA_{HH}$ = THA (at level n). At the level n of the whole society, the total amount of hours available per capita is 8760h/year. This total amount THA is divided as follows:

- level n, whole society: THA = 8760h/year;
- level n−1, production and consumption: HA_{PW} = 860h and HA_{HH} = 7900h;
- level n−2, within PW: HA_{PS*} = 320h; HA_{SG} = 530h and HA_{ES} = 10h.

Chosen set of flows of exosomatic energy

The flow exosomatic energy is made available by the energy sector (ES) to the whole society. Its total size (TET) refers to the whole society at level n. It is then

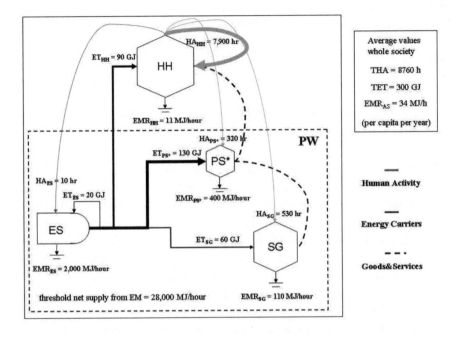

Figure 4.9 *Socio-economic metabolic pattern based on benchmark values*

split over various elements depending on the choice of the analyst. Obviously, we adopt here the same series of distinctions used in the previous analysis. That is a split between HH and PW (level n−1), and between PS*, ES and SG (level n−2). Total exosomatic energy consumption (TET) per capita in this hypothetical developed society is set at the benchmark value of 300GJ/yr, and is divided as follows.

- Level n, whole society: TET = 300GJ
- Level n−1, production vs consumption: ET_{HH} = 90GJ and ET_{PW} = 210GJ
- Level n−2, within PW: ET_{PS*} = 130GJ; ET_{SG} = 60GJ; and ET_{ES} = 20GJ

Resulting profile of energy intensities

The resulting profile of exosomatic metabolic rates (EMR_i), commonly referred to as energy intensities, reflects the power levels of the various compartments and can be related to the quantity of biophysical capital (exosomatic devices) used per hour of human activity in these sectors. As mentioned, exosomatic devices enhance the efficacy of human activity in performing the relative tasks and require exosomatic energy as their energy input. The benchmark values characterizing the various power levels in our typical developed society are as follows.

- Level n, whole society: $EMR_{AS} = 34MJ/h$
- Level n−1, production vs consumption: $EMR_{HH} = 11MJ/h$ and $EMR_{PW} = 240MJ/h$
- Level n−2, within the PW: $EMR_{PS*} = 400MJ/h$; $EMR_{SG} = 110MJ/h$ and $EMR_{ES} = 2000MJ/h$

The reader can recall the similar analysis proposed by H. T. Odum to analyse the forced relationship between parts and whole in the metabolism of ecosystems (Figure 2.4). Obviously, real values for specific countries will show a certain spread around these benchmark values for EMR_i. Actual power levels in the US and Canada are higher, while those in Spain, Italy and Japan are lower than the benchmark values shown above and in Figure 4.9 (an analysis of the spread of benchmarks is discussed in Appendix 2, Figure A2.13). For sure, all developed countries exhibit an extremely high power level for the EMR in the energy sector.

Thus, after implementing the chosen grammar using reliable local dictionaries, we can calculate values for the expected power levels required in a developed society to obtain a balanced dynamic budget of metabolism across different hierarchical levels.

Notes on the calculation of exosomatic energy flows

The values for energy consumption in energy statistics, even if referring to the various sectors of society, are expressed in tonnes of oil equivalent (TOE), a value associated with the virtual consumption of a reference primary energy source (discussed in detail in Chapter 5). These values therefore depend on the particular mix of energy carriers used by individual compartments of society. When getting into a more accurate and country-specific analysis of the pattern of exosomatic metabolism, it may thus be necessary to implement purpose-made 'dictionaries' that express the relationships between the specific identity of fund elements and the relative requirement of a given mix of exosomatic energy carriers. That is, according to national statistics the industrial sector – on the side of the end-uses – consumes a certain amount of energy calculated, by summing together joules of electricity, heat and liquid fuels. But the various conversion losses referring to the conversion of primary energy sources into the various flows of energy carriers consumed are accounted elsewhere. This implies that the joules of energy consumption of the final sector does not map onto joules of TOE. This is why one should have an analysis of the typologies of consumption of energy carriers of the different sectors. An example of this approach is provided in Figure 4.10 with regard to the household sector.

tasks / energy carriers	heating	home appliances	air conditioning	cooking	car driving
electricity	...	a_{12}	a_{13}	...	...
heat	a_{21}	...	...	...	...
liquid fuels	...	...	...	...	a_{35}
gas	a_{41}	...	...	a_{44}	...
coal	...	...	...	...	...
wood	a_{61}	...	...	...	...

Figure 4.10 *Characterization of the exosomatic metabolic pattern of the household sector in terms of tasks and energy carriers*

A 'dictionary' requires a specification of the quantitative characterization of a given semantic (e.g. heating, cooking) in relation to the specific context of application; that is, the set of technical tasks associated with the energetic metabolism of that sector and the specific requirement of exosomatic energy carriers. This is to say, the semantic concept of household consumption and the different tasks (the meaning of the words in terms of linguistic metaphor) is different in Bamaco, Mali than it is in New York City, US. The same is true for the semantic concept of the agricultural sector, the building and manufacturing sector, etc. Therefore, in order to quantify the overall set of semantic relations provided by a grammar specifying the relationship between energy conversions, one has to adopt an appropriate set of dictionaries. In this way, the various concepts included in the grammar can be expressed by using the appropriate 'words' (e.g. household sector, building and manufacturing sector) required to describe different 'realities' (the set of activities performed in these sectors) in different places.

Clearly, the use of this analytical approach implies acknowledging that we can find different technological coefficients (associated with the specific set of conversions of exosomatic energy carriers into useful energy) for the same semantic specification of a given task: cooking at home, say, or producing aluminium cans. And it underscores the need of always contextualizing the

98 The Biofuel Delusion

chosen grammar before getting into any quantitative characterization of the exosomatic energy pattern, such as the one shown in Figure 4.10.

Our second note concerns the energy consumption in the transportation sector. The definition of the transportation sector, commonly used and included in energy statistics, does not match our narrative of the metabolic pattern of self-organization. In fact, some of the energy in transportation is used for the operation of the paid work sector, and some is used for activity in the household sector. This choice is made in international statistics to focus on the category 'liquid fuels'. For this reason, in our assessments of the exosomatic metabolic pattern of societies, energy consumption in the transportation sector is allocated to either the household sector (energy consumed in the private use of cars) or the service and government sector (energy consumed by commercial and public transportation of goods and people). Hence the amount of energy allocated to the service and government sector in Figure 4.9 is higher than that reported in energy statistics.

The Correlation of Bio-economic Pressure with Indicators of Development

Can we provide evidence that economic development generates a convergence towards an expected pattern of metabolism, which can be characterized by using the grammar presented in Figure 4.9? This section provides a data set which backs up our claim.

In the previous sections we illustrated the existence of internal biophysical constraints determining a forced relationship between the relative size of funds and flows defined over a fund-flow grammar describing the metabolic pattern of a society. When applying the representation based on four angles discussed there, we can visualize the constraint associated with bio-economic pressure (BEP) as illustrated in Figure 4.11.

According to Figure 4.11, there is a forced relationship between the various factors determining economic development (indicated on the four quadrants of the figure), which can only be changed in an integrated way over the impredicative loop. As a result of this forced relationship, economic development translates into an increase in the value of BEP. A quantitative definition of BEP can be obtained in two non-equivalent ways – as illustrated at the bottom of Figure 4.11 – using two distinct relations of congruence. The first relation is used in this section to check the hypothesis that an increase in the value of BEP is actually correlated to indicators of social development. The second relation is used in the next section, to show that changes in the value of BEP require an integrated set of changes in the pattern of societal metabolism across different levels.

A Crash Course in Bio-economics   99

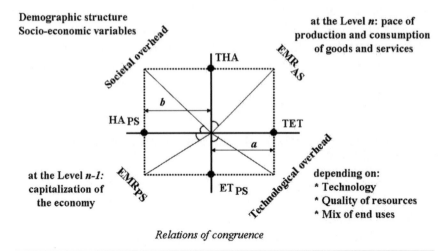

$$BEP = a/b = EMR_{AS} \times THA/HA_{PW} \times HA_{PW}/HA_{PS} = TET/HA_{PS}$$

$$BEP = TET/HA_{PS} = (MF \times ABM) \times (exo/endo) \times THA/HA_{PS}$$

Figure 4.11 *An overview of the characteristics of the exosomatic metabolic pattern determining the bio-economic pressure*

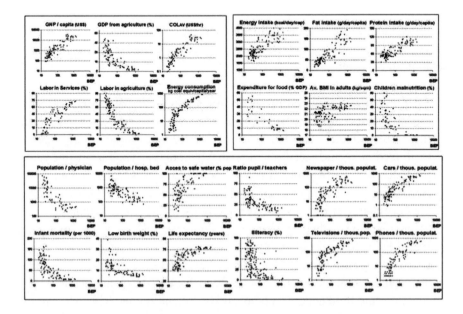

Figure 4.12 *The correlation between the BEP indicator and standard indicators of development*

The hypothesis that BEP should show a good correlation with conventional indicators of development has been confirmed in a study including more than 100 countries of the world, a sample including more than 90 per cent of world population. All the data in this and the next section are taken from Pastore et al (2000). An overview of the results obtained in this study is given in Figure 4.12, and clearly confirms our claim.

The existence of a pattern of exosomatic metabolism across hierarchical levels

The hypothesis of the convergence of different countries towards an expected exosomatic metabolic pattern can be explained by the effect of internal biophysical constraints related to the viability of the dynamic equilibrium. That is, can we detect the existence of a forced relationship between the relative sizes of funds and flows across levels?

The answer to this question is 'yes', but to prove it, we have to describe BEP using a different set of relationships that are non-equivalent to those indicated by the grammar of Figure 4.9. To obtain this non-equivalent set of relations, we express the ratio TET/HA_{PS} as the result of the product of three different factors determining a forced congruence of the resulting flows, using variables defined across different hierarchical levels:

$$BEP = TET/HA_{PS} = (ABM \times MF) \times (exo/endo) \times (THA/HA_{PS})$$

The three factors on the right of this relation represent three biophysical indicators of development, referring to different levels of analysis of the metabolism of society.

1. $ABM \times MF$ – average body mass × metabolic flow (MJ/hour) – refers to the endosomatic metabolism of human activity and addresses changes in the anthropometric characteristics of individuals. Average body mass is expressed in kg and metabolic flow in MJ/hour/kg. The higher this value, the better are the physiological conditions of humans living in the society. According to our database, $MF \times ABM$ has a minimum value (0.33; short life expectancy at birth, small average body mass) and a maximum value (0.43; plateau reached in developed countries). More explanations are available in Pastore et al (2000).
2. Exosomatic/endosomatic metabolic ratio (dimensionless); the ratio between exosomatic energy metabolized by the different sectors of the economy and food energy metabolized by human bodies. This refers to the level of technical capitalization of the economy. According to our database, 'exo/endo energy ratio' has a minimum value of around 5 (when exo-

somatic energy is basically in the form of traditional biomass, such as biomass and animal power). The maximum value is around 90 (for developed countries, based on an estimated endosomatic physiological metabolism of 10MJ/day per capita). The exo/endo ratio is a good indicator of economic activity; it is strongly correlated to the GNP per capita (as shown in Figure 4.13). The higher the exo/endo, the more goods and services are produced and consumed per capita.

3   THA/HA$_{PS}$ (dimensionless); an indicator referring to socio-economic changes, since it refers to the combined effect of two overheads: working/non-working, and working in SG/working in PS. This indicator determines the fraction of THA allocated in PS. When dealing with THA/HA$_{PS}$ it is difficult to define a low value, since in poor countries the distinction between the categories of human activity included within PW is weak, and the age limits for being considered as part of the the workforce are grey. An example of a low value for this ratio would be 9/1 in China. The maximum value we found is 45, in post-industrial societies with a large proportion of elderly people and a large proportion of the workforce absorbed by services. The value of this indicator reflects the social implications of development (longer education, greater proportion of elderly, more leisure time for workers, and an increased demand for activities in the services and government sector).

When considering each one of these indicators over the same sample of countries, we obtain the results illustrated in Figure 4.13.

These results confirm that each one of these three indicators can be used as a benchmark, since it is well correlated with both BEP and the set of indicators of development selected in the study by Pastore et al. This is to say that when considering different countries at different levels of development, the value of these factors changes in an integrated way over the impredicative loop. This is required in order to achieve a viable dynamic budget associated with the expression of the resulting exosomatic metabolic pattern. Talking of convergence, the block of graphs on the left of Figure 4.13 illustrates the correlation of conventional indicators of development with the exo/endo indicator. These graphs literally show the effect of the demographic transition, taking place after a given threshold value of exo/endo (around the value of 20/1). That is, after having reached a new viable state for the dynamic equilibrium, the value of the set of selected indicators reaches a plateau, and an additional increase in the rate of energy metabolism no longer affects the attributes of performance covered by the chosen set of indicators.

The relationship between these indicators is confirmed by the analysis given in Table 4.3, overleaf. In this table there is a good correlation between the three sets of values for:

1   BEP;
2   each one of the three indicators reflecting the changes taking place across the different levels of analysis (physiological/endosomatic metabolism, exosomatic metabolism, and the social profile of time allocation); and
3   the conventional indicators of development used by the World Bank and other international institutions.

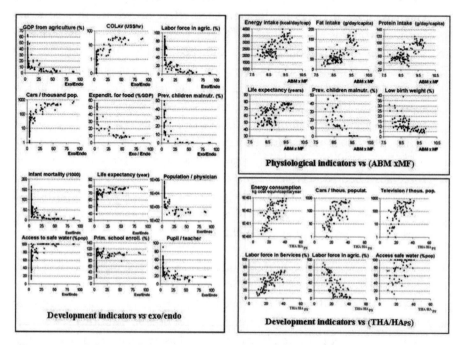

Figure 4.13 *The correlation between each one of the three factors determining the value of the BEP indicator and standard indicators of development*

Table 4.3 *The correlation between BEP, exo/endo ratio, THA/HA$_{PS}$ and ABM × MF over three different categories of indicators of development*

	Indicators	log (BEP)	log (exo/endo)	THA/HA$_{PS}$	ABM × MF
		r	r	r	r
Nutritional status and physiological well-being	Life expectancy	0.79	0.75	0.63	0.59
	Energy intake	0.82	0.81	0.55	0.73
	Fat intake	0.87	0.85	0.63	0.77
	Protein intake	0.85	0.85	0.57	0.72
	Child malnutrition	−0.71	−0.65	−0.63	−0.70
	Infant mortality	−0.76	−0.74	−0.57	−0.58
	Low birthweight	−0.65	−0.62	−0.49	−0.63
Economic and technological development	Log (GNP)	0.92	0.89	0.63	0.66
	% agriculture on GNP	−0.77	−0.73	−0.60	−0.54
	Log (COL$_{AV}$)	0.92	0.87	0.71	0.63
	% labour force in agriculture	−0.90	−0.81	−0.72	−0.66
	% labour force in services	0.90	0.83	0.76	0.56
	Energy cons/cap	0.92	0.95	0.53	0.67
	Food expenditure	−0.86	−0.87	−0.69	−0.78
Social development	TV/inhabitant	0.89	0.89	0.62	0.72
	Car/inhabitant	0.88	0.91	0.59	0.72
	Newspaper/inhabitant	0.77	0.80	0.47	0.60
	Phone/inhabitant	0.87	0.88	0.61	0.71
	Log (pop./physician)	−0.81	−0.76	−0.60	−0.67
	Log (pop./hospital bed)	−0.77	−0.78	−0.51	−0.70
	Pupil/teacher	−0.77	−0.76	−0.51	−0.62
	Illiteracy rate	−0.61	−0.58	−0.42	−0.44
	Primary school enrolment	0.44	0.39	0.38	0.36
	Access to safe water	0.78	0.77	0.53	0.59

Chapter 5

Assessment of the Quality of Alternative Energy Sources

The Treacherous Quicksand of Energy Analysis

The lack of clear consensus in the community of energy analysts about how to perform and use energy analysis is probably one of the explanations for the rapid decline of interest in energy analysis in the 1990s. If it is true that there is a general agreement about the existence of a link between energy and society, it is also true that so far, attempts to quantify this link have not delivered simple and easy-to-use analytical tools. 'Indeed, the scientists in this field ... were forced to admit that using energy as a numeraire to describe and analyse changes in the characteristics of ecological and socio-economic systems proved to be more complicated than one had anticipated' (Giampietro and Ulgiati, 2005).

We firmly believe that the failure to provide robust analytical procedures in the field of energy analysis can be ascribed to systemic problems associated with the conceptualization of the term 'energy', which is semantically very rich and therefore open to different criteria for quantification. This is a profound problem that affects elementary concepts (when dealing with aggregation – when and how to sum different energy forms) and quantitative analysis (when developing indicators based on the assessment of energy 'input/output'). In spite of this epistemological challenge, we are convinced that energy analysis has the potential to make a useful contribution to policy-making, especially when dealing with energy issues. It is for this reason that we make the effort to show the semantic problems that afflict and limit the methodological tools currently in use in the field of energy analysis – for example, the calculation of EROI in its present form – and how the framework of bio-economics could be used to improve these tools.

For this reason, we warn the reader that some of the following sections are written for those that already know basic elements of energy analysis. The reader can skip the technical parts of these sections without necessarily missing the main points, which are illustrated in the text with numerical examples. The rest of this chapter is structured as follows: the first section introduces in general terms the problems associated with the quantification of the semantically rich concept of energy. In particular, we illustrate the use of grammars for dealing with the analysis of the different energy forms found when analysing the metabolism of countries. The following section deals with the systemic problems encountered in energy analysis when attempting to perform a quantitative analysis of a network of energy transformation. The third section introduces the metaphor of heart transplant as an aid for the reader, and then moves on to the implementation of the concept of EROI using the framework of bio-economics (MuSIASEM) presented in Chapter 4. Finally, the chapter ends with a practical example of the process of assessing the quality of an energy input used for an economic purpose within an economic process. This example shows the relevance of the different dimensions and levels of analysis to be addressed.

Quantitative energy analysis depends on semantic pre-analytical choices

As discussed in Chapter 3, energy is a semantic concept which requires an appropriate set of pre-analytical choices for its quantification. In technical jargon, we say that the energy analyst has to define a proper grammar of the system of energy transformations before attempting any formalization. As defined earlier in Box 3.2, the definition of grammar refers to the pre-analytical phase in which the analysts must reach consensus about the meaning assigned to the numbers used in the model.

Why is it so important to use the concept of grammar, you might ask. It is because we deal with complex networks of energy transformations, and, as for all complex phenomena, we can always generate non-equivalent models to describe them. Each of these non-equivalent models allows us to focus on some of the characteristics of the investigated system, at the cost of ignoring others. This is beautifully explained by Box (1979, pp202–203), under the heading 'all models are wrong, but some are useful', when discussing the usefulness of quantitative models:

> *For such a model there is no need to ask the question 'is the model true?'. If 'truth' is to be the 'whole truth' the answer must be 'No'. The only question of interest is 'Is the model illuminating and useful?'*

Thus, when generating numbers in the field of energy analysis, what is really important is to obtain an agreement about the chain of semantic choices that leads to the final formalization. This requirement becomes crucial when these numbers are to be used in sustainability analysis and for decision-making (Giampietro et al, 2006a).

Problems related to use of mono-purpose grammar in energy analysis

Choosing just one quantitative protocol for the representations of a given situation is referred to as the adoption of a mono-purpose grammar. Clearly, this choice depends on the interests of the analyst, and on why one decides to generate the quantitative analysis in the first place. For example, the assessments of primary energy consumption found in international energy statistics – such as those of BP – are examples of quantitative assessments generated by a simple, mono-purpose grammar, characterized by the use of a limited lexicon and a limited set of production rules in the pre-analytical step (Figure 5.1).

As concerns the lexicon, the system of accounting used in international statistics does not deal with the entire set of energy transformations taking place in an economy. For example, solar energy, gravitational energy and food energy are not accounted for in these statistics. The grammar focuses only on a finite set of relevant categories of energy forms:

- primary energy sources (oil, coal, natural gas, nuclear, etc.);
- energy carriers (electricity, fuels); and
- end-uses/energy services (e.g. power generation, transport, residential).

Because of these choices, this grammar deals only with commercial energy.

As concerns the limited set of production rules, this simple grammar deals only with the correction of the accounting done in relation to the generation of electricity. This simple grammar has only one production rule consisting in a conversion factor, which transforms kilowatt-hours of electricity into a *reference* value of primary energy source, either joules of TOE or joules of tonnes of coal equivalent (TCE). This production rule is applied to any joule of electricity generated by using sources *other* than fossil energy. This production rule simply assumes a standard value for the conversion losses associated with the generation of electricity. In this particular example, the generation of 1J of electricity has a 'virtual cost' of 3J of oil equivalent. This implies that all the electricity produced by nuclear, hydroelectric, photovoltaic or wind sources is accounted as 3J of oil equivalent consumed in the category primary energy sources.

This mono-purpose grammar, therefore, is useful only in relation to the calculation of a standardized value of consumption of primary energy source – using fossil energy as the reference – for making international comparisons. It obtains this result by accounting 1J of electricity generated by alternative energy sources as a *virtual* energy consumption of oil, which has not actually been used in the generation of that 1J of electricity. The accounting of this virtual consumption makes it possible to compare the primary energy consumption of countries at different levels of electricity generation from non-fossil energy using a common standard. In this case, when applying the production rule, the amount of electricity generated is a *token* in the grammar (a data input) which determines the value to be assigned to the *name* (the assessment of 'primary energy consumption *equivalent*').

108 *The Biofuel Delusion*

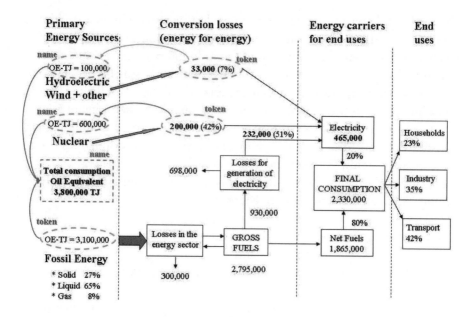

Figure 5.1 *Simple, mono-purpose grammar used to generate energy statistics: assessment of primary energy consumption in TOE of Spain in 1991*

Data are from EUROSTAT, 1995.

In this way, the overall assessment of primary energy consumption, used to compare different countries, is expressed either in TOE or TCE (exemplified in Figure 5.1 for Spain). In this grammar, the same standard conversion – in this example 3/1 – is applied to all countries included in the statistics, and hence country-specific variations are not accounted for. There are countries that produce electricity with inefficient coal power plants which may use 4MJ or even 5MJ to produce a 1MJ of electricity. Ironically, the use of a flat conversion factor should not be considered as an error within this grammar; systemic and easy comparison can only be obtained by *normalizing* apples and oranges in relation to a chosen criterion of equivalence: standard consumption of primary energy sources per kWh supplied. As a matter of fact, in recent years, the choice of how to account for this conversion factor has led to minor differences among data found in different statistics, for example EIA statistics and EUROSTAT.

The main point to be made about this example is that the bias associated with the given purpose of this grammar entails a limited applicability and usefulness of the numerical output. In this particular example, the overall assessment of primary energy consumption in TOE can *not* be related to the following.

- Final energy services (amount of goods and services obtained by the economy). This depends on how energy carriers are used on different end-uses.
- CO_2 emission. The overall assessment of primary energy consumption in TOE includes *virtual* TOE that does not correspond to any CO_2 emission (electricity produced by nuclear or hydroelectric power).
- Land requirement to obtain this energy supply or the pressure that the energy sector implies on land cover. These attributes are totally ignored in this grammar.
- Relative contribution of renewable and non-renewable primary energy sources.
- Total requirement of work hours in the energy sector.
- Total requirement of water.

An example of the general result of the application of this mono-purpose grammar to the characterization of the metabolic flow of a developed country is given in Figure 5.2. This is a standard map-flow of the energy consumption of a developed country. What is relevant in this figure is the clear adoption of a linear representation of the various flows. The energy flow changes its identity, and its numerical quantification, across the trajectory from left to right. It starts in the form of primary energy sources (on the left of the graph), then it becomes a flow of energy carriers (in the middle, because of transformations entailing energy losses). Finally, it ends (on the right of the graph) as a set of end-uses.

Within this representation, the correction factor adopted to account for the difference between primary energy sources and energy carriers (electricity) is used only to project the resulting correction on the left of the graph. It determines a comparable assessment of primary energy sources equivalent (the energy form used as reference: TOE) on the left of the graph. This choice entails an additional problem referring to the indication of the energy consumption of individual sectors (the arrows on the right of the map-flow). The consumption of energy – measured in the figure with numbers over the flows getting into the different sectors – no longer maps onto the relative consumption of joules of TOE. In fact, let's assume that two sectors – A and B – are consuming the same amount of energy carriers. If A is consuming 90 per cent of its energy in the form of electricity, whereas B is consuming only 10 per cent of electricity, we can conclude that A is responsible for a larger fraction of primary energy consumption than B. However, when using the method of accounting used for generating the quantitative analysis shown in Figure 5.2, the two sectors A and B would show as having the same consumption of energy. In relation to this point we proposed the use of dictionaries in multi-purpose grammars in Figure 4.10.

Clearly, it is possible to find quantitative representations, similar to the one

110   *The Biofuel Delusion*

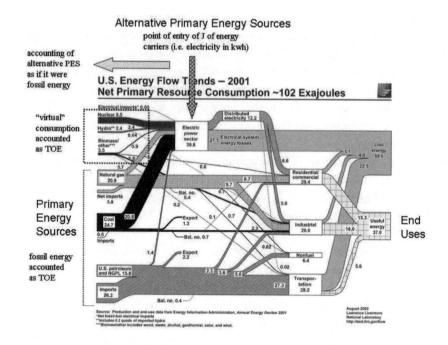

Figure 5.2 *The map-flow of energy consumption in the US, 2007*

Energy & Environment Directorate, Lawrence Livermore National Laboratory. Permission of the US government (via Wikimedia Commons); www.eia.doe.gov/emeu/aer/pdf/pages/sec1_3.pdf.

illustrated in Figure 5.2, which have been developed using other types of mono-purpose grammar, for example, to indicate the production of CO_2 associated with the energy consumption of a country or other relevant features. Examples can be found at www.eia.doe.gov/.

We believe that when dealing with the analysis and quantitative representation of the energy metabolism of societies, what is needed is a better-articulated or multi-purpose grammar: a grammar that, starting from a given data set, would allow us to compare the energy metabolism of different countries in relation to different benchmarks; and to analyse different scenarios, by switching 'names' with 'tokens' and/or imposing different production rules (for example, assuming technological changes).

In particular, this would make it possible to gain new information and understanding by adding new categories to the lexicon (tokens and names) of the grammar. Indeed, the conceptual tool of grammar is semantically open. This implies that additional categories and production rules can be added to the original data set, making it possible to enlarge the set of indicators to include some or all of the above-listed aspects. As discussed later on, this is essential when discussing scenarios of energetic metabolic patterns no longer based on fossil energy.

A general discussion of systemic epistemological problems associated with energy analysis applied to sustainability issues has been provided by the authors in several publications (Giampietro, 2003; 2006; Giampietro and Mayumi, 2004; Mayumi and Giampietro, 2004; 2006; Giampietro et al, 2006a; 2006b).

In the analysis of metabolic systems, no quantitative assessment of output/input energy ratio is ever substantive

Even though all energy forms can be expressed in the same standard SI unit of energy, joules (J), or in units that are reducible to joules (such as kcal, BTU, kWh), different energy forms may refer to logically independent narratives about change and in this sense they cannot be reduced to each other in a substantive way (see also Chapter 3). This implies that the validity and usefulness of a given conversion ratio, determining an energy equivalent for the conversion of one energy form into another, must always be checked in semantic terms.

Thus, to convert a quantitative assessment of a given energy form expressed in Joules into another, still expressed in Joules, one has to choose, first, a semantic criterion, for determining the equivalence class over the two energy forms; and, second, a protocol of formalization, to reduce the two to the same numeraire. Below we discuss three well-known problems in energy analysis that have their roots in this epistemological challenge: the standard problem associated with the summing of apples and oranges, the unavoidable arbitrariness entailed by the joint production dilemma, and the truncation problem.

Summing apples and oranges

Any aggregation procedure applied to a network of energy transformation must face the fact that different energy forms have different qualities. Often it is possible to find a particular perspective which makes it possible to consider them as 'belonging to the same equivalence class'. But one should be aware that they will be 'the same' only in relation to the particular perspective used for the comparison. The more we want to aggregate items that are different within the same equivalence class, the more the final assessment may be expected to lose relevance. In the end, it is the chosen protocol that will define the final number and its usefulness. Perhaps the metaphor of summing apples and oranges can help us to clarify the matter.

If we calculate the aggregate weight of a batch of apples and oranges, we find a number that is irrelevant for nutritionists concerned with a trace element present only in oranges, but very handy for the truck driver transporting them. On the other hand, if we sum the apples and oranges by using their aggregate content of vitamin C, we will end up with numbers that are useful for a nutritionist but irrelevant for either an economist or a truck driver. The more we aggregate items that can be described using different attributes (weight, price, nutritional content, colour, pesticide residues) using only a single category of equivalence, the more we increase the chance that the final number generated by this aggregation will be irrelevant for policy discussions (Giampietro, 2006).

This is a serious problem in energy analysis, since it is impossible to discuss the quantification of energy transformations without having agreed upon a useful accounting framework (Cottrell, 1955; Odum, 1971; 1996; Schneider and Kay, 1995; Kay, 2000; Fraser and Kay, 2002). For example, the quantitative assessment of 'the energy equivalent of a barrel of oil' cannot be calculated a priori, in substantive terms, without having specified first *how* that barrel will be used (its end-use) as a form of energy. Indeed, the same barrel of oil can have all of the following characteristics.

- A defined energy equivalent when burned as fuel in a tractor, but no energy equivalent when given as a drink to a mule, when using a narrative in which energy is assessed on the basis of its chemical characteristics which must match those of the converter.
- A defined, different energy equivalent when the barrel is used as a weight to hold a tent against the wind, when using a narrative in which energy is assessed by the combined effect of its mass and force of gravity within a given representation of contrasting forces.
- Yet another energy equivalent when the barrel is thrown against a locked door to break it, using a narrative in which energy is valued as the combined effect of its mass and the speed at which it is thrown, within a given representation of contrasting forces.

Joint production dilemma

We consider here as an example of a metabolic system a camel living in its natural habitat. This same camel can be considered as:

- a supplier of meat;
- a supplier of milk;
- a supplier of power;
- a supplier of wool;
- a supplier of blood to drink in emergencies in the desert; and
- a carrier of valuable genetic information.

How to associate with these different flows an energetic equivalent (MJ/kg) on the basis of their energy cost (Giampietro, 2006)? How to calculate the energy going into a camel? How to assign this energy input to the different outputs?

When dealing with multiple inputs and outputs – which are required and generated by any metabolic system – the choices made by the analyst will determine the relative importance (value/relevance) of end-products and by-products. When describing a complex metabolic system as a network of energy and material flows – linking different elements belonging to different hierarchical levels – it is always possible to generate multiple non-equivalent energetic representations. These different representations will reflect different choices about issue definition (narratives about the relevant changes to be investigated) and therefore quantitative assessments based on these different representations will be logically incoherent. Quantifications based on logically incoherent representations, even if referring to the same system, cannot be reduced in substantive ways to each other.

Truncation problem

When describing a system operating simultaneously on multiple scales, it is necessary to use several non-equivalent descriptions. By default, this entails the coexistence of different boundaries for the same 'entity' simultaneously perceived and represented at different scales. In particular, what should be considered as embodied in the inputs and/or outputs of a specific process depends on the choice of the scale (the choice of just one possible definition of boundaries) at which the assessment is performed. A famous example of the truncation problem is represented by the elusive assessment of the energy equivalent of one hour of human labour (see Box 5.1).

Stirling (1997) reported similar truncation problems for the economic accounting of the external costs of electricity generation. Costs reported in the studies monitored by Stirling varied from less than 0.05cents/kWh to more than 1000cents/kWh (a difference of 20,000 times!) depending on how the assessment was performed.

The truncation problem is highly relevant for the calculation of EROI (see the discussion of EROI in pre-industrial society in Figure 3.5).

Scale of representation in energy analysis

An important characteristic of metabolic systems, such as socio-economic systems, is that they auto-define the scale that should be used to represent their metabolism. For example, we cannot represent and quantify the energy input for a virus using the same descriptive domain as we do for representing and quantifying the energy input for a household or an entire society. In more general terms, we say that the very nature of a metabolic system entails a forced semantic interpretation of the categories that have to be used to represent their energy transformation (see Appendix 1).

> **Box 5.1** *Elusive assessment of the energetic equivalent of human labour*
>
> The literature on the energetics of human labour (reviewed by Fluck, 1981; 1992) shows many different methods to calculate the energy equivalent of one hour of labour. For example, the flow of energy embodied in one hour of labour can refer to:
>
> - the metabolic energy of the worker during the actual work alone, including (see Revelle, 1976) or excluding (see Norman, 1978) the resting metabolic rate – a scale of one hour;
> - the metabolic energy of the worker including non-working hours (see Batty et al, 1975; Hudson, 1975; Dekkers et al, 1978) – a scale of one day or one year over a single individual;
> - the metabolic energy of the worker and his dependents (see Williams et al, 1975) – a scale of one year over the whole family;
> - all embodied energy, including commercial energy, spent in the food system to provide an adequate food supply to the population (Giampietro and Pimentel, 1990) – a scale of one year over the food system of society;
> - all the energy consumed in societal activities (Fluck, 1981) – a scale of one year over the whole society; or, finally
> - H. T. Odum's EMergy analysis (1996), which includes in the accounting of the energy embodied in human labour a share of the solar energy spent by the biosphere in providing the environmental services needed for human survival – open in the definition of time and the whole biosphere.
>
> As a matter of fact, rigorous scientific assessments of the energy equivalent of 1 hour of labour found in literature vary from 0.2MJ to more than 20GJ, a factor of 100,000 (Giampietro et al, 2006b). This clearly shows that the quantification of an energy input required for a given process (or an energy output) in reality depends on the choice made when defining the boundary of that process.

This implies that metabolic systems can survive and reproduce only if they manage to gather what *they* define as energy inputs (negative entropy or *exergy* within a given well-defined system of accounting) and to discard what *they* consider waste (positive entropy or degraded energy). However, in this 'contextualized' definition, what is waste or positive entropy for one system (such as manure for a cow) may be seen as an energy input or negative entropy by another (such as soil insects) (Giampietro, 2006).

This key characteristic of metabolic systems entails that they have an expected identity, which translates into an expected range of values for the consumption rate of their specific energy input. Well-established typologies of metabolic systems (such as human beings, given models of car, developed countries) do have expected benchmark values in relation to their power level.

For example, humans cannot eat for long periods of time much less (100kcal/day or 0.42MJ/day) or much more (30,000kcal/day or 125MJ/day) than their expected daily food intake (about 1000–3000kcal/day or 4.2–12.5MJ/day); they would manifest health problems and eventually die.

Accepting that all metabolic systems are reproduced in the form of a metabolic pattern defined over a network of fund elements and flows entails accepting the existence of an admissible range for the pace of the various metabolized flows (of the whole and of the parts) associated with the identities of the various elements. This expected range of values for metabolized flows implies that the very same substance (such as a vitamin) can be beneficial or toxic for the system, depending on the congruence between the paces at which the flow is required and supplied. What is considered a resource when supplied at a given pace can become a problem when supplied at excessive pace. This is exemplified by the eutrophication of water bodies (too many nutrients to metabolize for the aquatic ecosystem) and by the disposal of human excrement. The latter can represent a valuable resource in rural areas (an energy gain for the system) but a serious waste problem in cities (an energy loss related to the construction and operation of treatment plants). A self-organizing metabolic network is defining a scale for its own metabolic pattern, when determining the size and the intensity of the metabolism of its parts and the whole.

The metaphor of a heart transplant

We propose here the metaphor of a heart transplant, illustrated in Figure 5.3, as an aid to check in semantic terms the feasibility and desirability of an alternative energy sector for society (such as an energy sector based on agro-biofuels rather than fossil fuels) (Giampietro and Ulgiati, 2005; Giampietro et al, 2006b). Indeed, the energy sector can be seen as the pulsing heart of modern society. We will refer to this metaphor throughout this and other chapters. Making use of this metaphor, we also introduce here the distinction between definitions based on functional and structural types (equivalence criteria), which is useful for the discussion of how to assess the quality of alternative energy sources.

We imagine the actual energy sector, based on fossil energy as primary energy source, as the heart (the organ or compartment) that keeps modern society (the individual or the whole) alive. Imagine now that we want to replace this heart with an alternative one (an energy sector based on agro-biofuels) which is supposedly better (such as zero GHG emissions from the energy sector and the fully renewable supply of energy carriers). Before actually performing the transplant, however, we may want to check whether or not the proposed substitution is feasible. Such a check must necessarily involve a comparison of the performance of the actual heart with that of the proposed alternative. The check should verify the congruence between the expected pace of blood flow (the required flow of energy carriers, determined by characteristics of society) and the pace of blood flow delivered by the heart we want to implant (the net supply of energy carriers which can be achieved by the 'alternative' energy sector).

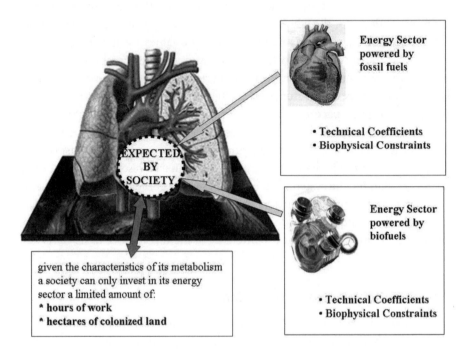

Figure 5.3 *The metaphor of heart transplant surgery*

Adapted from Giampietro, 2003.

The main point of our metaphor is that when dealing with the metabolic performance of a particular element (heart or energy sector) of a metabolic network (whole individual or whole society), we can define that particular element both in functional terms (it is interacting with the rest of the network (other organs or other sectors)) and in structural terms (it is expressing a given set of characteristics). Analysing the element in functional terms implies considering the mutual information regarding that element in relation to the function it performs. It represents the behaviour expected from that element by the rest of the network to which the element belongs. Analysing the element in structural terms implies considering the characteristics of its encoded identity; for example, the information contained in its blueprint and used in its production (see also the discussion in Appendix 1, Figure A1.1).

According to these two definitions, a heart transplant can be successful only if we implant a given realization of a structural type (an energy system generating a net supply of energy carriers), capable of expressing the functional type expected by the rest of the network (delivering the required supply of energy carriers at the right pace). In the case of an alternative system of the production of liquid fuels, we can use the expected characteristics of the functional type – what is working now – to check whether the characteristics of the new process generating the supply are similar to or better than the old one based on fossil fuels. In Chapter 4, we have illustrated a method of accounting which makes it possible to define expected/forced relations among the sizes of fund and flow elements used to represent a metabolic system.

Implementing the Concept of EROI within the Approach of Bio-economics

In the rest of this chapter, we discuss the problems found when trying to use the semantic concept of energy return on investment (EROI), often used in the field of energy analyses as an indicator to assess the 'quality' of alternative energy sources. In our discussion we build on theory, concepts and metaphors presented above and in previous chapters. In particular we attempt to integrate in our approach the following four strategies.

1. Addressing the multi-level nature of the dynamic budget of society and the peculiar characteristics of the autocatalytic loop of energy on which the dynamic budget is based. In relation to this task, it is important to develop a method for identifying functional and structural types within a pattern of metabolism, useful for defining parts versus whole.
2. Blending information about output/input energy ratios with information about power levels. A ratio of energy flows expressed in homogeneous units – the EROI – is dimensionless and out of scale. Recalling the discussion in Chapter 3, we can calculate a given EROI analysing the output/input energy ratio of a spider making a web. However, this assessment says nothing about the minimum threshold of power level required for the supply of the energy carriers associated to the exploitation of the primary energy source. To get out of this impasse, we can observe that the same EROI ratio can be obtained by dividing either two amounts of energy (ratio MJ/MJ), which is the classic approach, or two power levels (MJ/hour/MJ/hour). Using the MuSIASEM approach presented in the previous chapter, this latter method can be adopted, and the definition of

118 *The Biofuel Delusion*

    EROI can be scaled on the representation of a given pattern of societal metabolism by expressing forced relations over key flow/fund values.
3 Recognizing the need of making a distinction between quantitative assessments referring to different energy forms, such as primary energy sources, energy carriers and end-uses.
4 Making a distinction between the two types of biophysical constraints that affect the pattern of exosomatic metabolism: external and internal constraints.

Assessing the functional requirement of energy as defined by the rest of society

This section integrates, within a formal analysis of the dynamic exosomatic energy budget, the two concepts of EROI and bio-economic pressure (BEP). To this purpose we present a series of forced relations over fund and flow elements. These relations should not be seen as deterministic equations, but rather as a set of congruence constraints. As explained in Chapter 4, these relations refer to multi-scale analysis of the impredicative loop, which is behind the concept of dynamic energy budget (see mail delivery example in Figure 4.7). The labels used to indicate the identities of the various funds and corresponding flow elements must coincide ($ET_i \leftrightarrow HA_i$) in order to define relevant benchmarks. An example of the characterization of exosomatic metabolic pattern of developed society, obtained in this way, has been described in Figure 4.9.

The relation of congruence (1) below describes the requirement of exosomatic energy for the metabolism of the whole society:

$$TET = EMR_{AS} \times THA \qquad (1)$$

Where TET = total exosomatic throughput; THA = total hours of human activity; EMR_{AS} = average exosomatic metabolic rate (per hour) of society.

As noted in Chapter 4, EMR_{AS} is a benchmark very well correlated with GDP per capita. It provides an indication of the typology of the society we are dealing with. EMR_{AS} values < 5MJ/h are typical of developing countries, while values in the range 25–40MJ/h are typical of developed countries.

Relation (2) below describes the flows of exosomatic energy referring to the metabolism of the various fund elements making up the society and characterized at the level of the parts (n−1).

$$TET = ET_{HH} + ET_{SG} + ET_{PS^*} + ET_{ES} \qquad (2)$$

Where ET_i = exosomatic throughput relative to the fund element HA_i. The i compartments are: HH = household; SG = service and government; PS* =

productive sector (includes agriculture, building and manufacturing, and mining minus the energy sector); ES = energy sector.

In analogy with relation (1), the exosomatic throughputs in relation (2) can be written as follows:

$$\text{TET} = \text{EMR}_{AS} \times \text{THA} = \Sigma \, (\text{EMR}_i \times \text{HA}_i) \tag{3}$$

For example:

$$\text{TET} = (\text{EMR}_{HH} \times \text{HA}_{HH}) + (\text{EMR}_{SG} \times \text{HA}_{SG}) + (\text{EMR}_{PS^*} \times \text{HA}_{PS^*}) + (\text{EMR}_{ES} \times \text{HA}_{ES})$$

In Figure 4.9 we presented an overview of the pattern of exosomatic metabolism of developed countries analysed at the level of the parts, characterized by relation (3). This characterization entails:

- an expected set of relations of the parts over the graph;
- a set of expected benchmark values for the EMR_i of the various compartments; and
- a forced congruence over the resulting values of the flows (ET_i).

What is particularly relevant about this set of forced relations is that the profile of allocation of human activity does affect and is affected by the relative level of power (EMR_i) and energy flows (ET_i) of the various fund elements. In particular, recalling the dendrogram presented in Figure 4.8, we can write the effect of the series of three reductions in the form of the following relation (4):

$$\text{THA} \times (\text{HA}_{PW}/\text{THA} \times \text{HA}_{PS}/\text{HA}_{PW} \times \text{HA}_{ES}/\text{HA}_{PS}) = \text{HA}_{ES} \tag{4}$$
level n   level n/n−1   level n−1/n−2   level n−2/n−3   level n−3

This series of splits of human activity over lower level compartments (fund elements) must be congruent with the biophysical capability of the system to express the various exosomatic metabolic rates (EMR_i) typical of the fund elements considered. The approach that can be used to perform this check is presented in Appendix 2.

According to relation (4) and using the benchmarks used in Figure 4.9, we can write an *expected* relation between total human activity (THA) and work hours in the energy sector (HA_{ES}):

$$\text{THA}/\text{HA}_{ES} = (\text{THA}/\text{HA}_{PW} \times \text{HA}_{PW}/\text{HA}_{PS} \times \text{HA}_{PS}/\text{HA}_{ES}) \sim 1{,}000/1$$
                    1/ 0.1   1/ 0.35   1/ 0.03

Then, using relation (1) and the benchmark referring to EMR_{AS} (25–40MJ/h), we can calculate a benchmark value for the net supply of exosomatic energy per hour of labour in the energy sector required by a developed society. This value reflects the given bio-economic pressure (according to the analysis illustrated in Figure 4.4). The resulting range for this benchmark is:

$TET/HA_{ES} \sim 25{,}000\text{–}40{,}000 \text{MJ/h}$

These values indicate what is required by the rest of society (what is required by the rest of the body in the metaphor of the heart transplant) according to the expected set of relations and identities of the parts making up the dynamic budget of exosomatic energy of developed society.

Supply of exosomatic energy delivered by the energy sector

In general terms, we can write for the supply side of the dynamic budget of exosomatic energy the following relation (5):

$$TET = ET_{ES} \times EROI = ET_{ES} \times TET/ET_{ES} \qquad (5)$$

The supply of exosomatic energy delivered to society (TET) by the energy sector depends on the energy invested in the energy sector (ET_{ES}) and its return (EROI). This can be expressed as the ratio between the amount of energy supplied to society per amount of energy invested (TET/ET_{ES}). Please recall that even though TET and ET_{ES} are flows of energy, being defined in time (that is, on a yearly basis) they both have the dimension of power (GJ/year). One is the total flow of energy consumed by society in a year and the other the flow of energy consumed by the energy sector in that same year. The numbers used for this ratio must be expressed in the same unit and the same energy form, that is, they have to be expressed in joules of TOE. However, to have a better definition of the power level scaled to the size of society (measured in terms of hours of human activity) we can express these same two flows as exosomatic metabolic rates (flow/fund in MJ/h) of compartments of human activity (fund elements expressed in hours/year). That is, we can rewrite relation (5) (using relation (1)) as follows:

$$EMR_{AS} \times THA = (EMR_{ES} \times HA_{ES}) \times TET/ET_{ES} \qquad (6)$$

Relation (6) can also be formulated as:

$$EMR_{AS} \times THA/HA_{ES} = EMR_{ES} \times TET/ET_{ES}$$

Leading us to our key relation:

$$TET/HA_{ES} = EMR_{ES} \times TET/ET_{ES} \quad (7)$$

On the left side of relation (7) we have the term TET/HA_{ES}, which indicates the intensity of the supply of energy required (expected) by society per hour of labour in the energy sector. On the right side of the equation we have the product of two factors, $EMR_{ES} \times TET/ET_{ES}$, determining the *actual* supply of energy to the rest of society per hour of labour in the energy sector.

In semantic terms, the term TET/ET_{ES} (on the very right of the equation) can be regarded as the EROI: the energy return on the energy investment. However, when considering the simple numerical value of this ratio, we get a number which is out of scale; it does not provide any information about the size of the flows considered in the ratio. This is why it is essential to consider also the value of the other term: EMR_{ES}. In fact, by multiplying the value of EROI by the value of EMR_{ES} we can define a scale for the flow of the energy supplied to society within the given pattern of metabolism.

To make things more difficult, there is an additional complication. The semantic relation provided by relation (7) may be clear in its meaning, but when it comes to the quantification of this relation for specific societies, things become complicated. In fact, when using assessments for TET from energy statistics we obtain data expressed in primary energy source equivalents (for example, TOE). Recalling the map-flow given in Figure 5.2, the assessment refers to the numerical values indicated on the left of the graph. This entails that ET_{ES} should also by expressed in TOE. But if we do that, then, the ratio TET/ET_{ES} no longer coincides with the numerical assessment of the output/input of *energy carriers* going into and out of the energy sector: what is often measured and called EROI when considering the quality of energy sources. This discrepancy refers to the point made earlier that there is a difference between energy accounting based on energy carriers (for example, MJ of gasoline, MJ of ethanol) and that based on primary energy source equivalents (for example, the resulting value of these energy carriers when expressed in TOE or TCE). This distinction is negligible when dealing with the exploitation of different fossil energy sources – coal, oil or natural gas – but is very important when dealing with alternative primary energy sources, such as agro-biofuels. This point will be discussed later on.

Returning to relation (7), we reformulate it to write relation (8):

$$TET/ET_{ES} = (TET/HA_{ES}) / (ET_{ES}/HA_{ES}) \quad (8)$$

in which the concept of EROI (TET/ET_{ES}) is finally expressed in a way that makes it possible to frame it in relation to the analysis of the pattern of societal metabolism described in Chapter 4. On the left of relation (8) we have the ratio between two flows of energy in time, TET (GJ/year) and ET_{ES} (GJ/year), and on the right, the ratio between two key flow/fund ratios, characterized in relation to the scale of the societal metabolism: first, the required pace of net supply of energy to society per unit of work time in the energy sector (TET/HA_{ES}), a characteristic of society; and second, the power level which has to be expressed by the energy sector ($EMR_{ES} = ET_{ES}/HA_{ES}$) to achieve this goal; a characteristic of the process generating energy carriers through the exploitation of primary energy sources.

Both (7) and (8) point at the key role that EMR_{ES}, the power level expressed by the energy sector, plays in making possible a given exosomatic energy pattern. Indeed, the value of EMR_{ES} determines the viability of the dynamic energy budget. Moreover, EMR_{ES} can be linked to the analysis of economic factors relevant for assessing the economic viability of the process of generation of energy carriers. For example, EMR_{ES} is related to the requirement of capital and labour. Therefore, this value is also crucial for the economic accounting of fixed versus circulating costs.

Checking the implication of EROI against internal and external constraints

Having now a solid formal frame, we return to the basic rationale of EROI and its interpretation. To this purpose, we draw an analogy with the concept of EROI used in economics. In financial analysis, the concept of *economic* return on investment addresses at least three different relevant issues.

1 Ratio between return (output) and investment (input). How promising is the investment at first sight? In order to be useful, this first indicator must be contextualized using other indicators.
2 Absolute amount of investment required. Can we afford and handle such an investment? This has to do with the ability to generate and control the flow of input required to obtain the output, a check referring to *internal constraints*.
3 Payback time of the investment. How reliable is the information about the expected return? The longer the payback time, the more probable it is that original favourable conditions may change, thus altering the original estimate of output/input. With regard to energy, this has to do with stocks that may become depleted and funds that may deteriorate in time: a check referring to *external constraints*.

With regard to these three issues, relation (7) confirms that the biophysical viability of an energy sector (high net supply of exosomatic energy: the benchmark TET/HA$_{ES}$) depends not only on the output/input (TET/ET$_{ES}$) but also on the power level in the energy sector (EMR$_{ES}$: the feasibility in relation to internal constraint). As regards compatibility with external constraints, the integrated representation of EROI within the overall pattern of exosomatic metabolism can be used to calculate the overall consumption of primary energy sources associated with the generation of the given output of energy carriers required by society.

If the required supply of energy carriers is to be delivered using as little labour as possible, as is the case in developed society, society must invest in the capitalization of the energy sector. In this way, internal energy consumption increases (energy-for-energy) and so does energy consumption per hour of labour in the energy sector. What are the effects of this increased size of internal investment? They will depend on the quality of primary energy sources exploited. In the case of concentrated stock-flow energy sources, it is easy to increase power levels and obtain economies of scale by using increasingly powerful exosomatic devices (see Figure 5.6). The opposite is true when exploiting disperse primary energy sources (see Figure 5.7).

In conclusion, when facing high bio-economic pressure and hence the need for a very large supply of energy per hour of labour in the energy sector (TET/HA$_{ES}$ in the range 20,000–40,000MJ/h), we must have, beside a high output/input, a high power level per work hour in the energy sector. Even if assuming a pretty high output/input for energy carriers in the exploitation of given primary energy sources (for example, 20/1), this will not be enough if coupled to a low power level (for example, EMR$_{ES}$ < 500MJ/h) of labour in the energy sector. Actually, as discussed in Chapter 7, this is the case with ethanol produced from sugar cane in Brazil. By adopting labour-intensive practices one can keep the output/input of energy carriers high (for example, > 8/1) but this achievement is paid for by a low value of EMR in the production of ethanol from sugar cane (well below 100MJ/h).

Quality of Primary Energy Sources

It should be clear that when calculating the output/input energy ratio of energy carriers – often confused with EROI – we are *not* referring to assessments of primary energy sources. We saw in Chapter 3 that according to the first law of thermodynamics, it is not possible to have a closed process of energy conversion in which the overall output/input energy ratio is higher than one. Thus, whenever we have an output/input energy ratio larger than one – a greater-than-one return on the investment – we are dealing with an analysis based on investment (input) and gross return (output), measured in terms of energy carriers, which are used in the exploitation of primary energy sources. The energetic analysis of this exploitation generates a situation that is logically difficult to accept. The output/input ratio of energy carriers refers to, first, the output (gross supply of energy carriers (GSEC) resulting from this exploitation); and, second, the input (the fraction of this gross supply, the output, which is used for the exploitation). In this process of exploitation of primary energy sources, it is the output of energy carriers that generates the input of energy carriers! This point is illustrated in the left graph of Figure 5.4. But when considering the causality over the various flows of energy carriers in the opposite direction, we have to acknowledge that we cannot have the output without having the input. The analysis of this pattern proves that when dealing with energy analysis, it is necessary to learn how to deal with impredicative loops, in spite of the esoteric name! When analysing this pattern, the calculation of the output/input ratio of energy carriers is required to calculate the difference between the amount of primary energy sources (PES) consumed and net supply of energy carriers (NSEC) made available to society.

To clear up this confusion once and for all, it is crucial to consider the following point: any conversion of primary energy sources into a net supply of energy carriers to society entails two types of energetic costs referring to distinct types of losses.

Distinct types of energy losses in the conversion of primary energy sources into net supply of energy carriers

When studying the overall conversion of a flow of consumed PES into an NSEC, we have to study the combined effect of two types of losses, according to the overview given in Figure 5.4. The first loss refers to the conversion of one energy form into another. It is a 'pure loss'; it refers to the conversion of a given amount (in J) of primary energy source into the corresponding, but lesser amount (in J) of energy carrier, the difference being lost. The second loss refers to the consumption of energy carriers (input: GSEC−NSEC) involved in this conversion process (for example, fuels used in exosomatic devices to realize the conversion) in order to produce the energy carriers (output: GSEC). This is an internal loop of energy-for-energy in which the energy consumed is used to perform useful work and can be seen as an overhead over the gross output of energy carriers. That is, a certain fraction of the gross production is not available for the rest of society, since it has to be used inside the energy sector for its own production.

Thus, when exploiting a given primary energy source, it is necessary to invest a certain amount of energy carriers in the exploitation process in order to obtain a gross supply of energy carriers. This input (GSEC−NSEC) is used to generate the output (GSEC) of the exploitation process. The output/input ratio of the exploitation of primary energy sources therefore refers to the ratio GSEC/(GSEC−NSEC). This fact implies that PES > GSEC > NSEC.

Assessment of the Quality of Alternative Energy Sources

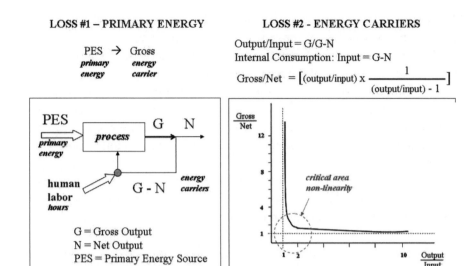

Figure 5.4 *The two factors determining the quality of energy sources*

The distinction between types of energy losses becomes crucial when different types of primary energy sources and energy carriers (EC) are involved in the same process, as is the case with the use of gasoline (EC) obtained from fossil energy (PES) for the production of agro-biofuels (EC).

When the output/input ratio GSEC/(GSEC−NSEC) becomes low (for example, <3/1), negative consequences of this loop amplify in a non-linear way according to relation (9):

$$\text{GSEC/NSEC} = (\text{output/input}) \times \frac{1}{(\text{output/input}) - 1} \qquad (9)$$

Where:

- GSEC/NSEC = calculated over energy carriers (dimensionless);
- output/input = output/input ratio calculated over energy carriers (dimensionless);
- output = GSEC (in J of energy carrier or PES); and
- input = GSEC−NSEC (in J in terms of EC or PES).

The values of the output and the input can be expressed in joules of PES by considering the first loss of conversion of PES into GSEC.

For this reason, when operating with an output/input of energy carriers below 3/1 the feasibility of the relative exploitation process becomes doubtful. When the ratio gets below 2/1 the exploitation process does not have any chance of being feasible (let alone desirable) as an exosomatic energy source. In fact, as illustrated in Figure 5.4, the curve gross/net supply of energy carriers entails a non-linear increase in internal energy consumption for low values of output/input.

The overview given in Figure 5.4 tells us that the quality of a process of conversion of a primary energy source (J/year) into a net supply of energy carriers to society (J/year) depends on the following.

1   The technical coefficients of the conversion of primary energy source into a gross supply of energy carriers (loss #1). When the primary energy source is oil, a fossil energy source, we deal, for example, with the conversion of crude oil (in J) into gasoline (J). When we deal with agro-biofuels, we have a series of two conversions: first, the conversion of solar energy and other energy forms required for biomass production (in J) into crop energy (in J), and second, the conversion of crop energy (in J) into ethanol or biodiesel (in J).
2   The output/input ratio of energy carriers for the process of generation of energy carriers (loss #2). When the primary energy source is fossil energy, then this loss could refer to, for example, fuels (in J) spent to generate fuels. In this case, the loss is moderate because the output/input ratio for gasoline production from oil is well above 10/1. When we deal with agro-biofuels, then this loss refers to the double overheads of energy carriers needed to perform the series of two conversions: solar energy to biomass and biomass to ethanol. Depending on the overall output/input ratio of energy carriers over these two conversions, this internal loop can imply very severe losses.
3   The net supply of energy carriers per hour of labour and per hectare of colonized land in the process, generating NSEC. When the primary energy source is fossil energy, the intensity of NSEC per hour of labour and the density of NSEC per hectare of colonized land is determined by both the intensity and density of the extraction and refinery of fossil energy sources, and the severity of the losses. When we deal with agro-biofuels, the intensity of NSEC per hour of labour and density of NSEC per hectare of colonized land is determined by, first, the combined intensity and density of the activities performed in the two steps of the agro-ecological transformation process (solar energy → crop energy → biofuels); and, second, the severity of the losses.

Quality of fossil energy as primary energy source

Having made these distinctions, we can make an important observation regarding the extremely high quality of fossil energy as a primary energy source. The quality of fossil energy is so high that even professional energy analysts have overlooked the distinction between the two types of losses described above, and consequently they have often confused the definition of EROI with the assessment of the output/input ratio of energy carriers. As observed by Georgescu-Roegen (1979), in the 1970s literature on energy analysis showed a certain confusion. The terms 'gross energy' and 'net energy' were used to indicate the distinction between primary energy sources and energy carriers respectively, *as well as* the distinction between gross supply and net supply of energy carriers, determined by the output/input ratio of energy carriers used to exploit primary energy sources. This confusion has gone largely unnoticed, and this is why we keep using the linear representation of energy flows, such as the one indicated in Figure 5.2. This is not at all surprising, because when dealing with fossil energy (and in particular oil), the first loss related to conversion of the energy form per se, for example crude oil to gasoline, is very low and difficult to detect since a part of the barrel of oil, used as input, goes into useful by-products. In the 1960s and 1970s, the second loss was very small as well, resulting in a very high output/input ratio of energy carriers, well above 40/1 and far away from the non-linearity indicated in the graph of Figure 5.4. This implied a very small combined loss in the overall conversion from PES to NSEC.

Moreover, the quality factor of 3/1 (referring to a virtual loss of PES equivalent) used in international statistics is used only for the accounting of energy carriers generated using alternative PES, for example electricity generated by nuclear power. This implies that, looking at the map-flow illustrated in Figure 5.2, the effect of this correction over the accounting of amount of energy flow goes only from the centre (from the entry point of energy carriers generated by alternative sources) to the left of the graph (the resulting assessment of virtual TOE of nuclear energy).

This linearization in the representation of impredicative loops of energy carriers generated by the exploitation of alternative PES implies losing a lot of valuable information. In fact, every time we add a new primary energy source to the exosomatic metabolic pattern of a society – photovoltaic, bio-energy, hydroelectric – we introduce a new typology of impredicative loop, with special characteristics, into the network of energy conversions. So far, the problem generated by alternative PES has been solved by pretending that alternative energy sources were TOE or TCE. This had the aim of keeping the linear representation of flows across different energy forms – as done in Figure 5.2 – which are then mapped onto amounts of input equivalent of fossil energy on the left. But the question is: how useful is this representation if we want to study the feasibility and desirability of an exosomatic metabolic pattern of a society in which a large fraction of PES will no longer be fossil energy? In relation to this point, we suggest that it would be much more useful to base the representation of exosomatic metabolic patterns on multi-purpose grammars based on the accounting of flows of energy carriers. In this way, it would be possible to address the existence and to study the implications of different typologies of impredicative loops referring to different typologies of primary energy sources.

Quality of agro-biofuel as primary energy source

If we consider the performance of agro-biofuels, we find a situation completely different from that of fossil energy fuels. Agro-biofuels have a pretty bad performance in relation to both types of losses indicated in Figure 5.4.

The first loss (conversion PES → EC) refers to two distinct reductions in the flow of energy: from solar energy to crop biomass, and from crop biomass into biofuel. The efficiency of the first conversion is low. Given the on-average 58,600GJ/ha of solar energy reaching the Earth's surface each year (Giampietro, 2002), crop biomass can fix only in the order of hundreds of GJ/ha. Hence, the first reduction in energy flow, from solar energy (PES) to crop biomass, is severe (hundreds of times smaller). The conversion from crop biomass energy to biofuel energy implies an additional reduction (data on ethanol production from corn and sugar cane are provided in Chapter 7).

This second loss is crucial when dealing with crop production for biofuel in temperate areas (for example, ethanol from corn, biodiesel from rapeseed or sunflowers) because the output/input ratio of energy carriers is critically low (smaller than 3/1), thus generating a non-linear increase in the GSEC/NSEC ratio.

Many energy analysts consider the loss associated with the first conversion (solar energy into biomass energy) irrelevant, since solar energy is free. On the other hand, this conversion is extremely relevant for the indirect economic and environmental costs that the large-scale exploitation of low-density energy flows implies; a large demand for labour, capital and land, and the loss of natural habitat for biodiversity. The additional requirement of primary energy sources associated with the second type of energy loss of biofuel production aggravates these indirect costs, provided of course that biofuel production is fuelled by its own production and not by fossil energy.

When dealing with agro-biofuels, it does not make much sense to use the same method of accounting based on TOE as for fossil fuels. Energy carriers in the form of biofuel would require selecting an ad hoc criterion for defining a class of equivalence. But agro-biofuel, when considered as a fully renewable and self-sufficient system (covering all its energetic costs), has nothing in common with fossil energy sources (fund-supply versus stock-supply), and therefore does not provide many criteria for defining an equivalence class related to TOE.

Full substitution of fossil energy with agro-biofuels is impossible!

The effect of the differences in quality of the primary energy sources fossil energy and fund-flow production of biomass is illustrated in Figure 5.5. This is a thought-experiment based on the heart transplant metaphor illustrated in Figure 5.3. In the top graph, we have the pattern of metabolism expressed by a modern society based on fossil energy (Figure 5.5a). The overall output/input of 15/1 refers (calculated over data expressed in J of TOE in relation to the ratio TET/ET_{ES}) to a power level of 300GJ/person/year and an EMR_{ES} of 2000MJ/h, assuming that only 10 hours of work per capita per year (HA_{ES}) are allocated to the energy sector.

Now, we will substitute the energy sector of this hypothetical developed society while keeping the energy consumption of 'the rest of society' at the original level ($TET-ET_{ES}$ = 280GJ/year per capita). This would be the case in a heart transplant. We further keep the same profile of human time allocation (10 hours per capita per year in the energy sector) and the same flows of energy consumption for all sectors other than the energy sector, as shown in Figure 5.5a, that is:

- household sector: ET_{HH} = 90GJ/year in final consumption (residential plus private transportation) and HA_{HH} = 7900h pc/year (from Figure 4.9);
- service and government sector: ET_{SG} = 60GJ/year and HA_{SG} = 530h pc/year;
- building and manufacturing minus energy sector: ET_{PS*} = 130GJ/year and HA_{PS*} = 320h pc/year.

We 'implant' in this system an energy sector powered by a primary energy source with an output/input ratio of energy carriers of 1.33/1; this is a typical value of agro-biofuel production in temperate areas. The lower part of Figure 5.5b shows that, when the output/input is 1.33/1, the supply of 1MJ of NSEC to society requires the generation of 4MJ of GSEC by the energy sector. Thus, the energy sector consumes 3MJ of energy carrier (the input defined as GSEC−NSEC) for each MJ of energy carrier delivered to society.

Thus, performing a naive analysis based on the assumption that all else is equal and an ambiguous definition of EROI (without getting into the distinction between output/input of energy carriers and EMR_{ES}), we would infer that this energy sector would have to consume for its own operations an ET_{ES} of

130 *The Biofuel Delusion*

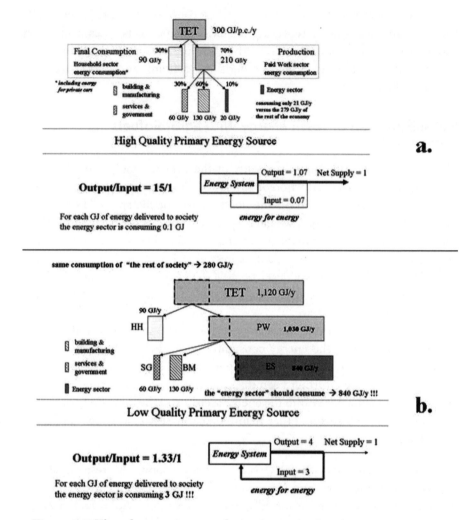

Figure 5.5 *Thought experiment: substituting society's primary energy source*

840GJ/year per capita. This would raise the total energy consumption of society (TET) to 1120GJ/year per capita, an increase of four times the original level of energy consumption per capita!

However, things are actually worse. If we perform an impredicative loop analysis over the dynamic budget of the exosomatic energy metabolism illustrated in Figure 5.5b, we find that this pattern is simply not feasible. We now integrate the analysis of EROI with the other two terms discussed earlier – the output/input of energy carriers in the energy sector, and the resulting EMR_{ES} –

in order to be able to perform the double check against external and internal biophysical constraints. By dividing the value of ET_{ES} = 840GJ/year per capita (estimated using the ceteris paribus hypothesis) by the hours of labour allocated in modern society to the energy sector (10 hours/year per capita) we obtain a EMR_{ES} of 84,000MJ/h. Thus, based on this hypothesis, operating the energy sector would require a power level more than 40 times higher than the power level found at this moment in the energy sector powered by fossil energy. Note that the power level actually achieved in the energy sector of developed countries is already the highest ever achieved and achievable so far by human technology.

This additional boost in power requirement would determine a clear biophysical impossibility. In fact, even if we agree that it would be possible to further boost the power level in the energy sector by 40 times, the consequent dramatic increase in the internal consumption of energy carriers and the relative requirement of exosomatic devices (technical capital) would increase even further the required investment of energy-for-energy, and therefore would generate additional non-linear increases in the gross/net ratio of the process, blowing any chance of a feasible metabolic pattern.

In conclusion, adopting a low-quality energy source for sustaining the exosomatic metabolism of developed society is simply not feasible, not even in a thought-experiment! The take-home message from the example discussed in Figure 5.5 is that any massive shift to a primary energy source with either a much lower output/input of energy carriers and/or a much lower EMR_{ES} than fossil energy would generate a collapse in the functional and structural organization of the economy.

Therefore, the real question to be answered in relation to the agro-biofuel idea is: if a massive move to agro-biofuels is not possible, can agro-biofuels supply at least a significant fraction of the liquid fuels consumed by modern society? This question will be answered using the same approach and empirical data in Chapter 7.

Additional Methodological Considerations

When integrating the analysis of the semantic concept of EROI with a bio-economic analysis of the dynamic budget of the pattern of the exosomatic metabolism of societies, we find that EROI is no longer a simple number calculated by dividing two others (an assessment of energy output over an assessment of energy input). Rather, we obtain a more robust insight into the underlying semantics and syntax. The fact that any representation of a given pattern of metabolism across multiple levels is not and cannot be substantive should no longer be considered a problem. On the contrary, it should be considered a plus of the proposed method. Since autocatalytic loops, by definition, escape the simplifications typical of reductionism, we can take advantage of this feature by forcing the analyst to explain and defend the choices made when selecting and implementing a given system of accounting. That is, the discussion over the assumptions to be used when defining the EROI in relation to a set of required 'power levels' associated with the expression of a metabolic pattern at a given level of bio-economic pressure can be a powerful tool for understanding the underlying issues.

A second advantage of a method of accounting based on the framing of the EROI concept within the bio-economic approach proposed in Chapter 4 is related to the use of the concept of grammar (see Box 3.2) for the formalization of semantic relations. In bio-economics, the definition of a grammar takes place when deciding about the identities used for characterizing the various funds and flows used in the quantitative representation. Once the grammar is defined, the set of expected relations between fund and flow elements entails that several congruence constraints can be used to check the feasibility of the resulting energy budget (recall the example of delivery of letters by the mail service in Figure 4.7). Therefore, any quantitative representation will depend on the decisions taken about how to account the role of the various fund elements (see the example of possible ways for calculating the EROI in Table 3.1) in the dynamic budget.

When adopting the concept of grammar, we can escape the problem of the infinite recursive loop. The choice of grammar defines, within a given multi-level matrix of funds, the dendrogram of splits over funds and flows (Figures 4.6 and 4.8). In this approach, the identities assigned to the various fund elements are then used to calculate the corresponding flows of energy and metabolic rates (Figure 4.9). Therefore, when addressing the whole metabolic pattern of society, any arbitrary choice by the analyst for calculating inputs and outputs on a chosen identity of a given fund element (reflecting the given definition of boundary) will also affect the accounting done for the remaining fund elements. That is, any decision about what should be included (or excluded) when calculating a particular element will also be reflected in what should be included (or excluded) in the calculation of the remaining elements. This is what the impredicative loop is about. Since the characteristics of all the fund elements (parts of the whole) must be accounted for in this integrated analysis of EROI, the arbitrariness of the choice referring to individual elements will be 'normalized', so to speak, over the entire system of accounting.

A practical example of the check over the amount of liquid fuels that can be covered with agro-biofuels, based on this approach, is given in Chapter 7; a few examples of analysis of the pattern of the exosomatic metabolism of modern society based on the adoption of grammars referring to the approach of bio-economics are provided in Appendix 2.

Recap of the integration of the concept of EROI and bio-economic pressure

We want to drive home the following points with regard to the concept of EROI:

- EROI is a semantic concept, which can be used to study the feasibility and desirability of a dynamic budget. The original concept of economic return on investment, which provided the inspiration for EROI, is an accounting valuation method developed in financial analysis and used to assess the feasibility and desirability of a financial investment. For this reason, it is essential that applications of this concept to energetic analysis address the crucial importance of critical power thresholds determining the feasibility and desirability ('convenience') of a given metabolic pattern.
- EROI, being a semantic concept, can be formalized in different legitimate ways, depending on the goal of the analysis. The final choice depends on the criteria considered relevant for the study.
- The existence of multiple legitimate formalizations of the concept of EROI implies that its quantification has to be based on the development of a specific grammar. This requires reaching an agreement on how to define and implement such a grammar.
- When dealing with metabolic systems, we can take advantage of the existence of benchmark values, determined by a typology of expected relations: for example, the values found when studying the metabolic patterns of developed countries.
- Defining EROI as a ratio between two flow/fund ratios (power level/power level) gives a more powerful characterization than defining EROI as a ratio of a simple amount of energy output over an amount of energy input. The latter ratio does not provide any information about the labour requirement, does not clarify the relationship between energy carriers and primary energy sources, and does not say anything about the impact on external constraints.
- Defining EROI as a ratio of two metabolic rates (expressed in MJ/h), we do not calculate just an output/input ratio, but we can also check the congruence over two different paces of flows: first, TET/HA_{ES} (the value of the net supply per hour of labour in the energy sector expected by society) versus the value that can be achieved by the proposed alternative; and, second, the level of capitalization of the energy sector ($EMR_{ES} = ET_{ES}/HA_{ES}$; the exosomatic metabolic rate of ES), as the value expected by society versus the value that can be achieved by the proposed alternative.

A picture is worth a thousand words

After this lengthy theoretical analysis, we conclude our discussion with some pictures, which we believe can help us to visualize what we mean by the high power levels (EMR_{ES}) achieved in the delivery of fossil energy in the energy sector. This refers to 'the ability of the investor to handle the required flow of investment' (in financial language) or 'the power level required to process the input of energy carriers to be used in the exploitation of the primary energy source' (in energetic language).

Figure 5.6 *Illustration of high power level associated with the exploitation of fossil energy: Bagger 288*

Source: photo courtesy of RWE Power AG, http://www.rwe.com/generator.aspx/rwe-power-icw/infotainment/baggertransporte/rueckblick-in-bildern/contextId=56674/language=de/id=134684/bild10-img.html.

The machine shown in Figure 5.6 (Bagger 288) is used for extracting coal in an open mine in Germany (in the photo, it is being transported). This exosomatic device extracts 240,000 tonnes of brown coal per day, and is operated by seven people. Instead of this picture, we could have used a photo of a super oil tanker operated by a small crew, a pipeline moving huge amounts of oil per day, or a power plant generating electricity at a pace of MW. In all these examples, the flow of energy per hour of paid work is orders of magnitude higher than the minimum threshold of 1500–2000MJ/h required for the energy sector.

In comparison, the pictures in Figure 5.7 of the activities taking place on an oil palm plantation provide a good illustration of the power level behind each worker in the production of agro-biofuels.

Figure 5.7 *Illustration of the power level achieved in an oil palm plantation*
www.flickr.com/photos/nagacocoa/sets/72157603186921140/, with permission.

Example of the Challenge of Energy Analysis when Integrated with Bio-economics across Hierarchical Levels and Scales

Distinction between energy input, energy carriers, applied power and useful work

In this chapter, we have seen that an analysis of the quality of primary energy source should be related to a quantitative analysis of, first, the requirement of energy services of a given society; and, second, the supply of energy services generated within that society by its energy sector. The problem faced when trying to perform such an integrated characterization is that it would require the calculation of at least four non-equivalent numerical assessments, related to four different key pieces of information (Figure 5.8):

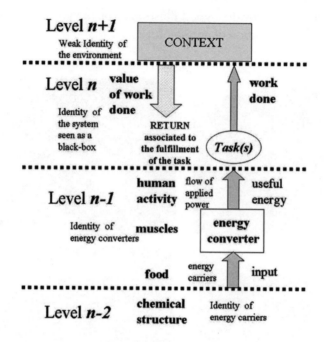

Figure 5.8 *Relevant qualities for characterizing the performance of a chain of energy conversions across hierarchical levels*

1. The flow of a given energy input required and consumed by the converter to generate applied power, but which can be measured in terms of the consumption of either primary energy sources or energy carriers. The efficiency of the conversion of primary energy sources into energy carriers will affect this assessment.
2. The power level at which useful energy is generated by the converter. The power level at which the various converters (defined at different levels) are operating.
3. The supply of applied power generated in the conversion process. Also in relation to this point, the amount of end-uses generated by the consumption of a given amount of energy input depends on the efficacy of the conversion of energy carriers into end-uses.
4. The usefulness of work done by the flow of applied power. This is elusive information, referring to how useful is 'that which is achieved by the particular profile of end-uses'. In fact, as in traditional physics, 'work' – what is achieved by the application of power – is an elusive quality that requires many assumptions to be measured and quantified in biophysical terms. For example, it is impossible to assess in biophysical terms the

difference between the work performed by a famous orchestra director and the work performed by a policeman directing traffic, when looking at their endosomatic energy consumption, let alone comparing the work performed by an orchestra director and a refrigerator.

Since this discussion may appear too theoretical, especially in relation to the energetic assessment of agro-biofuels, we provide in the next section an example of an analysis that addresses these various issues in practical terms. Our example is about answering a simple question: Are crop residues a feasible and desirable energy input for feeding cows? The analysis used to answer this question will help us clarify the type of information required for answering more important questions, like: Are agricultural crops a feasible and desirable energy input for feeding the exosomatic metabolism of modern societies?

Example: Are crop residues a feasible and desirable energy input for feeding cows?

After the bombardment of theoretical discussions and examples in the previous sections of this book, we hope the reader is able to come up with the right answer to the question posed in this section: that is, 'It depends'. In fact, there are situations in which crop residues are a valuable resource for raising cows, and other situations in which they are neither a feasible nor a desirable energy input for raising cows. Understanding the mechanism that makes a given energy flow either a resource or a problem is crucial to understanding why, at the moment, agro-biofuels are not a feasible or desirable alternative to oil.

In order to answer this question, we have first of all to:

- adopt an integrated analysis of the process of production of agro-biofuels, which can describe its performance across different levels and dimensions;
- specify the type of constraints that the metabolic pattern of a developed society entails on the production of energy carriers, such as the effect of bio-economic pressure on the expected performance of the energy sector; and
- define the type of ecological constraints determining the stability of the fund element (ecological processes), guaranteeing favourable boundary conditions. This constraint determines whether or not the production is really renewable and ecologically friendly.

In relation to this integrated analysis, the analysis of complex phenomena entails looking for *emergent properties* of the whole – seen at a given level – that cannot be detected when looking at the set of energy transformations associated with the activity of individual converters (parts) operating within the black box (at a lower level).

To clarify this point, we provide an example based on an analysis of the feasibility and desirability of feeds of different nutritional quality in a system of animal production. This analysis is based on the work of Zemmelink (1995) and illustrated in Figure 5.9.

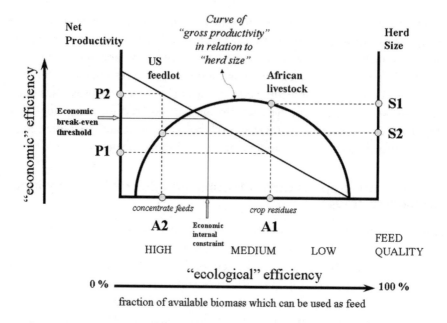

Figure 5.9 *Analytical scheme checking the possibility of using feeds of different qualities in meat production*

Based on data from Zemmelink, 1995

In the graph of Figure 5.9, the horizontal axis (A1, A2 and so on) represents differences in the quality of feed. This quality is defined by the nutrient and energy content of the feed per unit of mass. Looking at the feed types possibly available in a given agro-ecosystem (the given lexicon of possible energy sources), we can have, for example:

- feed obtained from dedicated crops or valuable by-products – high quality feed;
- feed obtained from tree leaves and other fleshy biomass – medium quality feed; and
- feed obtained from rice straw – low quality feed.

Therefore, moving on the horizontal axis implies changing the mix of possible feed types. 'Very high quality feed' implies that only dedicated crops or valuable by-products are used (a small fraction of the total biomass produced in the ecosystem); 'very low quality feed' implies that rice straw can be used in the mix, increasing the amount of feed available.

The points on the curve refer to possible herd sizes (S1, S2 and so on) which can be maintained as a stock in the system, in relation to the chosen quality of feed. The diagonal line indicates the relationship between the productivity (pace of output) of animal products, such as beef (P1, P2 and so on on the vertical axis on the left), and differences in the quality of feed (A1, A2 and so on). Different levels of productivity (analogous to power level) entail a difference in the quality of feed used as inputs for animal production (for example, the points A1 and A2 on the horizontal axis).

According to this graph, when using only animal feed of high quality one can get a high level of productivity (boost the output). However, by doing so, one can only use a small fraction of the total primary productivity of a given agro-ecosystem (low ecological efficiency). Another consequence is that the total stock of animals in the system will be small. Thus, this analysis describes an expected relation among:

- productivity in time (power level – left vertical axis);
- ecological efficiency (utilization of available biomass – horizontal axis); and
- stocks in the system (herd size – right vertical axis).

However, with this analytical output, obtained by applying a grammar to the set of possible energetic transformations, we cannot yet define the feasibility and desirability of different feed types (energy inputs for the system). In fact, the feasibility and desirability depend on the required productivity and characteristics of the agro-ecosystem. The required productivity is determined by the socio-economic context. In quantitative terms, a proxy variable would be the result of an economic analysis, in which the break-even point defines a minimum threshold of productivity. This is illustrated on the left vertical axis. The characteristics of the agro-ecosystem refer to the set of biological conversions and the ecological context that are required to guarantee the productivity of the agro-ecosystem providing the biomass used as feed.

Also in this example, looking at the feasibility and desirability of feed for animal production, we find the expected trade-off between power level and energy consumption that is so familiar for the mileage of cars. The need to operate at a high level of productivity – a high bio-economic pressure – translates into a reduction of ecological efficiency in using available biomass resources (high speed entails larger consumption). When the socio-economic constraints force operation at a high level of productivity, a large fraction of the available biomass, such as tree leaves and rice straw, can no longer be considered a useful input for the system (feed), and will result in waste.

This analysis clearly demonstrates the need for contextualization in biophysical analysis. If we look only at biophysical variables we can only characterize whether or not a feed input of quality A1 is an input of 'adequate quality' for a system of beef production operating at productivity level P1. But the ultimate decision about whether or not productivity level P1 is feasible and desirable for the owner of the beef feed-lot cannot be made from within this biophysical analysis. The viability and desirability of productivity level P1 depends on the constraints faced on the interface between the beef feed-lot and the rest of society. The evaluation of desirability has to be done considering a different dimension of analysis. In this case, the acceptability of P1 has to be checked using a socio-economic dimension (the position of the economic break-even point on the left vertical axis): an evaluation of the pace of generation of added value, linked to productivity level P1 and required for the viability of the production system.

In conclusion, the very same feed input of quality A1 can be either perfectly adequate for that system of animal production in a given social context (for example, when the production takes place in a developing country) or unacceptable (when moving the same biophysical process from a developing country to a developed country). That is, a change in socio-economic context can make level P1 no longer acceptable. The same owner of the feed-lot, forced to operate at higher productivity (for example, P2) in order to remain economically viable, would judge feed input of quality A1 no longer viable or desirable. On the other hand, when considering this case only in biophysical terms, the feed input A1 would remain of adequate quality for sustaining a given population of cows. But this information is irrelevant to whether or not this input has an adequate quality for sustaining, in economic terms, the threshold of productivity required by the owner for his feed-lot to survive in the actual context.

The set of relations described in Figure 5.9 is based on a series of well-known biological processes for which it is possible to perform an accurate analysis (as shown by Zemmelink) referring to physiological conversions associated with animal production. Put another way, we can easily develop an ad hoc grammar and even perform an exergy analysis (that is, a sophisticated evaluation based on thermodynamic reading). Yet, due to the complexity of the metabolic pattern of a system of animal production operating across multiple scales, and due to the different dimensions of analysis which have to be considered, the concept of 'quality of the energy input to the whole system' will always depend on, first, the hierarchical level at which we decide to describe the system (for example, cow versus the whole beef feed-lot), and, second, the context within which the system is operating, when considering the economic side of the animal production system.

That is, if we want to consider socio-economic interactions, as required by sustainability analysis, then there are 'emergent properties of the whole' that will affect the viability or desirability of an energy input: the minimum admissible feed quality for achieving an economic break-even point. These emergent properties can affect the admissibility of the metabolic pattern of the whole, and therefore induce an internal biophysical constraint within a particular conversion process, such as the need of reaching a certain threshold of power level. For example, in the feed-lot case, the pace of transformation of feed (plant energy) into beef (animal energy) at the hierarchical level of the whole production system has to be compatible with the bio-economic pressure (supply of products per unit of labour) typical of the socio-economic context.

This can imply that what is an effective energy input, when operating at a lower power level – in this example, the mix of feed of quality A1 in Mali – is no longer a viable or desirable energy input when the feed-lot is operating in the US. That is, even when the biophysical parameters of the system remain completely unchanged across level n/level n−1/level n−2 (that is, keeping the same cows, the same set of potential energy inputs for the feed, the same techniques of production), if we couple the same system with a different external context across level n/level n+1/level n+2 (the interaction between the beef feed-lot and the rest of society) we will obtain a different biophysical definition of quality for what should be considered as a feasible and desirable energy input; see the scheme in Figure 5.8.

In conclusion, the question about the viability and desirability of crop residues as alternative feed cannot be answered just by looking at one particular dimension (biophysical accounting) and one scale of analysis (considering only a series of three contiguous levels of conversions). Crop residues may or may not provide quality nutrition to cows, but their viability and desirability as a feed depends on the severity of the biophysical constraints determined by the pattern of societal metabolism expressed by the whole, which in turn depend on the interaction of the whole with its context.

A self-explanatory overview of the different hierarchical levels required to properly describe funds and flows over this integrated analysis is given in Figure 5.10.

An overview of the resulting non-equivalent descriptive domains, which have to be used simultaneously across different hierarchical levels to properly describe funds and flows in this integrated analysis, is given in Figure 5.11.

As shown by Figure 5.11, the descriptive domain referring to the interface between level n−2 and level n−1 (lower-right quadrant) deals with physiological transformations, the relation between animal physiology and typology of organisms (types of organisms in herd). On the interface between level n-1 and level n (lower-left quadrant), we deal with the emergent property of the herd determined by lower-level characteristics (the relative presence of adults, calves bulls and cows will define the function of the herd).

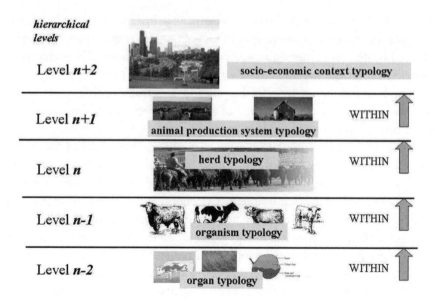

Figure 5.10 *Five distinct, contiguous hierarchical levels (n−2, ... , n+2) used in the narrative of the beef production system*

The pictures included in this figure are from Wikimedia Commons and used under Creative Commons Attribution ShareAlike 3.0 license. Seattle panorama: Joe Mabel, http://commons.wikimedia.org/wiki/File:Seattle_-_I-5_%26_downtown_from_Beacon_Hill.jpg; US feedlot: Dereck Balsley, http://commons.wikimedia.org/wiki/File:Cattle_Lot.jpg.

The two upper quadrants in Figure 5.11 reflect the socio-economic dimension; they provide meaning or contextualization to the biophysical readings of the lower quadrants. In this figure we can appreciate the implications of the different socio-economic contexts. At the interface between levels n and (n+1) (upper-right quadrant), we perceive the performance of the economic agent in relation to local boundary conditions, while at the interface between levels (n+1) and (n+2) (upper-left quadrant) we perceive the performance of the typical local production system in relation to the broader socio-economic context.

For example, in the US this larger socio-economic context implies:

- the high cost opportunity of labour (high bio-economic pressure);
- ample availability and use of capital;
- strong constraints on economic viability in relation to maintaining an adequate level of exosomatic metabolism (the need for high economic labour productivity and maximizing of profit in monetary terms); and
- weak constraints in relation to maintaining an adequate level of endosomatic metabolism (food security guaranteed by society).

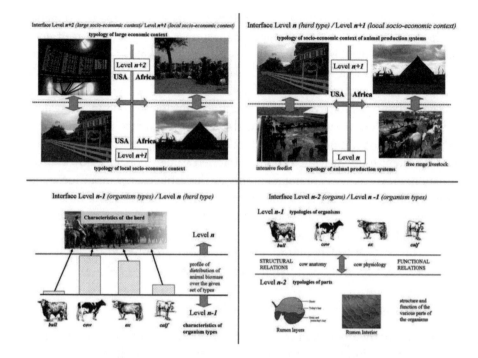

Figure 5.11 *Non-equivalent descriptive domains related to the beef production system*

The pictures included in this figure are from Wikimedia Commons and used under the following license agreements.

GNU Free – Livestock Africa: http://members.virtualtourist.com/m/tp/10d610/; House USA: Daniel Case, http://commons.wikimedia.org/wiki/File:Fury_Brook_Farm.jpg.

Creative Commons Attribution ShareAlike 3.0 license – House Africa: http://flickr.com/photos/mvvermeulen/2921924279/; US feedlot: Dereck Balsley, http://commons.wikimedia.org/wiki/File:Cattle_Lot.jpg; Market Africa: www.flickr.com/photos/41099823@N00/224362521/sizes/l/in/set-72157594389619477/.

Public domain – e-ticker: klip game –http://commons.wikimedia.org/wiki/File:E-ticker.jpg.

This leads to a strategy of the maximization of profit.

In contrast, in Mali the larger socio-economic context implies:

- the low cost-opportunity of labour (low bio-economic pressure);
- a systemic shortage of capital;
- weak constraints on economic viability (the maximization of profit in monetary terms is not a priority); and
- strong constraints in relation to maintaining an adequate level of endo-somatic metabolism (the presence of threats to the physical survival of family members of the economic agent).

Insufficient food security pushes for priority being given to the reproduction of the fund human activity (the people first) over the reproduction or accumulation of the fund exosomatic devices (the biophysical capitalization of the farm). This leads to a strategy of minimization of risk.

These differences define a set of constraints affecting the option space of the animal production system. For example, considering the interface level n/level n+1, US farmers must have a cash flow comparable to those enjoyed by other workers operating in other sectors of society. Since livestock production in this system must provide a source of added value (monetary flow), in this situation, crop residues are not a feasible or desirable energy input for animal production. Considering the same interface, Malian farmers must be resilient and flexible against perturbations, since they are on their own in case of economic troubles. Since livestock is used in this system also as a form of savings investment (safety buffer), in this situation, crop residues are a feasible and desirable energy input for animal production.

Conclusions

After this lengthy discussion of the animal production system, our conclusion is surprisingly simple and straightforward: it is impossible to collapse the complexity of multi-scale metabolic systems into a single numerical assessment. The epistemological predicament associated with complexity in energy analysis deals with the impossibility of reducing to a single quantitative assessment – an output/input energy ratio – two types of representation: first, the representation of events taking place simultaneously across different scales; and second, the representation of events that require non-equivalent narratives to be adopted. This predicament implies that we should abandon the idea that a single index or number can be used to characterize, compare or evaluate the performance of the metabolism of complex energy systems.

Discussing the trade-off between energy efficiency and power delivered, Odum and Pinkerton (1955) write: 'One of the vivid realities of the natural world is that living and also man-made processes do not operate at the highest efficiencies that might be expected from them', meaning that the idea that the output/input energy ratio (a measure of efficiency of a conversion) should be maximized or considered as a very relevant characteristic for defining the performance of an energy system, is not validated by the observation of the natural world.

The same explicit call for the adoption of a more integrated analysis based on multiple criteria and wisdom (addressing and acknowledging the pre-analytical semantic step) was given by Carnot, the father of thermodynamics, almost two centuries ago. Carnot (1824) stated in the closing paragraph of his book *Thoughts on the Motive Power of Fire and on Machines Suitable for Developing that Power*:

We should not expect ever to utilize in practice all the motive power of combustibles. The attempts made to attain this result would be far more harmful than useful if they caused other important considerations to be neglected. The economy of the combustible (efficiency) is only one of the conditions to be fulfilled in heat-engines. In many cases it is only secondary. It should often give precedence to safety, to strength, to the durability of the engine, to the small space which it must occupy, to small cost of installation, etc. To know how to appreciate in each case, at their true value, the considerations of convenience and economy which may present themselves; to know how to discern the more important of those which are only secondary; to balance them properly against each other; in order to attain the best results by the simplest means; such should be the leading characteristics of the man called to direct, to coordinate the labours of his fellow men, to make them cooperate towards a useful end, whatsoever it may be.

Chapter 6

Neglect of Available Wisdom

In the previous chapter we have illustrated, in theoretical terms and with practical examples, the key characteristics to be considered when analysing the quality of alternative energy sources. In this chapter, we contextualize this analysis of the quality of alternative energy sources in relation to, first, the wisdom accumulated in the field of energetics over the last two centuries; and, second, the recent debate provoked by a paper published in *Science* on the assessment of agro-biofuels. The chapter closes by proposing an alternative grammar for the integrated analysis of agro-biofuels, starting out from the controversial study published in *Science*, and building on the concepts developed by the great pioneers in energetics.

Brief History of Energetics and Bio-economics

Attempts to integrate economic analysis with biophysical analysis, with the aim to improve our understanding of the functioning and evolution of human society, have a long history. We feel it is appropriate here to pay tribute to the scientists whose ideas and efforts have shaped a unique interdisciplinary scientific approach. We pinpoint here those pioneers who have proposed theories and concepts relevant for the development of our approach and for the subjects covered in this book. A more exhaustive overview has been provided by Joan Martinez-Alier (1987) in his book *Ecological Economics*.

Pioneers of energetics and bio-economics

William Stanley Jevons (1835–1882)
Jevons is one of the founders of neoclassical economics, together with Hermann Heinrich Gossen, Carl Menger and Leon Walras. Jevons was very interested in the biophysical basis of the economic process. In his famous book *The Coal Question* (1865), Jevons analysed the key role that coal played as a high-quality energy source in the performance of the UK economy, and discussed evolutionary trends in relation to possible exhaustion of fossil energy (a subject which has been discussed ever since!). In reference to the continuous increase in energy consumption associated with economic growth, Jevons pointed at an interesting feature of the evolutionary pattern of metabolic systems: an increase in 'efficiency' in using a resource (increase in

output/input ratio) leads to an increase rather than a decrease in the use of that resource in the medium to long term. This phenomenon has been known since then as 'the Jevons paradox'.

At that time, Jevons was discussing possible trends of future coal consumption and reacting to those urging to dramatically increase the efficiency of steam engines in order to reduce coal consumption. In the face of such claims, Jevons correctly indicated that more efficient engines would have expanded the possible uses of coal for human activities and argued that increased efficiency would have boosted the rate of consumption of existing coal reserves rather than reducing it.

An overview on the history, theoretical analysis and empirical evidence of the Jevons paradox has been provided by Polimeni et al (2008) in the book *Jevons's Paradox: The Myth of Resources Efficiency Improvements*.

Sergei Podolinsky (1850–1891)

Podolinsky was a pioneer in the field of agricultural energetics. According to Joan Martinez-Alier (1987), he was the first to calculate the EROI for agricultural systems. By using his energetic reading of the surplus of energy produced by agricultural labour, he provided a critical appraisal of the labour theory of value developed by Marx and Engels. His biophysical analysis was pointing at the fact that the Marxist theory was neglecting the key role provided by nature in making available high-quality resources (Martinez-Alier, 1987). It is interesting that several of his ideas (Podolinsky, 1883) became, later on, major lines of research in the field of energetics: the systemic input/output analysis of energy flows in agricultural production; the analysis of the relationship between increases in labour productivity and increases in energy inputs supporting a worker; and net energy analysis over processes producing energy carriers.

Wilhelm Ostwald (1853–1932)

A professor of physical chemistry who, after winning the Nobel Prize in Chemistry in 1909, became obsessed by the idea of establishing a new scientific discipline of energetics (*der energetische imperativ* or the energetic imperative). Ostwald presented a manifesto of his vision of energetics in 1887 in Leipzig, at the inaugural address of the first chair of physical chemistry in Germany, entitled 'Energy and its Transformations'. His vision of an organized society was that of a well-functioning body that coordinated individual organs to maximize its energetic efficiency (Ostwald, 1907). Therefore, for him the measure of cultural progress was the increased ability to harness energy to boost the efficacy of human activity. 'The progress of science is characterized by the fact that more and more energy is utilized for human purposes, and that the transformation of the raw energies ... is attended by ever-increased efficiency' (Ostwald, 1911, p870).

Vladimir Vernadsky (1863–1945)

Vernadsky was an important pioneer in the field of environmental sciences and father of the systemic view of the cascade of interactions based on energy and material flows taking place in the biosphere. He developed the key concept of the bio-geochemical cycle, which connects the metabolism of individual components of ecosystems to whole ecosystems, and individual ecosystems to the larger-scale metabolism of the whole planet (Vernadsky, 1926). He also addressed the existence of a hierarchical relation between the different spheres at which self-organization operates on this planet: the geosphere (dissipative systems with no informed autocatalytic loops) is needed in order to support the activity of the biosphere (living metabolic systems capable of processing information), which in turn is required in order to support the activity of the noosphere (the last stage of evolution, the expression of human civilization, in which consciousness and new forms of informed autocatalytic cycles have been introduced).

Alfred Lotka (1880–1949)

A statistician and mathematician, Lotka was a pioneer in mathematical biology. He introduced the principles of energy analysis into theoretical biology. We discuss here two of his major contributions relevant to the subjects covered in this book.

First, Lotka proposed a general principle of evolution framed in energetic terms and related to the ability of achieving a high power level in a given set of expressed activities. In particular, his 'law of maximum energy flow' for biological systems is based on the concept that the survival of biological systems is enhanced when the activities of growth, reproduction and maintenance are boosted. These three activities all require the ability of converting an energy input into useful work at an increasing rate. Therefore, the greater the ability of biological systems (that is, individual organisms, species and ecosystems) to gather energy inputs and to convert them into growth, reproduction and maintenance, the higher will be their survival rate. In his words: 'in the struggle for existence, the advantage must go to those organisms whose energy capturing devices are most efficient in directing available energies into channels favourable to the preservation of the species' (Lotka, 1922, p147). In relation to this concept, in a famous paper Odum and Pinkerton make explicit reference to the fact that this law implies an increase in power level at which fund elements use flows of energy: 'In other words, we are taking the "survival of the fittest" to mean the persistence of those forms which can command the greatest useful energy per unit time (power output)' (Odum and Pinkerton, 1955, p332).

The second major contribution of Lotka refers to the introduction of the concept of bio-economics by extending the general energetic approach also to the analysis of socio-economic systems (Lotka, 1956). Lotka proposed the establishment of a new field of economics, which would have to be based on the integrated analysis of the metabolism of energy and material flows in relation to economic performance. In respect to this idea, he proposed the distinction between exosomatic and endosomatic energy, to make the point that technology is nothing but an extension of the natural pattern of self-organization, obtained using non-physiological energy converters (exosomatic devices).

Friederick Soddy (1877–1956)

Like Ostwald, Soddy was a Nobel Prize winner in Chemistry who turned his attention to the role of energy in economic systems. In particular, he criticized the excessive focus of economists on the analysis of monetary flows to study economic changes (Soddy, 1926). In his view, the characterization of economic flows quantified in terms of money referred to the existence of a debt. That is, the very idea associated with money is that society will have to provide to the money holder either a product or a service of an equivalent price. Tracking money means tracking debt, and therefore those studying monetary flows are studying only a reflection (as in the Plato metaphor of the cave) of real wealth. Real wealth is the biophysical process of production and consumption of actual goods and services. In this analysis, 'real' wealth is generated by the use of energy to transform materials into physical goods and services. The generation of real wealth, therefore, is affected by the typical constraints associated with thermodynamic transformations. Like Ostwald, Soddy was very aware of the fact that the big boost in economic progress experienced in the 20th century was due to the massive switch to fossil energy as the primary energy source for powering the economic process.

Leslie White (1900–1975)

White was a famous US anthropologist who, from his disciplinary perspectives, supported the energetic basis of cultural evolution. For White, 'the primary function of culture' and determining the level of progress is its ability to 'harness and control energy' (White, 1943). White (1949) proposed 'the basic law of cultural evolution', which states:

> *culture evolves as the amount of energy harnessed per capita per year is increased, or as the efficiency of the instrumental means of putting the energy to work is increased ... we find that progress and development are effected by the improvement of the mechanical means with which energy is harnessed and put to work as well as by increasing the amounts of energy employed.*

Here, White refers both to exosomatic devices (able to increase the power level) and the total amount of energy consumed (the energy input consumed), echoing the basic ideas of Lotka that the development of technology represents a move towards an increased exosomatic metabolism for human societies.

In White's view, humans develop technology to tackle their daily problems of survival. In relation to this goal, the more tasks they can perform, the more problems they can solve. In order to carry out more effectively a larger number of tasks (solving more problems) they require a larger amount of capital and a larger flow of energy. Therefore, those societies able to harness more energy and use it more efficiently to perform more tasks will get an edge over other societies. As with other pioneers in the field of energetics, White saw the trajectory of technical progress as a move through a ladder of different energy converters and different energy sources in order to get higher power levels. First, human muscles alone are used as converters, then animal and wind power, and finally technical power based on fossil energy, which boosted the ability of human beings to harness larger and larger quantities of exosomatic energy.

George Kingsley Zipf (1902–1950)

Zipf was a truly creative thinker who provided a lot of interesting new concepts, some of which are in fashion today in complex systems theory (Zipf, 1941; 1949). In particular, in his analysis of the organizational pattern of societies, seen as biosocial organisms, Zipf (1941) introduced for the first time the notion of critical organization (the expression of power laws in organizational patterns). This is now an expected feature of complex dissipative systems operating across different levels of organization (Bak, 1996).

Another important concept introduced by Zipf is the explicit acknowledgement of the key role that the reproduction of the fund of human activity plays in the exosomatic autocatalytic loop of energy. When explaining the restructuring of the US economy associated with the Great Depression, he says: 'Expressed differently, in 1929, the United States discovered a new "raw material": leisure time, which in a way, is just as much a "raw material" as coal, oil, steel or anything else, because for many types of human activity, leisure time is an essential prerequisite' (Zipf, 1941, p324). As we observed in Chapter 4, the total amount of human activity has to be wisely invested in a balance of activities that guarantee both a requirement and a supply of goods and services. In relation to the need of readjusting simultaneously the whole set of compartments operating over the entire impredicative loop, Zipf says: 'any change in kind or amount of goods or processes within a social economy will necessitate a restriation within the social economy itself' (Zipf, 1941, p324). By 'restriation', Zipf meant that in order to establish a different pattern of exosomatic metabolism, the various fund and flow elements making up an impredicative loop must be rearranged in a coordinated way in order to obtain a balanced dynamic budget of the various critical flows.

William Frederick Cottrell (1903–1979)

Cottrell was a US sociologist who has provided an extraordinary analysis of the relation between socio-economic changes and changes in the metabolic pattern associated with societal structure and function. In particular, he focused on changes that can take place in relation to both the mix of energy sources used to generate energy carriers, and the mix of converters used to generate useful energy. Following the tradition of this field, he described the evolution of human culture in terms of increased control over energy transformation. According to Cleveland (1999), Cottrell was the first social scientist to identify the importance of the net energy return, or the energy surplus, delivered by the energy sector of a society. Cottrell also emphasized the importance of energy transitions, such as the shift from animate energy sources (human labour and draught animals) to inanimate energy sources and their associated converters (fossil fuels, steam and the internal combustion engine).

Interestingly, Cottrell (1955) clearly identified several crucial points in relation to performing a sound energetic analysis of the quality of energy sources:

- it is the identity of the converter which defines the identity of the energy input;
- it is the net surplus of energy made available to society that matters;
- the productivity of labour depends on the amount of energy used per hour of work, to make the activity of the workers more effective; and
- the Industrial Revolution is the result of a massive injection of exosomatic devices and fossil energy to boost the productivity of labour in the economy.

In particular, in relation to the definition of quality for an energy source:

> *Cottrell believed that the most important quality of an energy source was the surplus of energy it delivered. Cottrell observed that, in general, societies adopted a new energy technology only if it delivered a greater energy surplus, and hence a greater potential to produce goods and services.* (Cleveland, 1999)

This point is directly related to the need to adopt an accounting of the net supply of energy carriers from the energy sector to the rest of society. If the energy sector is producing energy carriers and consuming energy for its own operation, this does not provide any potential for producing and consuming goods and services in the other economic sectors!

Howard T. Odum (1924–2002)

A pioneer in the development of ecological theory, H. T. Odum can be considered one of the founders of systems ecology, together with his brother

Eugene P. Odum and Ramon Margalef. H. T. Odum has also provided a crucial contribution to the theoretical analysis of sustainability issues by applying basic principles derived from ecological theory to the analysis of the metabolic pattern of socio-economic systems (Odum, 1971; 1983; 1996). We have presented some of his ideas in Chapter 2, so we mention here only his crucial theoretical concept for the development of the field of energetics and bio-economics.

Any system belonging to the class of 'metabolic networks self-organizing through informed autocatalytic loops' (his definition) expresses a set of expected systemic properties. In particular, we can expect that any member of this class of systems will evolve towards an increased consumption of energy as a whole (if favourable boundary conditions are available) and towards a more efficient use of energy at the level of the parts: the combination of maximum energy flow and minimum entropy production principles (Giampietro and Mayumi, 2004; Mayumi and Giampietro, 2004; Polimeni et al, 2008). Moreover, when dealing with the analysis of these systems we can expect to find typical patterns when looking at the relation between parts and the whole, and typical changes in the systemic relations across parts and the whole during their evolution: for example, growth versus senescence. It should be noted that the definition chosen by H. T. Odum is generic enough to include both ecosystems and socio-economic systems. According to these ideas, it is possible to find useful benchmarks referring to the typical food metabolic rate of a person, the average gas consumption of a model of a car, or the typical GDP per capita of a developed country. These benchmarks are useful for characterizing typologies of fund elements, which are stabilized within complex metabolic systems (see Appendix 1).

Nicholas Georgescu-Roegen (1906–1994)

A very creative and profound thinker in economics, Georgescu-Roegen provided a crucial contribution to the theoretical discussion of sustainability by developing a theoretical framework for bio-economics (Georgescu-Roegen, 1971; 1975; 1976). His ideas and theories have enormous relevance for those working in the field of sustainability analysis, and for this reason there is ever-growing interest in his work (for an overview, see Mayumi and Gowdy, 1999; Mayumi, 2001). We already presented some of the concepts developed by Georgescu-Roegen relevant to applied energetics and bio-economics in Chapters 2–4. Besides being a great economist, Georgescu-Roegen was also a great epistemologist. In particular, he addressed the epistemological challenges associated with the quantitative analysis of the evolution of economic systems. Two key issues, relevant for our methodological approach of bio-economics presented in Chapter 4, are discussed in Box.6.1.

Box 6.1 *Two epistemological issues that are key to the development of a multi-scale analysis of the exosomatic metabolic pattern of modern economies*

Addressing the semantics behind the chosen quantitative representation

Many of those studying the functioning of socio-economic processes seem to be confused by what is produced by the economic process. According to Georgescu-Roegen, the economic process does not produce goods and services, but it produces an autopoietic system which is able to reproduce itself, via the establishment of an integrated process of the production and consumption of goods and services. When dealing with the analysis of the economic sectors – those producing added value – they do not produce goods and services, but rather they produce those processes required to produce goods and services (there is a discussion of this point in Appendix 1). When considering the whole socio-economic system, it is the integrated action of the productive economic sector and the sector of final consumption which has to be considered. Using Georgescu-Roegen's terminology, the economic process has the goal and function of reproducing various fund elements, defined simultaneously across different levels and scales by using disposable flows (see Figure 6.1).

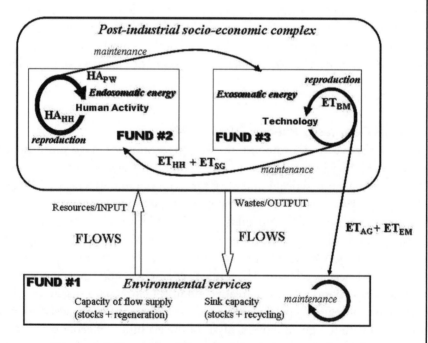

Figure 6.1 *Network of flows and funds in post-industrial socio-economic systems*

Procedural issue: the Roegenian method

For the reasons discussed so far, Georgescu-Roegen proposed a procedural approach for the study of the economic process. Two key points behind his approach are (Dogaru, 2008):

1. it is only by assigning a purpose to the analysis of an economic process that it becomes possible to take a reference point; and
2. it is important to be aware that any *arithmomorphic* representation (that is, a quantitative characterization of the economic process based on numbers and therefore simplifications) entails a dialectical tension.

Something is gained by the power of analytical representation, and something is lost by the simplification required by having the arithmomorphic model in the first place. For this reason, Georgescu-Roegen suggested a procedure which explicitly acknowledges the need of adjusting the semantic and the syntax to each other in an iterative process.

According to Dogaru (2008) the procedure suggested by Georgescu-Roegen is based on three steps:

1. the definition of various elements of the process, based on a definition of borders in time and space (parts and the whole);
2. the identification of some main features of the economic process at the general level on which to focus (whole and context); and
3. the development of a simple analytical model.

Then there are three steps to guarantee the validity of the relative semantic:

1. a semantic description of the various elements of the analytical model;
2. an elimination of pseudo-measures, together with a check over the presence of inconsistency in the representation of the economic process; and
3. the identification of the purpose of the analysis for each economic process considered.

In relation to this point, we believe that the idea of performing an impredicative loop analysis starting with a definition of a grammar based on a fund-flow model represents an implementation of the suggestion of Georgescu-Roegen.

The flourishing of energetics and bio-economics in the 1970s and 1980s

It was not until the 1970s that energetics and bio-economics started to flourish as a scientific field, probably due to two parallel events. First, the contributions to the theory by H. T. and E. P. Odum and Georgescu-Roegen undoubtedly provided a solid basis; and second, energy analysis got a major boost in practical terms because of the first oil crisis in the 1970s.

The supply of oil suddenly became more expensive and less reliable. For this reason, many governments became interested in studying the dependence of their economies on oil. The availability of research funds made it possible to involve academic institutions in energy analysis focusing, from different angles, on the relation between economic performance and the energetic metabolism of society. The adoption of the basic rationale of net energy analysis (Gilliland, 1978) resulted in the development of quantitative approaches aiming at the calculation of output/input energy ratios. The term 'energy analysis', rather than 'energy accounting', was officially coined at the IFIAS workshop of 1974 (IFIAS, 1974). Building on the ideas of the pioneers of energetics, in the 1970s and 1980s the field of energy analysis experienced an explosion of research activity in relation to several different applications. The most important are listed below:

- the energy analysis of agricultural production and, in general, the relation between energy consumption and food production (Steinhart and Steinhart, 1974; Leach, 1976; Slesser, 1978; Pimentel and Pimentel, 1979; Smil, 1983; 1988; Stout, 1991; 1992);
- anthropological studies of the link between energy and society (Cipolla, 1965; Rappaport, 1968; 1971; Adams, 1988; Debeir et al, 1991; Fischer-Kowalski and Haberl, 2007);
- the analysis of economic performance based on the concept of embodied energy (Herendeen and Bullard, 1976; Costanza, 1980; 1981; Hannon, 1981; 1982; Herendeen, 1981; 1998; Slesser and King, 2003);
- the analysis of economic performance in relation to the different qualities of energy forms and natural resources (Cleveland et al, 1984; 2000; Hall et al, 1986; Gever et al, 1991; Kaufmann, 1992; Hall, 2000; Ayres et al, 2003; Ayres and Warr, 2005); and
- the integrated analysis of the issue of sustainability (Tsuchida and Murota, 1987; Watt, 1989; 1991; Smil, 1991; 2001; 2003; 2008; Allen and Hoekstra, 1992; Kay and Schneider, 1992; Schneider and Kay, 1994; Kay, 2000; Allen et al, 2003).

Quite remarkably, after the boom in the 1970s and 1980s, the long and glorious story of energy analysis came to a rather abrupt end. With the return of cheap oil, the issue of energy analysis in relation to sustainability quickly lost priority and consequently funding. The lack of research funds translated into a progressive disappearing of all the groups and departments studying energy analysis within the academic universe. Interest in theoretical discussions over the issue of how to perform energy analysis in the context of sustainability quickly declined in the 1990s outside the original circle of concerned scientists. (We will return to this argument in Chapter 9.)

Now, in the third millennium, the two challenges of climate change and peak oil have returned the topic of 'energy and society' to the political and scientific agenda. The re-opening of the tap of research funds has generated a revival of interest in this field. But who are our new generation of energy analysts? Has this new generation of scientists, encouraged by funding for alternative energy sources, learned anything from the body of knowledge already available in the field of energetics? Did they bother to study the enormous amount of work done on this topic, for more than a century, by a remarkable group of outstanding scientists?

Recent Debate on the Energy Analysis of Agro-biofuels

In this section we scrutinize a recent study by Farrell et al (2006) published in *Science* and providing an energy analysis of agro-biofuels. The study compares various assessments of output/input energy ratio for the production of ethanol from corn, and then provides a comparative analysis of the CO_2 emissions of two processes: gasoline production from oil, and ethanol production from corn. This paper triggered a heated debate in the academic community concerned with energy analysis and, we believe, for good reasons (Cleveland et al, 2006; Hagens et al, 2006; Kaufmann, 2006; Patzek, 2006). In this section we follow up on the discussion initiated by this study, especially because we believe that the quantitative analysis provided by Farrell et al is representative of the confusion found in the use of energy analysis by the new generation of energy analysts in the debate over agro-biofuels. In particular, the paper is a good case study for the following reasons:

- The journal in which it is published, *Science,* is supposedly a prestigious journal, which implies a thorough review procedure (especially when dealing with controversial and highly relevant issues). *Science* is supposed to guarantee that what it publishes is the state of the art in the field.
- The goal of the study; this was a report prepared for *Science* in order to settle a controversy. It had the explicit aim of cutting through the confusion of contrasting assessments and providing the 'last word' on data and accounting procedures.
- The degree of sophistication of the quantitative analysis; the study was carried out by the prestigious Energy and Resource Group at Berkeley, using a pretty elaborate protocol. The authors analysed published and grey literature to check the results presented in six studies illustrating the range of assumptions and data found in the case of the corn-based production of ethanol.

158   The Biofuel Delusion

According to the concepts introduced so far, we provide below a critical appraisal of this paper, from the choice of the grammar used for quantification (the pre-analytical definition of the set of energy forms considered and their transformation) to a discussion of the quantitative results provided.

Critical appraisal of the grammar used by Farrell et al

The grammar used by Farrell et al (2006) is illustrated in Figure 6.2. According to this grammar, they calculate an output/input of ethanol greater than 1 but smaller than 2/1 (1.26/1). Recalling that it is impossible to have a conversion of energy forms generating an output of energy larger than the input, we can explain this result by acknowledging that what is accounted as 'the input' in this analysis is only the flow of energy carriers spent in the internal loop (loss #2 in Figure 5.4). That is, this positive result is obtained by ignoring the initial input of primary energy sources into the process (loss #1 in Figure 5.4) of ethanol generation. On the other hand, when looking at the output/input of gasoline in this same study, we find a value smaller than 1 (0.79/1). This lower value does not reflect a larger use of energy carriers in the internal loop (loss #2), but rather the choice made by the authors of including in the accounting of the input not only energy carriers but also the *primary energy source*: crude oil (loss #1). This simple observation indicates that something is not right in this method of accounting. In our view, there are three very disturbing problems related to the definition of this grammar.

First, there is no distinction between primary energy sources and energy carriers. We saw in the previous chapter that the assessment of primary energy sources is required to check the compatibility of the process generating energy carriers with boundary conditions (when looking at the interface of black box and context); the assessment of energy carriers is required to check the compatibility of the process generating energy carriers with internal biophysical constraints (when looking at the interface of parts and black box). For this reason it is important to be clear about how to account for different forms of energy flows and why.

Second, there is no reference whatsoever to power levels or power densities. Hence, there is no way to establish a relation between the numerical values, used to characterize the various energy flows, and the relative funds; that is the requirement of human labour, capital and colonized land. The numbers generated using this grammar can *not* answer questions such as: What is the labour demand associated with the supply of these two energy carriers? What is the land demand associated with the supply of these two energy carriers? What is the requirement of capital, water, soil or other key elements associated with the supply of these two energy carriers? Therefore, the analysis performed by Farrell et al does not provide any relevant information to

The chosen grammar (energy graph)

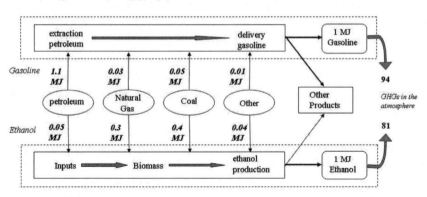

The quantitative results

GASOLINE	Inputs → 1.1 + 0.03 + 0.05 + 0.01 = 1.19 MJ	Output/Input = 0.84
ETHANOL	Inputs → 0.05 + 0.3 + 0.4 + 0.04 = 0.79 MJ	Output/Input = 1.26

Figure 6.2 *The grammar used in the analysis of Farrell et al, 2006*

study the feasibility and desirability of the process of the generation of energy carriers for society, either in relation to environmental constraints or in relation to biophysical constraints associated with bio-economic pressure. Indeed, the semantic relevance of this analysis reminds us of the proposal to produce biodiesel from human fat.

The third problem with this grammar is that it does not allow a distinction between the net supply of energy carriers and the gross supply of energy carriers. Since the category 'energy carrier' is not defined in the lexicon as an item distinct from the category 'primary energy source', it is impossible to calculate the effect of the internal loop of energy carriers which must be consumed to make available a net supply of energy carriers.

Finally, there is another systemic/logical error in the chosen system of accounting. The implementation of the accounting over this grammar neglects the issue of scale. The authors introduce the concept of co-production credits (for example, for dried distiller grains), which are assessed as a net output of the energy carrier for the ethanol production system. However, this concept can be justified only when the production of agro-biofuels is performed in limited quantities. This point will be discussed in more detail, with numerical examples, below.

160   *The Biofuel Delusion*

Critical appraisal of the formalization used in this study

We now go more in detail through the numerical results provided by Farrell et al following the explanations of the numbers found in their paper. We want to remind the reader that in this comparison we are dealing with two completely different types of fuels. Gasoline is an energy carrier derived from oilfields (stocks) and thus oil is the relative primary energy source. The process of oil exploitation to obtain gasoline represents a stock-flow supply of energy carriers. On the other hand, ethanol is an energy carrier derived from corn. Corn is *not* a primary energy source but a type of energy carrier. Corn can carry either food energy for humans or biomass energy that can be burned directly (corn stove) or indirectly by converting it into alcohol. Corn biomass, in turn, is derived from a set of primary energy sources including solar energy, and other energy forms such as rain and biota activity. The set of natural processes leading to the production of corn is *supposed* to represent a fund-flow supply of energy carriers, at least in the good intentions of the proponents of agro-biofuels.

The overall conversion of primary energy sources into a net supply of energy carriers; external constraints

As concerns gasoline, we need information about the overall consumption of the primary energy source in order to know the speed of stock depletion on the supply side and the amount of CO_2 emissions on the sink side. When dealing with ethanol, we need information about the overall consumption of the primary energy source in order to know:

- on the supply side, the overall size of the fund, which is required to guarantee a given net flow of energy carriers (land and water requirement); and
- on the sink side, the amount of CO_2 emissions, and the emissions of other pollutants such as pesticides, phosphorus and nitrogen leakages.

However, in order to be able to calculate the overall consumption of primary energy sources per MJ of net supply of energy carriers, we also need a detailed analysis of the internal loop of energy carriers required to make available the energy carriers (see Figure 5.4).

Accounting for primary energy flows

As concerns ethanol production, Farrell et al state (in footnote 6): 'by convention photosynthetic energy is ignored in this calculation'. The note does not explain who decided such a convention, and why this should be considered a wise choice. Maybe this is a remnant of the document produced by IFIAS in 1974. As a result of this choice, the proposed grammar and system of accounting totally ignores the interface of the process of production of ethanol with natural processes. This choice is difficult to understand in semantic terms, because the main concern about the large-scale production of agro-biofuels is related to excessive land demand, which can compete with food production, imply the destruction of habitats, and generate pollution in the agricultural phase. So ignoring the interface with nature does not seem a particularly wise choice for generating a useful grammar to study the performance of agro-biofuels (see Box 6.2).

Box 6.2 *The need to address the interface between biomass energy production and natural processes*

In a review of 17 comprehensive studies about the possible contribution of biomass to the future global energy supply, Berndes et al (2003) complain that these studies tend to concentrate on only two parameters: land availability and yield levels in energy crop production, neglecting other important aspects of the problem. The authors note that:

> the question [of] how an expanding bio-energy sector would interact with other land uses, such as food production, biodiversity, soil and nature conservation, and carbon sequestration has been insufficiently analysed in the studies. It is therefore difficult to establish to what extent bio-energy is an attractive option for climate change mitigation in the energy sector. (Berndes et al, 2003, p1)

The grammar used in Farrell's study does not even consider the land requirement!

Concerning the production of gasoline from oil, we learn (in another note): 'The metric for petroleum use include crude oil used as a feedstock for gasoline, and the metric for GHGs' (Farrell et al, 2006, note 19). Thus, we have a clear logical inconsistency: the authors adopt different conventions for gasoline and corn-ethanol. For the former, they consider the energy associated to the primary energy source; for the latter, they do not. The reason for this choice is obvious: in this way the authors can compare the overall emissions of CO_2 over the two processes. For assessing CO_2 emission it is essential to address the overall consumption of primary energy sources for generating the net supply of a MJ of gasoline and the fossil energy fuels used in the production of ethanol. So this is another example of a mono-purpose grammar. But there are many other attributes of performance that matter beside the emission of CO_2! The authors consider the negative effects associated with the consumption of the primary energy source (fossil energy) for the making of gasoline, but they ignore the negative effects associated with the consumption of primary energy sources, such as the destruction of habitat for biomass production, in the making of ethanol.

We observe a similar inconsistency in relation to the adoption of the 'displacement method'. This method is adopted by the authors to give energy credits to ethanol, in relation to the possible use of co-products of the process of ethanol production. However, this method is *not* applied to the process of gasoline production. Both processes, however, generate co-products (Figure 6.2). The production process of gasoline generates many co-products that are used in our daily life, from bitumen for paving roads, to key chemicals used in industrial production, to plastic materials used for a virtually infinite number of applications. As a matter of fact, about 17 per cent of the material in a barrel of crude oil ends up as other petroleum products, not fuel (Hall et al, 2009).

This example demonstrates that in energy analysis, the choice of the convention made by the authors is arbitrary. Given that pre-analytical choices are never neutral in relation to the final outcome of energy analysis, they should be justified by robust semantic explanations, and not just mentioned in footnotes to the text. The choice of including the category 'primary energy source' in the grammar when dealing with gasoline entails focusing on the problem that gasoline production from oil has on the sink side: CO_2 emission. Whereas the choice of excluding the category 'primary energy source' from the grammar when dealing with ethanol from agricultural crops (ignoring the photosynthetic step) entails neglecting the problem that ethanol production from corn has on the supply side: the huge differential in the demand of land and environmental services, which is clearly favourable to gasoline. By giving energy credits to by-products of ethanol, the authors increase the overall output/input ratio of corn-ethanol. By ignoring the co-products of gasoline production from oil, the authors reduce the overall output/input ratio of gasoline.

Accounting for energy carriers flows

The flows of energy represented in vertical lines in Figure 6.2 refer to the expenditure of energy carriers, which are corrected for quality factors in the chosen system of accounting. However, looking at the choices made by the authors, the chosen grammar seems to have the goal of providing a meaningful comparison over the two systems of production only in relation to the CO_2 emission associated with the gross production of 1MJ of energy carrier. No discussion of the gross/net ratio is provided. The assessment of the gross production is in any case muddled by the arbitrary choice of how to calculate energy credits to co-products. We emphasize again that CO_2 emission is *only one of the many problems* that the generation of energy implies on the sink side. There are other problems that matter. What about nitrogen and phosphorus – associated with the agricultural production of crops – leaking in the water table and inducing dead zones in the ocean? What about SO_2 in the emissions of fossil energy leading to acid rain? What about soil erosion? What about the loss of habitat and biodiversity because of the expansion of either crop monocultures or palm-oil plantations?

To make things worse, the calculation of the output/input ratios over energy carriers, such as 1MJ output/0.79MJ input, is dubious since it is based on the accounting of *virtual* amounts of net supply of energy carriers, which are introduced by adopting the displacement method (with energy credits for co-products). This method is controversial in its basic logic (since it is arbitrarily applied to ethanol but not to gasoline), but, above all, it is not applicable to the large-scale production of agro-biofuels, as discussed below.

Credit for by-products and the issue of scale

In this section, we discuss the validity of the idea of assigning energy credits to co-products of ethanol using the displacement method. We use here the explanation of the rationale given by Farrell et al (2006) in their article in *Science*.

The production of ethanol generates a flow of material, the so-called dried distiller grains (DDG), containing solubles such as corn gluten feed and corn oil. The displacement method of accounting consists of giving an energy bonus – the so-called energy credit – to the output of ethanol (as done by Shapouri et al, 2002). The explanation given for this choice is that the co-products of ethanol from corn can be used as feed. Therefore, according to this rationale, the amount of fossil fuels that would be required to generate the same amount of feed obtained as co-products has to be included in the calculation as if it were an actual net supply of energy carriers: an amount of energy carrier equivalent to the amount saved.

Critics of this assessment (such as Pimentel et al, 2007) argue that the energy credit given to DDG is too high because the quality of the feed based on DDG is much lower than the quality of the feed they are supposed to replace. Besides the discussion over the *quantity* of energy to be accounted for as credit, there is also the logical inconsistency of giving by-product credits to ethanol but not to gasoline. The same rationale could be applied to the accounting of gasoline production, which entails many more useful co-products.

However, we do not want to enter any further into this debate here. What we do want to address is another serious problem relative to the validity of this accounting convention, which is that the rationale backing up the energy credit for by-product feeds does not address the issue of scale (Giampietro et al, 1997; Giampietro and Ulgiati, 2005). If the production of biofuels were implemented on a large scale, the amount of DDG generated would exceed the existing demand for feed by several times. This implies that DDG would represent a serious environmental problem, a pollutant on the sink side, to which analysts should associate an energetic and economic cost rather than a positive return (Giampietro et al, 1997).

Let's perform some back-of-the-envelope calculations to support our point. Using the assessment of output/input of energy carriers calculated by Shapouri et al (2002) – which is slightly higher than that calculated by Farrell et al (2006) – we obtain an output/input ratio of 1.3/1. That is, Shapouri et al, following the displacement method, assign a generous positive energy credit for by-products in their calculation of the ratio gross/net supply. Now, we calculate the amount of distilled grain generated if we want to cover 10 per cent of transportation fuel in the US (that is, 3 per cent of total US exosomatic energy consumption) with corn-ethanol production (replacing fossil-energy-based gasoline).

As discussed earlier, in order to be fully renewable and with zero emission, this biofuel system – operating at 1.3/1 output/input of energy carriers – should produce a gross supply of 4 litres of ethanol to generate a net supply of 1 litre of ethanol. This means that the net supply of 3EJ of ethanol (about 140 billion litres of ethanol) – the flow F5 in the grammar of Figure 6.3, discussed below – translates into a gross production of 12EJ of ethanol, or 558 billion litres (the flow F1 in the grammar of Figure 6.3).

We can now translate this requirement of gross ethanol production into the corresponding requirement of corn production. The production of 558 billion litres of ethanol would require, first, the gross production of 1500 million tonnes of corn (six times the whole production of corn in the US in 2003; USDA, 2006); and, second, the generation of 500 million tonnes of DDG by-products. This amount of DDG co-products is *ten times the total US consumption* of high protein commercial feeds – 51 million tonnes – recorded in 2003 (USDA, 2006).

At this point, the negative effect generated by an enlargement of scale becomes crystal clear. When reaching a scale of production of zero-emission ethanol able to cover 3 per cent of total exosomatic energy consumption of US, the co-products of ethanol will cause a serious environmental problem (and an energetic cost!), not deserve a credit for the displacement of fossil energy.

The consequences of this point are crucial. If we drop the bonus associated with the displacement method from the accounting of the output/input of energy carriers in the production of ethanol from corn, then the overall output/input of the corn-ethanol systems goes down to 1.1/1. This last assessment shows that the emperor is actually naked; currently, the production of ethanol from corn in the US is a process that is basically converting oil into ethanol. Or, put in a different way, if the US could really produce ethanol as a fully renewable energy source with zero emissions, then it would have to face a further increase in the overall demand of land, water and labour by a factor of 11.

Is the poor quality of the paper in *Science* an isolated case?

The editor in chief of *Biofuels, Bioresources and Biorefining*, in a commentary in his own journal, says: 'Net energy is a (mostly) irrelevant, misleading and dangerous metric' for discussing biofuels as an alternative energy source (Dale, 2008a). Rather surprised by such a blunt remark, and not knowing Dale personally, we checked his work and found another paper, also in his journal, entitled: 'Thinking clearly about biofuels: ending the irrelevant "net energy" debate and developing better performance metrics for alternative fuels' (Dale, 2008b). In this paper, Dale states: 'The net energy argument revolving around fuel ethanol offers a textbook example of how not to think about biofuels and other alternative transportation fuels' (Dale, 2008b, p15). Halfway through this paper, Dale uses the data presented by Farrell et al (reproducing the figure illustrated in Figure 6.2) and compares the production of gasoline from oil with the production of ethanol from corn. Dale then claims that the net energy accounting for gasoline is negative (minus 18 per cent), whereas the net energy from ethanol is positive (plus 26 per cent). In relation to this claim, we already observed that the grammar adopted in the study by Farrell et al (and used as a basis by Dale) lacks any semantic/logical coherence.

Thus, it is obvious that Dale's claim has the sole goal of generating additional confusion to convince the reader that the concept of net energy analysis should not be used. In fact, the paper ends with the following statement: 'Useless, misleading and irrelevant metrics such as net energy must be eliminated from our discourse on fuel alternatives' (Dale, 2008b, p17). In this way, Dale uses nonsense formalizations to discredit a sound semantic concept.

This is an important point, since generating confusion over the energetic assessment of agro-biofuels may become an effective strategy for avoiding criticisms of biofuel policies. For this reason, we believe it important to be fussy and explain in detail, in theory and with practical examples, the various problems found both in the semantic and in the formalization of energy analysis applied to the issue of agro-biofuels.

We do agree with Dale on the point that energy analysis has proved to be particularly slippery regarding the definition of agreed protocols of calculations. This applies also to the use of net energy analysis and EROI. However, the existence of confusion about the calculation of the energetic performance of agro-biofuels does not imply that anything goes. It does not imply that we should stop using our common sense.

Resuming the findings presented in this section, we can say that there is great confusion in the literature regarding the assessment of the energetic performance of biofuels (such as Farrell et al, 2006; Shapouri et al, 2002; Patzek, 2004; Pimentel et al, 2007). This confusion is not due to bad data or sloppy calculations. It is due to the lack of agreement on how to generate a useful multi-purpose grammar capable of providing an integrated analysis of the energetics behind the generation of a net energy supply of biofuel from energy crops.

Proposing an Alternative Approach for Dealing with the Analysis of Agro-biofuels

As discussed in the introduction, the great appeal of agro-biofuels in the public perception is due to its alleged ability to kill two birds with a stone. The claim is that agro-biofuels represent a self-sufficient system that is both fully renewable – they come from a fund-flow supply situation, and therefore will never run out – and has zero emissions (presumedly the biomass used as feedstock is fixing CO_2 from the atmosphere, thus compensating for the CO_2 released when burning the agro-biofuel.) It seems that not everybody is aware that the actual production of agro-biofuel is possible only because it heavily depends on fossil energy inputs in the production process.

In the face of these claims, it is important to check the overall performance of the system producing energy carriers (agro-biofuels) from a fund-flow process, which has to be able to sustain itself with a positive balance in terms of the production and consumption of energy carriers. As observed in Figure 5.4, when the ratios between the output of energy carriers and the input of energy carriers are pretty low (e.g. < 2/1), we face a dramatic increase in the internal consumption of the energy sector, due to an internal loop of energy-for-energy. What are the consequences of this internal loop in terms of labour requirement, land requirement, water requirement and capital requirement per MJ of net supply of energy carriers to society?

Obviously, the grammar proposed by Farrell et al (and used by Dale) cannot answer any of these questions, since it does not even make the fundamental distinction between primary energy sources (a category needed in order to be able to study external constraints, such as the total requirement of colonized land) and energy carriers (a category needed in order to be able to study internal constraints, such as the total labour and capital required in the process). Actually, their analysis does not even mention that the generation of a net supply of energy carriers for society requires labour and/or colonized land. This is a crucial omission. Using the grammar illustrated in Figure 6.2, it is impossible to deal with the crucial issue of threshold values of energy intensity (per hour of human activity in different funds: TET/HA_{ES} and EMR_{ES}) and of energy density (per hectare of colonized land in different funds: the density of the net flow (W/m^2) of energy carriers) required for assessing the feasibility and desirability of the agro-biofuel option. In conclusion, the semantic behind this grammar ignores the most fundamental principles of energetics and bio-economics. When dealing with the analysis of the feasibility and desirability of agro-biofuels, the grammar shown in Figure 6.2 is simply totally irrelevant for an informed discussion.

Multi-purpose grammar

We now propose an alternative grammar, shown in Figure 6.3, for the assessment of ethanol production from agricultural processes and studying the performance of agro-biofuel systems. This grammar makes it possible to check whether or not the production of ethanol is fully renewable (the system must be able to cover its own consumption of energy carriers), and to verify the zero-emission claim (all the carbon in the CO_2 emissions associated with the use of energy carriers has to be fixed by the biomass used in the production of the energy carriers). With this grammar, we can distinguish between gross supply and net supply of energy carriers, and therefore calculate the overall demand of labour, land and other inputs per MJ of net supply of ethanol to society. This makes it possible to calculate key benchmarks: TET/HA_{ES}, EMR_{ES} and densities of flows per unit of land. In this way, we can also calculate how much land has to be kept in agricultural production to obtain the required net supply of agro-biofuels to society.

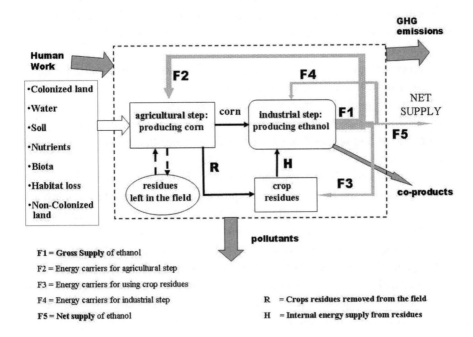

Figure 6.3 *Multi-purpose grammar useful for the energetic analysis of agro-biofuels*

The grammar presented in Figure 6.3 addresses explicitly the existence of an internal loop of energy carriers that are required for the process of generating a net supply of biofuels. That is, when dealing with scenarios in which biofuels are expected to be a significant source of energy carriers for society – that is, more than 5 per cent of the actual consumption of liquid fuels – to replace fossil energy carriers, then one has to deal with the unavoidable consequences that the internal loop of energy-for-energy entails. Its relative system of accounting has been presented in detail in Giampietro et al (1997) and Ulgiati (2001).

In the middle of Figure 6.3, we have a box (with a dotted line) in which we represent the internal flows of energy carriers associated with the production of agro-biofuel. In this example we deal with ethanol from corn.

If we look inside the black box, there are five flows of energy carriers:

- F1 is the amount of gross biofuel output generated in the production process;
- F2, F3 and F4 are the various flows of energy carriers required/consumed by the process in the form of fuel energy; and
- F5 is the amount of net biofuel supply generated in this production process.

Looking at the interface between black box and context (Figure 6.3), there are several flows which have to be considered in order to calculate the performance of this system in relation to its context, in an integrated way:

- GHG emissions – the overall emissions of the system (referring to the aggregate consumption of primary energy sources);
- F5 – the net supply of energy carriers to society;
- co-products – which have to be considered either as a resource or as a pollutant, depending on their quantity and density in relation to the demand of the society;
- pollutants – flows of matters that are harmful for the environment;
- human work – the hours of work required for the various phases of ethanol production. This is a relevant variable in relation to the bio-economic pressure operating in the society; and
- the set of environmental services guaranteeing the reproduction of the fund that generates the supply of biomass, starting with the requirement of colonized land. This represents the flow of primary energy sources required for the conversion of solar energy into the energy associated with chemical bonds contained in biomass.

Having all this information, we can establish a relation between the internal technical characteristics of the process (the internal loop of energy carriers required to make available energy carriers to society) and its feasibility and desirability in relation to ecological and socio-economic constraints. In particular, in relation to the interface with the environment we can define both the overall requirement of land, water and other environmental services on the supply side, and the overall environmental loading associated with GHG emissions and other pollutants on the sink side. In relation to the interface with the socio-economic process, we can define the competition that the energy sector creates with other socio-economic compartments in terms of:

- labour demand;
- capital demand;
- energy demand;
- key material flows demand, for example water and mineral resources; and
- colonized land demand, for example, competing with other necessary land uses such as food production.

In our view, it is essential to develop this multi-purpose grammar in order to establish a link between the internal characteristics of the energy system and its overall performance. In fact, as discussed earlier, when dealing with corn-ethanol production systems, several of the criteria of performance described above depend in a non-linear way on the value taken on by the ratio between the gross output of biofuel (F1) and the aggregate amount of fuel consumed in the process (F2 + F3 + F4). The internal loop of energy-for-energy must be *sufficiently small* to prevent an excessive demand of land and labour per unit of net fuel delivered to society. The output/input must be larger than 5/1. Recall that the output/input value of corn-ethanol in the study of Farrell et al is 1.26/1, which falls exactly in the most critical part of the curve shown in Figure 5.4. To make things worse, if one corrects the assessment of the output for the dubious energy credits applied, this ratio becomes even lower.

Two meta-grammars for the integrated analysis of the quality of energy sources

In our opinion, an effective semantic problem structuring – the choice of an appropriate grammar – must be the crucial starting point of any discussion of alternative fuels in general. In fact, when dealing with energy systems generating a net supply of energy carriers there are always several relevant attributes of performance which must be considered.

- There are energy carriers that are produced (for example, in the form of ethanol or oils).
- There are energy carriers that are consumed (for example, in the form of

electricity and fossil fuels during the production of the energy crop, transport, and in the conversion of biomass into the final biofuel). These energy carriers can have different 'qualities', for example, 1J of electricity versus 1J of liquid fuel.
- There are primary energy sources that are depleted (when considering stock-flows).
- There are primary energy sources that entail a requirement of environmental services and a competition of land use, and these can be damaged (when considering fund-flows).

Either stock-flows or fund-flows must be assessed against biophysical constraints coming from the metabolic pattern of both the socio-economic system and ecological systems. To be considered as an energy source, the output of energy carriers of this process needs to exceed the input of energy carriers, or – according to the formalization given in Figure 6.3 – $F1 > (F2 + F3 + F4)$. But even more important, in relation to its feasibility and desirability, the requirement of land, labour and capital for generating a net supply of biofuels should not imply a serious interference with the actual functioning of the whole socio-economic system, or the actual functioning of the ecosystems embedding the society.

When it comes to the decision of how to allocate money for developing alternative fuels, it would be wise to decide priorities on the basis of an integrated set of criteria characterizing their feasibility and desirability. To this end, we provide below a meta-grammar for the analysis of the compatibility of a system generating a net supply of energy carriers with the exosomatic metabolism of the society that should adopt such a system. This meta-grammar can be tailored to the specific problem to be analysed – for either stock-flow energy inputs or fund-flow energy inputs – using the approach illustrated in Chapters 4 and 5.

Multi-purpose grammar to analyse the exosomatic metabolic pattern of a society based on PES providing stock-flow energy carriers

The grammar presented in Figure 6.4 is based on the graph language proposed by H. T. Odum. The elements in the graph have the following meanings.

- Hexagons represent elements consuming energy carriers: household (HH), agricultural (AG), building and manufacturing (BM) and service and government (SG) sectors.
- The rectangle with a semicircle on the right represents the element generating the supply of energy carrier: the energy and mining sector (EM). However, this element is also consuming a part of the flow of the exosomatic energy carriers it produces.

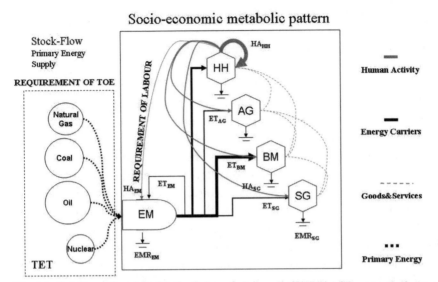

The set of relations determining biophysical constraints on the FLOWs of *Exosomatic Energy*
(1) internal constraint - Bio-Economic Pressure
(2) external constraint - Environmental Loading

Figure 6.4 *Meta-grammar useful for the energy analysis of fossil fuels*

- The circles represent the amount of primary energy sources available to the energy sector. These primary energy sources are not produced by the energy sector. It is important to track their consumption, since they map onto emissions and rate of stock depletion.
- The earth symbol represents sinks. In the case of fossil energy, emissions go into the atmosphere and do not require a specific localization. This is another major advantage of fossil energy as a primary energy source; its waste disposal does not require locally a large additional expenditure of energy carriers as compared to, for example, the generation of nuclear energy.

Internal interactions between the elements in Figure 6.4 are represented in the following ways:

- dotted lines are flows of exosomatic energy from primary energy sources to the energy and mining sector (EM);
- black lines are flows of exosomatic energy carriers from the EM sector to all the other elements;
- grey lines are investments of human activity generated by the household sector (HH) to all the other elements; and

- grey broken lines are exchanges of economic goods and services across different elements, excluding energy products. This exchange can be characterized in terms of an input/output matrix.

In this example, we have not considered the role of monetary flows, which is to define the profile of values of the various benchmarks over the metabolic pattern. This has to be addressed in a more comprehensive analysis, such as the MuSIASEM approach presented in Appendix 2. What is relevant in this grammar is the ability to establish a mapping of the flows of primary energy sources and energy carriers, and their effect on internal and external constraints. With regard to the examples of bio-economic analysis discussed in Chapter 4 and the overall view of the exosomatic metabolic pattern given in Figure 4.9, we see that the fund human activity is generated by the household sector, and then distributed to the other sectors. The ratio between the amount of fund human activity and the flow of exosomatic energy determines the power level of each element (EMR_i – exosomatic metabolic rate), which can be used as a specific benchmark to define the characteristics of the various elements.

By implementing this meta-grammar into a specific grammar for a specific situation (defining local dictionaries), it becomes possible to build an integrated system of indicators capable of analysing the importance of internal constraints (the implications of the bio-economic pressure) and external constraints, related to critical values of environmental loading. In relation to this point, it should be noted that when dealing with the exploitation of fossil energy sources (natural gas, goal, oil) and nuclear energy, humans have dealt with very concentrated forms of PES. This made it possible to reach very high levels of power: EMR_i. For this reason, so far, the analysis of the quality of energy sources has focused only on the value of the output/input ratio of energy carriers (confused with EROI). The overall performance of the energy sector of modern societies depends on the mix of PES, and the key characteristics (TET/HA_i, EMR_i, the spatial density of the supply) of each one of the various PESs used in the mix. The situation is completely different if such a mix includes an important fraction of alternative energy sources such as photovoltaic cells, bio-energy or wind power.

Multi-purpose grammar to analyse the exosomatic metabolic pattern of a society based on PES providing fund-flow energy carriers

The grammar presented in Figure 6.5 is again based on the graph language proposed by H. T. Odum, and the meanings of symbols and interactions are identical to those described above for Figure 6.4.

The relevant interactions between the fund elements in Figure 6.5 can be summarized as follows. The overall requirement of energy carriers of the various compartments is determined by their relative size in terms of human

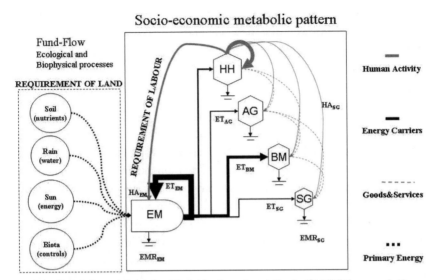

The set of relations determining biophysical constraints for the FLOW *Exosomatic Energy*:
(1) *internal constraint - Bio-Economic Pressure*
(2) *external constraint - Environmental Loading*

Figure 6.5 *Meta-grammar useful for the energetic analysis of agro-biofuels*

activity (HA_i) and requirement of power level (EMR_i). This will define an aggregate internal requirement of a net supply of energy carriers, which in turn will translate into an aggregate requirement of a gross production of energy carriers within the energy sector, depending on its gross/net ratio. After determining the gross production of energy carriers, we can check the interference with ecological processes that a given level of production implies. That is, by studying the particular biophysical processes of production of the energy carrier – in this case we consider biofuel – we can define a family of indicators of environmental impact related to such a production. Again, the formalization of this calculation has to be tailored to the specific method of biofuel production considered – ethanol from corn, ethanol from sugar cane, biodiesel from sunflowers, ethanol from switchgrass or oil from palm-oil – using specific dictionaries.

In this case, the low density in the production of the gross flow of energy carriers entails the need to invest an important fraction of the output in the exploitation process (a very high gross/net ratio). The alternative is to increase the fraction of working time to be allocated to the energy sector. This trade-off is extremely clear for agro-biofuels and it is discussed in detail with empirical data in Chapter 7. Before concluding, we want to note that it would also be important to perform this type of analysis to assess the performance of other alternative energy sources – for example, other applications of bio-energy, photovoltaics, wind power, the new generation of nuclear energy – when considering the whole life cycle.

Concluding remarks

In this chapter, we illustrated a few examples of the pitfalls of applying energy analysis to the study of the feasibility and desirability of agro-biofuels. These pitfalls can be contextualized within the dramatic swing that took place in the field of energy analysis in the last century. The pioneers of energetics provided a very strong semantic for the analysis using weak methods of formalization. The last generation of energy analysts applied a strong formalization to a very weak semantic. In the last section of this chapter we proposed a constructive approach capable of retaining the original semantic richness of the pioneers, while adopting new methods of formalization (multi-purpose grammars to be implemented with an ad hoc set of dictionaries), which are nowadays possible because of the computational power of computers. We believe that this middle way is possible, and that therefore energy analysis – if properly implemented – can represent an invaluable tool for a comprehensive and critical appraisal of proposed solutions to our energy problems.

Chapter 7

A Reality Check on the Feasibility and Desirability of Agro-biofuels

Performing a Reality Check on the Feasibility and Desirability of Renewable and Zero-Emission Agro-biofuels

In Chapter 5, we proposed the metaphor of a heart transplant as a conceptual tool for analysing the feasibility of an alternative energy sector expected to deliver a net supply of energy carriers to society at a required pace. Then we illustrated that this feasibility depends on three key characteristics:

1   the output/input of energy carriers associated with the exploitation of a primary energy source to generate a net supply of energy carriers;
2   the power level that can be achieved in the energy sector in order to process an adequate amount of energy input per unit of time (the level of EMR_{ES}); and
3   the compatibility of the resulting overall demand of primary energy sources with existing boundary conditions.

The high power level required to guarantee an adequate net supply of energy carriers to society per hour of labour in the energy sector refers to the compatibility with internal constraints. An adequate availability of primary energy sources refers to the compatibility with external constraints.

The typical pattern of exosomatic metabolism in a developed country, based on fossil energy as primary energy source, has been discussed in Chapter 4 (see Figure 4.9). For a developed country, the average power level of exosomatic energy used to produce and consume goods and services – at the level of the whole society – entails an EMR_{AS} of 25–40MJ/h. When looking for the internal constraints determining the feasibility of the relative dynamic budget, the benchmark values for the two key characteristics are: an output/input around 14/1, and EMR_{ES} of about 2000MJ/h. However, this particular quantification (the values of the benchmarks) depends on the conventions of energy accounting used in international energy statistics. They are based on the definition of a reference energy form as the primary energy source. When considering fossil energy as primary energy source, the energy form used as reference can be TOE in joules.

176 *The Biofuel Delusion*

As explained in Chapter 5, an operationalization of the elusive concept of EROI in quantitative terms requires reaching an agreement on the choice of a protocol: the selection of an appropriate grammar and dictionaries. This clearly generates a problem when dealing with alternative primary energy sources. In fact, the quantitative assessment of the exosomatic energy consumption of countries based on international statistics is based on a system of accounting tailored to and calibrated on fossil-energy-based exosomatic metabolic patterns.

In Chapter 5 (Figure 5.5), we illustrated (as a mental experiment) the impossibility of powering a modern society if agro-biofuels *entirely* replaced fossil energy as the primary energy source. Of course, the metabolism of a country being sustained by just one type of primary energy source and one type of energy carrier is an extreme assumption. The analysis presented in Figure 5.5 would become more difficult had we considered a combination of different primary energy sources used to generate a mix of different energy carriers. However, such a complex analysis can be performed by implementing the bio-economic principles based on the MuSIASEM approach and by developing a system of grammars and dictionaries (see Chapter 5 and Appendix 2). We do not get involved in this endeavour here!

In this chapter, we simply check whether it would be possible to cover with agro-biofuels – a fully renewable and zero-emission production system – a small fraction of the liquid fuels consumed in modern society. For this type of analysis we do not need complex calculations. We can study the feasibility and desirability of the idea by focusing on two critical thresholds:

1 The intensity of the net supply of exosomatic energy from the energy sector per hour of labour. This can be assessed using available data.
2 The density of the net supply of exosomatic energy from the energy sector per hectare (referring to the area of land used by the energy sector). This can also be assessed using available data.

Threshold of energy intensity of the supply per hour of labour in the energy sector

According to the analysis based on the benchmarks given in Figure 4.9 and the formal analysis provided in Chapter 5, the range for the intensity of the net supply of exosomatic energy from the energy sector per hour of labour in developed societies is: TET/HA_{ES} ~ 25,000–40,000MJ/h.

In Chapter 4 we looked at two countries, Italy and the US, which have a value for this parameter at the lower and higher extreme, respectively, of the above-mentioned range: Italy, with a low consumption of exosomatic energy, at 23,000MJ/h; and the US, with a high consumption of exosomatic energy, at 47,000MJ/h.

Threshold of energy density of the supply per hectare of colonized land

This value can be extrapolated by the comparative reading of the two graphs shown in Figure 7.1. This method of visualization, taken from the work of Vaclav Smil (2003), is based on the analysis of the density of flows of energy carriers (produced and consumed) in relation to the category of PES. It provides a comparison of the power density ranges of different typologies of supply of exosomatic energy as related to different primary energy sources (left graph in Figure 7.1) and a comparison of the power density ranges of different typologies of land use associated with the pattern of metabolism of developed societies (right graph in Figure 7.1). Differences in power densities between typologies of supply and consumption of exosomatic energy are so big – by orders of magnitude – that we need to use a logarithmic scale.

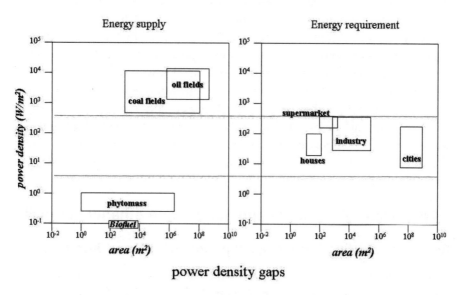

Figure 7.1 *The density of typologies of supply and requirement of exosomatic energy*

Based on data in Smil, 2003, pp242–243.

Figure 7.1 clearly illustrates the existence of a marked gap between the density of energy consumption – on the vertical axes measured in W/m^2 – of typologies of land uses expressed in developed societies (on the right) and the density of production of energy carriers based on biomass as a primary energy source (on the left). Indeed, when relying on biomass as a primary energy source, the power density in production is much lower than the power density in consumption typical of urban areas. Obviously, in such a situation it is impossible to have a large fraction of colonized land invested in land-use categories with a high density of energy consumption per hectare (see also Chapter 2). Due to this gradient in energy density, the number and size of cities (and hence the urbanized population) in pre-industrial societies were much smaller than in societies powered by fossil energy.

In this regard it is important to recall that if there is significant internal consumption of energy carriers to make energy carriers (a high ratio of gross/net biofuel), the required land in production for each self-sufficient process of energy-carrier generation will be amplified, and the density of the net supply of energy carriers per square metre will be further reduced. Thus, if one wants to use bio-energy, it would make more sense to directly burn the biomass – for heating or electricity – rather than converting it, to avoid a further reduction of its already low density. In fact, the conversion of agricultural biomass into biofuels makes things much worse. As illustrated in Chapter 5, the density of net power supply is markedly reduced by the internal loop of energy carriers consumed within the process generating biofuel. Starting from an energy carrier that can be burned (biomass), the production process of agro-biofuels uses this input to make beer, an intermediate product with a water content of more than 80 per cent. This step does not seem to be a smart move in terms of energy efficiency. From an energetic point of view, if one wants to use food energy in exosomatic way, a stove directly burning corn (as shown in Figure 3.1) would be better than using corn to make ethanol!

Corn-ethanol in the US and Sugar Cane-ethanol in Brazil

Results of existing large-scale experiments: US and Brazil

As observed, the production of the US and Brazil combined covers almost 90 per cent of the world production of biofuels. In 2006, the US produced 18.4 billion litres (46 per cent of the world's total) and Brazil 16 billion litres of ethanol (42 per cent of the world's total; World Bank, 2008). Even though this still does not represent a large-scale production at world level (in 2005, ethanol comprised only 1.2 per cent of the world's liquid fuel supply), it is large enough to allow an assessment of the technological coefficients from the analysis of aggregated values referring to the whole sector. This method is much more robust than inferring aggregate values from measurements taken at the level of individual plants, a method which implies a risky scaling-up.

The total output and total labour demand of ethanol production, both in the US and Brazil, used in the following assessment is based on data provided by the ethanol industry for the whole sector. Using these data we assessed the external and internal constraints (using the approach presented in Figure 5.4).

In relation to the external constraint, we focus here only on the requirement of land for the energy sector by using:

- technical coefficients (in Box 7.1 and 7.2) referring to the conversion of crops into biofuels: PES → GSEC (yield per hectare and litres of biofuels per kg of crop); and
- the GSEC/GNEC ratio (the consumption of energy carriers for making energy carriers) using the formula shown in Figure 5.4.

In relation to internal constraints, we focus here only on the requirement of labour in the energy sector by using an evaluation of the EMR_{AB}, the power level of exosomatic energy consumed per worker in the agro-biofuel production process. This evaluation depends on the output/input ratio, since the energy consumed in the process of generating GSEC coincides with the input of energy carriers (GSEC−NSEC) that has to be consumed in order to generate NSEC. The power level is calculated by dividing GSEC−NSEC by the hours of work in the process.

Finally, in this assessment we used the output/input calculated by Farrell et al (2006) for ethanol from corn in the US, but corrected the value by eliminating the energy credits for by-products. This follows the discussion in Chapter 6 regarding the credit for by-products and the issue of scale. In fact, even at a 3 per cent substitution of liquid fuel with the net production of ethanol – fully renewable and with zero emissions – the displacement method is not justified.

Benchmarks for the net supply of ethanol from corn in the US

There is a well-established data set for corn-ethanol production in the US, without major differences in the physical assessment of biophysical inputs and outputs among different studies. The differences found in the overall assessment of the output/input energy ratio are basically generated by different choices about how to convert the various inputs and outputs into energy equivalents, but not by the initial accounting of the quantities of biophysical inputs and outputs. Details of our calculations are given in Box. 7.1; no energy credit is given to the by-products in the form of energy carriers.

From this data set we obtain two benchmarks:

1. net supply of ethanol per hour of labour in the corn-ethanol production system: 224MJ/h;
2. net supply of ethanol per unit of land in the corn-ethanol production system: 6GJ/ha/year or $0.6MJ/m^2$/year or $0.02W/m^2$.

Box 7.1 *Technical coefficients of ethanol production from corn in the US: data on a yearly basis and referring to 2004*

Gross output (per hectare of crop production)

Corn production: 8000kg/ha
Ethanol production: 3076L/ha (2.69kg of corn to produce 1L of ethanol)
Gross supply of energy carriers: 66.13GJ of ethanol/ha (1L of ethanol = 21.5MJ)

Input in step 1: agricultural production of corn

Labour: 12h/ha/year
Land: 1ha
Fossil energy: 29.3GJ/ha

Input in step 2: fermentation/distillation of ethanol

Labour: 14.76h/ha/year
Land: negligible (in addition to the hectare in crop production)
Fossil energy: 31.9GJ/ha

Total input (over gross supply)

Total fossil energy carrier input: 61.2GJ/ha
Total labour demand: 8.8 hours/1000L or 114L/h

Output/input

Output/input in energy carriers: 1.1/1

Net supply

11L of GSEC ethanol → 1L NSEC ethanol
Net supply of energy carriers: 9 per cent of gross production of ethanol
Net supply per hour: 10.4L/h (9 per cent of the gross)
Net supply per hectare: 277L/ha (9 per cent of the gross)

Note on labour input

The assessment of labour demand for the phase of agricultural production is from Pimentel et al (2007), whereas the labour requirement for fermentation/distillation is based on two different assessments (USDA, 2007; USDE 2008). USDA (2007) suggests 17, 000 full-time jobs at the plant per billion gallons of ethanol. This is equivalent to an input of 33 million hours for 3780 million litres (8.7 hours per 1000 litres). USDE (2008) suggests 238,000 jobs in all sectors of the economy in 2007. This is equivalent to an input of 476 million labour hours.

> Since it is unclear whether the hours of agricultural production are already included in these assessments, for safety (in favour of the biofuel option) we took out the four hours of agricultural labour (per 1000 litres) from the most favourable of the two assessments.

Please note that when considering the requirement of fossil energy for a two-step process – the agricultural production of corn, and the conversion of the corn into ethanol – we assumed as valid the pro-ethanol claim that the by-products of agricultural production provide the entire heat energy consumption of the step of distillation. This is not always the case, but again we went for the most favourable assumption. Therefore, the requirement of fossil energy refers only to the consumption of energy carriers both for the phase of agricultural production (transportation, production of fertilizers, pesticides, irrigation, the making of steels and technical infrastructures) and only transportation and technical infrastructures for the phase of fermentation-distillation.

Benchmarks for the net supply of ethanol from sugar cane in Brazil

For this assessment we used official data and technical coefficients provided by a very detailed and informative study published by the Sugar Cane Agroindustry Union (UNICA) in Brazil (De Carvalho Macedo, 2005). These data have been checked against the assessment of ethanol production from sugar cane in Brazil provided by Patzek and Pimentel (2005) and Pimentel et al (2007), which were reporting a much worse performance. Therefore, in Box 7.2, we used two different assessments of the energy inputs used in the production of ethanol from sugar cane. These discrepancies in the assessments of technical coefficients (different output/input ratios) are relevant for the calculation of the gross/net ratio of ethanol production.

In relation to phase I – inputs for the production of agricultural biomass – Pimentel et al (2007) indicate an input of 40GJ/ha/year (this assessment is labelled with a H, standing for high), while De Carvalho Macedo provides an input of 15GJ/ha/year (this assessment is labelled with a L, standing for low). In relation to phase II – fermentation and distillation for producing ethanol – Pimentel et al (2007) indicate an input of 48GJ/ha/year (labelled H) versus an assessment of De Carvalho Macedo of only 4GJ/ha/year (labelled L).

The discrepancies between these two assessments have been investigated by Boddey et al (2008), who report that Pimentel and Patzek assume:

- a larger input for the agricultural production, especially for nitrogen fertilizer (probably because they are sceptical about the possibility of preserving the long-term health of soil producing sugar cane with low inputs of fertilizers) and irrigation; and
- a much larger input for high-tech, heavily mechanized industrial production.

Boddey et al (2008) make the following case:

- Brazilian cane varieties use much less fertilizer than the varieties cultivated in other areas of the world because of the peculiar association found in local varieties with N_2-fixing bacteria.
- High-tech production typical of the US has very little to do with the labour-intensive production (implying a large recycling of biomass for fertilization) in Brazil.
- The abundant supply of rain implies that in Brazil there is almost no requirement of energy for irrigation.
- The use of bagasse (a by-product of the process) makes it possible to cover not only the requirement of heat, but also the generation of electricity consumed in the production of ethanol.
- Some of the energy conversion factors used by Pimentel et al for phase II are much higher than those used by De Carvalho Macedo.

Because of this combination of fortunate conditions found in Brazil, we can fully appreciate the poor quality of biomass as primary energy source for making liquid fuels, when compared with the quality of fossil energy. In fact, a low use of energy carriers in the production of ethanol does not translate only into a high output/input, but also into a low power level per worker operating in the process of ethanol production (a low EMR). The opposite is true for corn-ethanol production in the US. Due to the low density of the primary energy input, it uses a lot of technical capital per worker – determining a high EMR – but this translates into a low output/input energy ratio.

In conclusion, the net supply of ethanol from sugar cane determined by a combination of the intensity per hour of labour and the density per hectare of colonized land remains very low, either when considering an overall output/input of energy carriers of 1.5/1 (according to the high-input assessment) or an output/input of energy carriers of 7/1 (according to the low-input assessment). A more detailed discussion of this issue is given further on.

Details on the data set used to generate the benchmarks for sugar cane production are given in Box 7.2. From this data set we obtain two sets of benchmarks. Using high input (H) estimates from Pimentel et al (2007), we obtain:

- a net supply of ethanol per hour of labour in the sugar cane-ethanol production system of 150MJ/h; and
- a net supply of ethanol per unit of land in the sugar cane-ethanol production system of 45GJ/ha/year, or 4MJ/year/m^2, or 0.1W/m^2.

Box 7.2 *Technical coefficients of Brazilian production of ethanol from sugar cane; data on a yearly basis and referring to 2003*

Gross output (per 1 million tonnes of sugar cane)

Total ethanol production: 83.33 million litres of ethanol
Gross supply of energy carrier (GSEC): 1,766,000GJ of ethanol

Gross input (per 1 million tonnes of sugar cane)

Labour: 2200 full-time jobs or 4 million hours/year (of which 73 per cent in agriculture)
Land in production: 13,333ha

Values per hectare

Sugar cane production per hectare: 75,000kg/ha
Ethanol production per hectare: 6250Ll/ha (12kg of sugar cane to produce 1L of ethanol)
Gross supply of ethanol per hectare: 134GJ of ethanol/ha (1L of ethanol = 21.5MJ)

Total inputs (from UNICA study)

Input in step 1: agricultural production of sugar cane
Labour: 210hours/ha/year
Land: 1 hectare
Fossil energy (H): 40GJ/ha
Fossil energy (L): 15GJ/ha

Input in step 2: fermentation/distillation of ethanol
Labour: 90hours/ha/year
Land: negligible
Fossil energy (H): 48GJ/ha
Fossil energy (L): 4GJ/ha

Total input over gross production
Total labour input: 300 hours/ha/year or 48 hours/1000 litres
Total fossil energy carrier input (H): 88GJ/ha
Total fossil energy carrier input (L): 19GJ/ha

Output/input

Output/input in energy carriers (H): 1.5/1
Output/input in energy carriers (L): 7/1

Net supply (H) (per 1 million tonnes of sugar cane)

Based on output/input 1.5/1 (Pimentel et al, 2007):
Net supply of energy carriers: 33 per cent of gross production of ethanol
3L of GSEC ethanol → 1L NSEC ethanol
Net supply: 27.7 million litres (33 per cent of the gross)
NSEC: 588,000GJ
Net supply per hour: 7L/h
NSEC per hour: 150MJ/h
Net supply per hectare: 2065L/ha (33 per cent of the gross)
NSEC per hectare: 44.3GJ/ha

Net supply (L) (per 1 million tonnes of sugar cane)

Based on output/input 7/1 (De Carvalho Macedo, 2005):
Net supply of energy carriers: 87 per cent of gross production of ethanol
1.15L of GSEC ethanol → 1L NSEC ethanol
Net supply: 72.2 million litres (87 per cent of the gross)
NSEC: 1,536,000GJ
Net supply per hour: 18.4L/h
NSEC per hour: 395MJ/h
Net supply per hectare: 5437L/ha (87 per cent of the gross)
NSEC per hectare: 117GJ/ha

Note on labour input

We could not find an assessment of the labour demand in the phase of agricultural production in the UNICA study. In fact, the statistics on employment given in that study refer to the entire production of sugar cane, which is used not only for ethanol production, but also for sugar production. In that study, however, there is an estimate of the employment for the sugar-ethanol sector referring to the 1990s for the area of Sao Paulo. The sector was employing about 30 per cent of the Brazilian workers in sugar cane-ethanol production. In the same section, the study says that, in the 1990s, the production of ethanol in the rest of Brazil was employing about 70 per cent of the workers in sugar cane-ethanol production. In the rest of Brazil, the production was much more labour-intensive: up to three times the value found in Sao Paulo. The same study says that in the last decade, the use of human labour in the process of sugar cane production has been decreasing: from 674,000 in 1991 to 448,000 workers in 2003 (De Carvalho Macedo, 2005, p212, Table 6). For this reason, we decided to use the evaluation done for the state of Sao Paulo in the 1990s as the average value for labour intensity of the production of sugar cane-ethanol for Brazil in the year 2003. This is probably a generous assumption.

Using low input estimates (L) from De Carvalho Macedo (2005), we obtain:

- a net supply of ethanol per hour of labour in the sugar cane-ethanol production system of 395MJ/h; and
- a net supply of ethanol per unit of land in the sugar cane-ethanol production system of 117GJ/ha/year, or 11.7MJ/year/m², or 0.4W/m².

A critical evaluation of fully renewable and zero-emissions ethanol production

The expected benchmarks and characteristics of the net supply of a fully renewable and zero-emissions agro-biofuel system are represented in Figure 7.2.

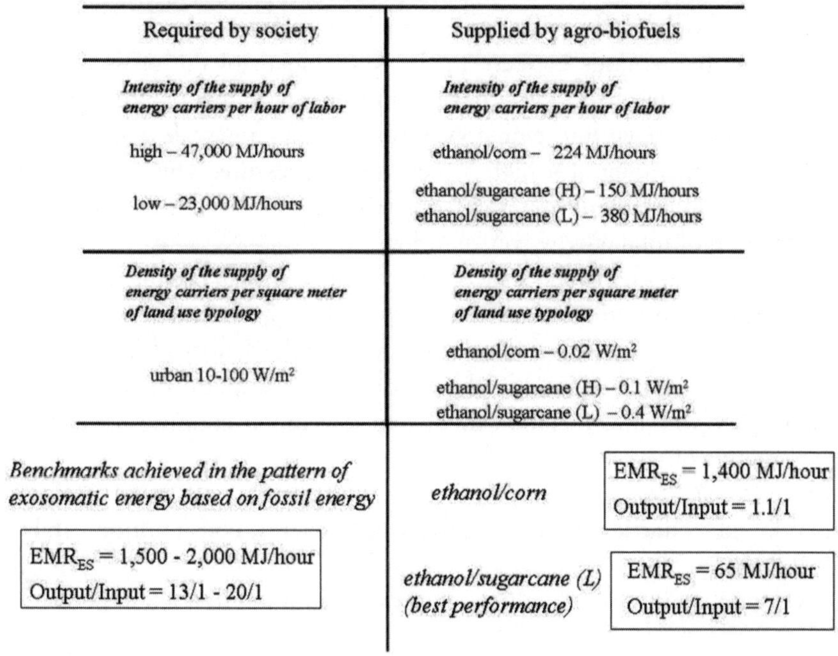

Required by society	Supplied by agro-biofuels		
Intensity of the supply of energy carriers per hour of labor	*Intensity of the supply of energy carriers per hour of labor*		
high – 47,000 MJ/hours	ethanol/corn – 224 MJ/hours		
low – 23,000 MJ/hours	ethanol/sugarcane (H) – 150 MJ/hours ethanol/sugarcane (L) – 380 MJ/hours		
Density of the supply of energy carriers per square meter of land use typology	*Density of the supply of energy carriers per square meter of land use typology*		
urban 10-100 W/m²	ethanol/corn – 0.02 W/m² ethanol/sugarcane (H) – 0.1 W/m² ethanol/sugarcane (L) – 0.4 W/m²		
Benchmarks achieved in the pattern of exosomatic energy based on fossil energy	ethanol/corn	EMR_{ES} = 1,400 MJ/hour	
		Output/Input = 1.1/1	
EMR_{ES} = 1,500 - 2,000 MJ/hour	ethanol/sugarcane (L) *(best performance)*	EMR_{ES} = 65 MJ/hour	
Output/Input = 13/1 - 20/1		Output/Input = 7/1	

Figure 7.2 *Overview of the benchmarks characterizing the requirement of society and the production process of ethanol*

In order to be completely renewable and have zero emissions, the ethanol-corn system, experimented with on a large-scale in the US, must use a lot of technical capital (exosomatic energy consumption per hour of work, relevant for internal constraint) and land (to gather the large amount of PES, relevant for external constraint). In fact, to provide a net supply of 224MJ/hour it should produce 11 times this amount of gross ethanol, multiplying by 11 all the activities required for the gross production of ethanol. At the moment, the power level of the workers in the corn-ethanol system is pretty high. The exosomatic metabolic rate per hour of labour (EMR) is 1400MJ/h. However, this value is obtained at present using fossil energy carriers. This is a value comparable to that of the workers operating in the energy sector of developed societies. If this energy input were taken out by the gross production of ethanol, then this high EMR would translate into a very low output/input of 1.1/1. As discussed in Figure 5.4, since corn-ethanol production is operating in the critical area of the output/input energy carriers ratio (< 2/1), the solution of using exosomatic energy to boost labour productivity (to match the internal constraint) entails a non-linear increase in the ratio of gross/net production of ethanol. This would generate a tremendous increase in the requirement of land and labour per unit of net supply delivered to society. If the production of corn-ethanol were self-sufficient, it would clash against an external constraint.

It is the tremendous increase in the requirement of the gross supply of energy carriers, due to the internal loop of energy-for-energy, that makes it impossible to produce zero-emissions biofuels on a large scale within a developed society operating at a high bio-economic pressure. This system would crash against internal and external constraints even if trying to cover only a small fraction of total energy consumption.

Just to give an idea of the level of impossibility, we will use the benchmarks characterizing the performance of the corn-ethanol system in terms of the intensity of the net supply per hour of labour, and the density of the net supply per hectare of colonized land in the energy sector, to perform an analysis of compatibility with the metabolism of two developed countries, selected as representative of:

- developed countries with a relatively low energy consumption (Italy in 1999); and
- developed countries with a high energy consumption (the US in 2006).

The case of a developed country with a low energy consumption
When dealing with Italy, a developed country with a moderate level of energy consumption (121GJ/year per person in 1999 vs more than 300GJ/year in the US), we check the feasibility of an ambitious scenario in which agro-biofuels are used to cover 30 per cent of the energy spent in transportation. This implies that starting from a total consumption of 7EJ/year (1EJ = 10^{18}J) and considering that 30 per cent of this amount goes in transportation (2.1EJ/year), we want to cover one-third of this consumption – 10 per cent of the total energy consumption of Italy (0.7EJ/year) – with fully renewable and zero-emissions agro-biofuels. According to the previous calculations, the agricultural sector

should provide *a net supply* of 32.5 billion litres of ethanol. When assuming a system that is fully renewable and capable of capturing the CO_2 emitted (ratio 11 gross/1 net), this amount of net ethanol would require the gross production of 358 billion litres of ethanol.

Using the benchmarks presented in Figure 7.2 for ethanol from corn, Italy would require, first, 34 Ghours of labour in biofuel production – this is 94 per cent of the work hours provided by the Italian workforce in 1999 (at the moment the energy sector absorbs less than 1 per cent) – and, second, 117 millions hectares of agricultural land – this would be more than 7 times the 15.8 million hectares of agricultural area in production in Italy in 1999.

Had we performed the same type of analysis with data referring to the production of biodiesel from sunflower using technical coefficients referring to this production in Italy (a detailed description of this calculation is in Giampietro and Ulgiati, 2005) we would have found the same type of results: a very large fraction of the workforce in the energy sector, more than 50 times higher than the current one, and a requirement of land several times larger than the current one.

This is the reason why it really is not necessary to invest time in debating over the accuracy of the assessments of output/input of the energy carriers in the agro-biofuel process of production. It really does not make any difference if the output/input (determining the ratio gross/net) is 1.2 or 1.5.

Moreover, there are other reasons reinforcing this incompatibility:

- few Europeans want to be farmers anymore, and in Italy it is at present difficult to find enough farmers to produce food;
- Italy does not have any surplus food production (since the food consumed in Italy would already require double the arable land in production; Giampietro et al, 1998);
- an expansion of agricultural production in marginal areas would dramatically increase the requirement of technical inputs – such as fertilizers – further reducing the overall output/input energy ratio, and generating an additional non-linear increase in the internal loop of energy-for-energy; and
- the environmental impact of agriculture (soil erosion, alteration of the water cycle, loss of habitats and biodiversity, accumulation of pesticides and other pollutants in the environment and the water table) is already serious. Any expansion in marginal areas would make it much worse.

This check indicates that in a densely populated developed country, it is impossible to cover a significant proportion of liquid fuels using fully renewable/zero-emissions agro-biofuels. It is impossible even when we are dealing with a country with low energy consumption per capita. What about a country, like the US, with higher consumption, but also with much more land available?

The case of a developed country with high energy consumption

When performing a check for the US we adopt a less ambitious goal for our scenario. Rather than covering 10 per cent of the energy consumption of the country, we want to cover just 10 per cent of the fuels used in transportation in the US (about 30EJ). Since the energy in transportation is less than 30 per cent of the total, we want to cover 10 per cent of 30 per cent, which is 3 per cent of the total energy consumption of the US. According to this target, the agricultural sector of the US should generate a net supply of 3EJ of ethanol per year, or a net flow of 140 billion litres.

Using the 11/1 ratio between gross and net energy supply of ethanol, we calculated that the net supply of 3EJ of ethanol – a net flow of 140 billion litres – would translate into a requirement of a gross production of 33EJ – 1540 billion litres. This gross production of ethanol would require:

- 148 Ghours of labour in biofuel production (almost 48 per cent of the labour supply that could be provided by the US workforce after absorbing all the unemployed!). This would require that half of the US workforce agree to go back into farming and rural activities; and
- 5500 million hectares of arable land (more than 31 times the 175 million hectares of arable land in production in the US in 2005).

Also in this case, we find a total lack of feasibility in using a self-sufficient corn-ethanol system aimed at reducing the dependence on fossil energy – 10 per cent of the transportation fuels – and generating zero CO_2 emission. This lack of feasibility clearly indicates that the actual production of ethanol in the US is possible only because such a production *is powered by fossil energy fuels!*

But if we drop the motivation of independence from fossil energy and the zero emission, then common sense suggests that for a developed country it is unwise to pay for importing a barrel of oil, then add a lot of capital, land and significant labour inputs – additional production factors that also have to be paid for – and, to top it off, consume natural resources and stress the environment (soil erosion, nitrogen and phosphorus in the water table generating dead zones in the ocean, pesticides in the environment, freshwater consumption) to produce 1.1 barrel of oil equivalent in the form of ethanol. Whether the price of oil is high or low, this solution will never be economically competitive.

The case of a developing country with low energy consumption per capita

The example of ethanol from sugar cane in Brazil is important, since it illustrates the best possible situation, so far known, for agro-biofuels. As observed earlier, in Brazil we deal with the combination of the highest output/input ratio of energy carriers achieved so far in the production of biofuels, and a country with enough land to produce sugar cane for energy (a semi-tropical agriculture) and food, because of low demographic pressure.

When considering the assessment provided by De Carvalho Macedo (2005), looking at the data in Figure 7.2, we observe that even if ethanol from sugar cane does much better than ethanol from corn in terms of output/input ratio of energy carriers, the differences in value from the performance of the sugar cane-ethanol system (the intensity of the net supply per hour of labour and the net supply per colonized land) and the performance that would be required to run the metabolism of a developed country is very large: in the order of hundreds of times.

In relation to this point, we can observe that the study by Boddey et al (2008) provides an even higher assessment of the output/input ratio of energy carriers for ethanol from sugar cane in Brazil: a value of around 9/1. However, using the value of 9/1 or 7/1 would not make a big difference in the final calculation of the two benchmarks. Looking at the curve given in Figure 5.4, we see that above a certain threshold value (that is, an output/input of greater than 5/1), the reduction over the gross/net ratio becomes negligible. In this case, moving from 7/1 to 9/1 would imply inducing a reduction of 11 per cent rather than a reduction of 14 per cent over the gross density per hectare and intensity per hour of labour. Actually, reducing the consumption of fossil energy on the input side is clearly not an improvement! In fact the problem with the production of ethanol in Brazil is not with the low value of the output/input ratio of energy carriers, but with the low value of EMR.

In general terms, we can say that the problem of sugar cane-ethanol depends on the low density of the flow of energy input coming from the fund-flow supply biomass associated with the high output/input of energy carriers. If we increased the EMR in order to collect more energy per hour (as done by the US corn-ethanol system), then we would get in trouble for the consequent reduction in the output/input ratio of energy carriers.

When dealing with the case of sugar cane-ethanol in Brazil, we can better focus on the issue of desirability. That is, after assuming that a fully renewable ethanol production from sugar cane is feasible in Brazil, how desirable is this choice for Brazilians? Is this providing an adequate income to workers? Is it delivering enough net supply of energy carrier (per hour of labour) to support the economic activity of other workers operating in different economic sectors? Would this choice increase the material standards of living of the Brazilian people, moving it closer to the level enjoyed by more developed countries? Or rather would it slow down the movement of the Brazilian economy over time?

Countries such as China, India and South Korea, which compete with

Brazil in the international markets, are relying on fossil energy as a primary energy source and are at the moment using their workforce, their capital, their finances and resources to boost other economic activities: to produce added value linked to the production of goods and services. On the contrary, Brazil, with the top priority given to agro-biofuel development, is using an increasing fraction of its economic resources and workforce just to produce an oil-equivalent ethanol supply; an input for economic activities, but not an economic activity in itself (unless the energy carrier is exported to other countries). Looking at the data on the qualitative composition of the workforce in the sugar cane-ethanol sector in Brazil given in the UNICA study (De Carvalho Macedo, 2005, p212, Table 6) we find that over the last decade, the number of permanent jobs has decreased, whereas the number of temporary jobs has increased, and that only 9100 college graduates are employed in the sector out of the 450,000 jobs directly related to the sugar cane-ethanol process (p210, Table 4). In the long term, this choice may very well imply a decrease in the ability to maintain a high level of bio-economic pressure within the Brazilian economy.

Thus, when dealing with the desirability of the agro-biofuel option, one should give due consideration to the following two observations. First, it is unclear whether or not it is economically convenient for Brazil to invest a large amount of economic resources (capital, land and labour) in the production of ethanol from sugar cane. In fact, these resources definitely do have an opportunity cost. Since Brazil is competitive in agricultural production, it may consider the option of moving the large flow of investments – presently allocated to biofuels production – into the implementation of policies and development of infrastructures enhancing the added value obtained from the export of food products.

Second, a net supply of energy carriers based on a significant amount of labour and land supply tends to generate a dangerous situation. The negative non-linear effect generated by the requirement of energy carriers in the internal loop of energy-for-energy implies that the use of energy carriers in biofuel production would be better kept to a minimum. This translates into:

- a low level of technology supporting the activity of workers (in Brazil a large fraction of sugar cane is still cut manually) and therefore a low labour productivity, resulting in a low salary for the workers; and
- a low level of technical inputs – such as fertilizers and irrigation – supporting agricultural production, and a low level of respect for the ecological integrity of the environment in which the production takes place.

To put it another way, in order to boost the efficiency of the double conversion (solar energy → sugar cane; sugar cane → ethanol) it is necessary to squeeze, as much as possible, both the workers in the plantations and the soil and biodiversity. Unless very strong regulation and monitoring (implying a lowering of the output/input ratio and higher economic costs) is continuously applied, it is unlikely that high standards of sustainability can be achieved and maintained. In general terms, it is possible to say that, so far, the massive transformation of traditional farming systems into intensive monoculture according to the paradigm of industrial agriculture – a phenomenon that is taking place in many countries in South America – did not bring about the rural development promised by the proponents of agro-biofuels (this point is discussed in Chapter 8).

As explained at the beginning, we are neither able nor willing to provide any normative judgement about the desirability of policy options in specific cases. We believe that decisions about what is desirable within a feasible option space have to be taken by those who will bear the consequences of the choice. The Brazilians are the only ones who can decide whether or not it is convenient for Brazil to invest in agro-biofuels. What our methodological approach can provide is a richer narrative to characterize the various pros and cons.

Linking Bio-economic Pressure to the Quality of Energy Sources

In Chapter 4, we focused on the effects of BEP, as an internal biophysical constraint, on the feasibility of the exosomatic metabolic pattern of developed societies. Indeed, BEP defines a biophysical constraint affecting the productive sector of society, which must be capable of delivering the expected supply of all the material and energy flows metabolized by society while using only the hours allocated to the productive sector (HA_{PS}). This aggregate flow of products includes the net supply of energy carriers from the energy sector (ES), the required supply of food from agriculture (AG), the required supply of exosomatic devices and infrastructures from building and manufacturing (B&M), and the required supply of minerals from mining (M). In practical terms, the productive sector (PS) is the sector guaranteeing the supply of all the goods required for stabilizing the exosomatic metabolism of society (and the endosomatic input sustaining human activity). The internal biophysical constraint consists of achieving the delivery of all the goods included in this list while using only 330 hours/year of work in the productive sector (PS) per capita (see also Figure 4.9).

Given this definition of BEP, the quality of an energy source can be assessed in relation to its contribution to maintaining, diminishing or raising the bio-economic pressure in society. A high-quality energy source should make it possible to achieve a high net supply of energy carriers per hour of labour in the energy sector and, more in general, per hour of labour in the productive sector (PS). Obviously, also the reverse is true; when substituting energy sources (such as fossil fuel) guaranteeing a net supply of around 28,000MJ of energy carriers per hour of labour in the energy sector with energy sources providing a net supply of energy carriers in the order of only 250–400MJ per hour of labour in the energy sector, we should expect a substantial involution of the process of economic development. In the latter case, society will have to allocate the vast majority of its labour force to the energy sector and/or dramatically reduce the energy consumption in the household and service sectors. The most probable solution is that both adjustments will take place simultaneously, moving the society 'backwards' toward a pattern typical of pre-industrial society.

Below we formalize this idea, given that the concept of bio-economic pressure can be quantified using a set of expected relations over funds and flows (see Giampietro and Mayumi, 2000a; 2000b). In particular, when looking for a numerical benchmark we can use a simple ratio, which can easily be calculated using empirical data:

$$BEP = TET/HA_{PS}$$

Where TET is the total exosomatic throughput (the total exosomatic energy expressed in terms of primary energy sources, for example TOE) and HA_{PS} is the amount of working hours in the productive sector, which includes the agricultural sector, building and manufacturing, and energy and mining. This simple ratio can be expressed (and calculated) in several non-equivalent ways. Using the grammar described in Figure 4.9, we can for example write:

$$BEP = \frac{\{(HA_{HH} \times EMR_{HH}) + (HA_{PS^*} \times EMR_{PS^*}) + (HA_{SG} \times EMR_{SG}) + (HA_{ES} \times EMR_{ES})\}}{(HA_{PS^*} + HA_{ES})} = \frac{TET}{HA_{PS}}$$

Note that in the above relation, the productive sector (PS) is disaggregated into PS* (defined as PS − ES) and ES, to 'isolate' the energy sector (ES) for our purpose.

When expressing the value of the benchmark BEP per person per year, we obtain:

$$BEP = TET/(HA_{PS^*} + HA_{ES}) = 300GJ/(320 \text{ hours} + 10 \text{ hours}) \sim 900MJ/h$$

Due to the required hours of work in the compartments of building and manufacturing and agriculture, the bio-economic pressure (BEP) is necessarily much lower than the threshold of net supply of energy carriers per hour of labour in the energy sector expected (i.e. required) by society (TNS_{ES}). In fact:

$TNS_{ES} = TET/HA_{ES} = BEP \times (HA_{PS}/HA_{ES})$
$TNS_{ES} = TET/HA_{ES} \gg BEP = TET/(HA_{PS^*} + HA_{ES})$
$(HA_{PS^*} + HA_{ES}) \gg HA_{ES}$ since $HA_{PS^*} \gg HA_{ES}$

Using available benchmarks we find that the *expected* intensity of the supply of energy per hour of labour from the energy sector is about 28,000MJ/h. That is, TNS_{ES} is more than 30 times the numerical value of BEP.

$TNS_{ES} \sim 28,000MJ/h \gg BEP \sim 900MJ/h$

We now consider the benchmark for the intensity of the supply of energy per hour of labour from the energy sector referring to the net production of ethanol. This value is 224MJ/hour for production taking place in temperate areas (ethanol from corn) and 380MJ/hour for production taking place in tropical areas (ethanol from sugar cane, most favourable assessment) (see Figure 7.2).

Then, let's imagine a miracle happens: the production of all the exosomatic devices (in manufacturing), the building of all infrastructures (in building), the production of all food (in agriculture) and the extraction of all minerals (in mining) consumed by a given developed society is now performed, free of charge, by friendly aliens willing to help. This miracle would make it possible to allocate all hours of human work available in the productive sector to the production of fully renewable and zero-emissions agro-biofuels. Would this be enough to balance the dynamic budget?

In spite of the miracle, due to the low net supply of energy carriers from the agro-biofuel system, that lucky society would not be able to operate at a level of bio-economic pressure of 900MJ/hour! Actually, it would not be able to cover even half of its energy consumption, even when using *all the available workers* in the productive sector (HA_{PS}) in the production of agro-biofuels.

The Implications of Peak Oil for Those Looking for Alternative Energy Sources

Peak oil in relation to exosomatic metabolic patterns

In this section, we briefly illustrate the basic concept of peak oil and why this is relevant for an informed discussion about alternative energy sources. In order to illustrate the concept, we will describe an analogous situation related to the metabolism of endosomatic energy flows. That is, we imagine a case of peak grain for a pre-industrial society that is harvesting the food it consumes from gigantic stocks located below the ground. This situation is illustrated in Figure 7.3.

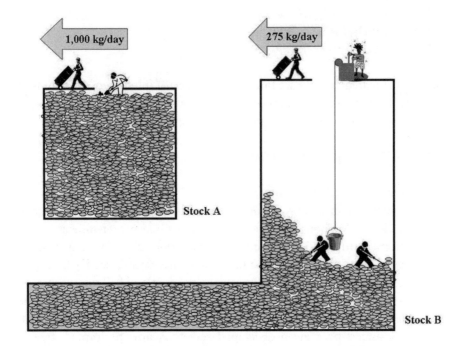

Figure 7.3 *Illustration of the conceptual difference between peak oil and stock depletion using the metaphor of peak grain*

The two underground stocks illustrated in Figure 7.3 contain different amounts of grain: stock A (on the left) contains 20 million kg of grain, while stock B (on the right) contains 30 million kg. This information refers to the size of the stock, but it does not say anything about how much grain we can get out of the stock per day. Because of their characteristics, the two stocks entail a different requirement of work and a different net output when exploited. This different relevant attribute calls for another set of specifications.

- Stock A can provide up to 1000kg of grain per day, depending on the investment; the possible output depends on the amount of workers that the society wants to use to gather grains.
- Stock B can only provide up to 275kg of grain per capita, no matter how many workers are used. That is, after having reached a given number of workers in the underground reservoir fully dedicated to moving the grain up to the surface, it is not possible to increase the net supply of grain per day by putting more workers into it. Additional workers will only interfere with each other.

Now, let's see the implications of the different characteristics of the two stocks for a society willing to use them for its daily food production, rather than producing grain using agriculture. In this mental experiment let's use the pre-industrial society described in Chapter 3 (Figure 3.5), with a size of 385 people divided in two classes: 25 rulers and 360 ruled. Recall the situation of that society when producing its food (required flow of grain) with a fund/flow supply (with agricultural production).

The pre-industrial society using agricultural production (fund/flow supply)

According to the assumptions provided in Table 3.1 (generation of food obtained through agricultural production), the endosomatic metabolic pattern of production and consumption of this society is determined by a forced relation over the characteristics (relative size and metabolic rate) of the three compartments (through an impredicative loop definition of constraints):

1 the K sector of rulers = 25 people consuming the equivalent of 1.1kg of grain per capita (partially used as feed for animal products);
2 the W sector of 50 workers forced to work in grain production in order to generate the net supply of grain required by society, consuming 0.7kg of grain per capita; and
3 the FF sector of farmer families = 310 people, the overhead of the W sector, also consuming 0.7kg of grain per capita.

As explained in Chapter 3, this situation is forced by the special characteristics of the fund/flow supply of grain, determining a given requirement of working hours in grain production (agriculture) and a given requirement of land.

Let's now see what happens when this society exploits either stock A or stock B.

Stock A

The society has a lot of room for adjusting the characteristic of the endosomatic metabolic pattern by changing the different characteristics of the loop. For instance, it can reduce the number of workers in agriculture from 50 to 10, and increase the consumption of grain per capita to 2kg of grain per capita for each member of the society, abolishing the classes. This will bring the consumption of grain of the society to a metabolic rate of 770kg of grain per day. Obviously, that this flow can be generated using only ten workers is due to the extremely favourable characteristics of stock A: the grain is very close to the surface. However, the more stock A is depleted, the more workers will be needed for extraction. This would call for a choice of either reducing the consumption per capita or increasing the number of workers in the reservoir (workers who would have to be taken away from other occupations). Looking at the lifespan of stock A, we can say that depending on the choice of the society, this stock could last for 71 years (at 770kg of grain per day). However, if the particularly favourable situation of this society (compared with neighbouring societies) generates a growth in the population (either by an increase in local fertility or by immigration), the lifespan of the stock would be reduced. In any case, in the short term this society has the option of expanding in size, by increasing either the population or the consumption per capita, or both.

Stock B

The society will have the same typology of endosomatic metabolic pattern it used to have when cultivating grain (before extracting it from underground reservoirs). In fact:

- the amount of grain required per day – previously produced by agriculture – is already equal to the maximum amount of grain that can be taken out from the stock; and
- the society has to invest the same number of workers in the underground reservoir as were employed in agriculture before, assuming that 50 workers are needed to guarantee the supply of 275kg of grain per day.

For this reason, even if the society has moved to a stock-flow it faces the same set of internal constraints that it had to face when it was operating in fund-flow. That is, it will be impossible to abolish classes, associated with two different levels of consumption of grain per capita, since the maximum amount of grain consumption cannot be increased. On the positive side, the generation of a net supply of grain to society will not imply the loss of habitat or the requirement of land, since the space occupied by the operations in the underground stock is negligible when compared with the former requirement of space for agricultural land. On the negative side, given the extremely bad characteristics of stock B – the grain is very distant from the surface and hard to transport to the surface – it is possible that in the future it might not even be possible to maintain the existing supply of 275kg of grain per day. However, good news arrives from the estimate of the lifespan of this stock. Assuming a maximum of 275kg of grain per day, and a progressive reduction of this metabolic rate in time, we can calculate that the lifespan of this stock is much more than 300 years. Peak grain and a lowering quality of the stock, associated with an increase in the cost of extraction, translate into a longer lifespan for the stock.

The lesson about peak oil

Peak oil is not about the fact that we will soon run out of fossil energy. The situation of stock B is already in a clear peak grain state, but it will last much longer than stock A. Stock A has a much shorter lifespan than stock B, but will not reach a situation of peak grain. That is, peak oil is not about the availability of the energy input, but the achievability of power level: the maximum pace of consumption of energy. Keeping this example in mind, we can discuss now a few implications of peak oil for an informed discussion about the search for alternatives to fossil energy.

First implication of peak oil: the end of the fairy tale of perpetual growth?

Peak oil can have a tremendous impact by shattering the set of ideological axioms which have been used so far in the perception of development strategies. For almost a century, Western society has been permeated by an uncontested perception that there is a direct relation between technological progress, levels of consumption and economic and social development. This is due to the fact that the energy supply was linked to stock exploitation, and the pace of exploitation never reached a peak. The internal constraint – not having enough capital – was the limiting factor to increased consumption.

Looking at the example of the society exploiting stock A, we can clearly say that by using better technology its citizens can make more food accessible, attract immigrants to do unwanted work or buy products they don't want to produce. By plundering more and more grain, they can try to eliminate all type of tensions. This is the magic of an expanding pie; if this strategy can be realized, then everyone, sooner or later would be better off. This situation is represented in the upper part of Figure 7.4a.

Within this uncontested perception of an expanding pie, the vision of sustainable development is associated with the very naive assumption that the typology of exosomatic metabolic pattern enjoyed now by developed countries could be extended to the entire planet, in spite of the size of a world population expected to reach 8–10 billion. According to this ideological view, a free market in a completely globalized economy will deliver the required win–win solutions. This fairy tale of perpetual growth is essential for two important reasons.

1 Tensions associated with the uneven distribution of resources (the equity issue) can be lessened by expectations about a better future. If the American dream can be extended to the entire planet, then the present generation can accept living in bad conditions. Sacrifices can be accepted if they deliver a better future for the children.
2 The economic performance of the society can be boosted by increasing the level of debt of a society.

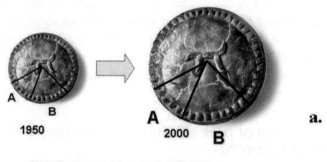

Globalization and free-trade will fix it . . .

Globalization and free trade enhance conflicts . . .

Figure 7.4 *Globalization and trade in an expanding pie and in a zero-sum game*

Soddy, an incredibly accurate prophet, made reference to this point as briefly discussed in Chapter 6. Loans make it possible to spend money now that refers to goods and products which have not yet been generated. The money that is spent now is based on the promise – e.g. made by someone who buys a house – that s/he will pay back the credit. However, the buyers can pay back the debt only if they are able to earn money. To do that, they will have to be able to take part in economic activities that will be performed in the future. Here, the discrepancy between the real economy and the virtual economy indicated by Soddy becomes relevant. If the pie keeps expanding, then these economic activities will keep increasing, also in biophysical terms. But when the wealth is generated by stock depletion (extracting and consuming the grain), we end up in the paradox indicated by Soddy: the virtual economy, based on the accounting of monetary flows, will perceive as economic growth any massive increase of debt – which many will confuse with an increase of wealth – at the very moment at which the resources are depleted. The real economy is reducing its original endowment of wealth.

This is to say that the biophysical reality check that should be associated with the issue of peak oil may induce a dramatic change in the way humans perceive and describe wealth and growth. We do not know whether peak oil has already been reached or not, but there is an important consensus that humankind is getting close to that point (plus or minus 20 years). According to the experts (www.aspousa.org/), it is not reasonable to expect that it will be possible to surpass the mark of 90/100 millions of barrel of oil produced per day; and it is reasonable to expect that the current pace of production will slowly decrease in time, due to the progressive reduction of the quality of the reserves. The consumption of oil, in the year 2008, has been fluctuating around 85 million barrels per day, making the market very nervous about a possible future mismatch between demand and supply. This explains the sudden increase in price to more than US$100 per barrel over a few months. So, even if we do not know when peak oil has been/will be reached, we can safely say that we are close to the situation illustrated by the exploitation of stock B in Figure 7.3. We may remain with oil in the ground for a long period, but we cannot expect major increases in the levels of consumption experienced in the year 2008.

If this is true, then, the narrative of an expanding pie is no longer a valid one (Figure 7.4b). This implies that humankind, as a whole, is getting into what can be characterized as a zero-sum game: if someone eats a larger share of the oil pie, someone else will have to eat less. When getting into this situation, it becomes more and more difficult to find win–win solutions capable of leaving everybody happy.

On the contrary, there is a concrete risk that the expansion of markets through uncontrolled globalization will increase the connectedness of socio-economic interactions to a point that will accelerate conflicts over limited resources. Natural ecosystems have developed wise strategies over millions of years for coping with the interaction of different communities sharing limiting resources. Ecosystems learned that niche differentiation – reducing the level of connectedness of interacting systems – is a powerful strategy for boosting the stability of complex metabolic networks.

In relation to this point, we want to touch on just two crucial implications of peak oil, which are illustrated in Figure 7.4.

Should we return to the population issue in discussions about sustainability?

In the last two decades, the concept of a population bomb – which was very popular in the 1970s – has been removed from the agenda of politically correct discussions about sustainability. However, peak oil could provide a clear reason for reconsidering this decision. Let's represent the total exosomatic energy consumption at the level of the whole world – as illustrated in Figure 7.5a – as the area of a rectangle characterized by two sides.

- Side A: the fund – the overall hours of human activity per year (population × 8760) – is proportional to the length of the horizontal axis. It would be a proxy of endosomatic energy.
- Side B: the pace of the flow – the exosomatic metabolic rate per hour of human activity (average over the whole planet) – is proportional to the height of the vertical axis. (It would be a proxy of the rate of exosomatic consumption per hour or the exo/endo energy ratio).

Therefore, the total consumption of energy is represented in this way (on the left side of Figure 7.5a) by the area of the rectangle.

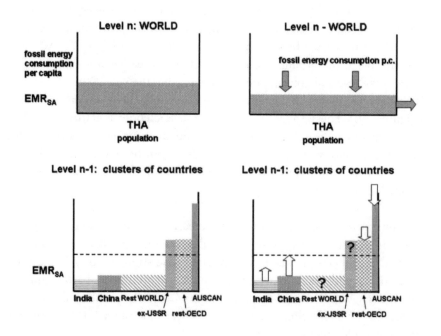

Figure 7.5 *After reaching peak oil: an increase in the size of human population entails reducing the pace of oil consumption per capita; an increase in the pace of energy consumption of a country entails reducing the pace of oil consumption in other countries*

By adopting this representation and by assuming a situation of peak oil, it becomes evident that since the area of the rectangle cannot be increased (it is impossible to increase the peak threshold/area of, let's say, 95 million barrels a day), any increase in population – an enlargement of the base of the rectangle

– will directly translate into a reduction in the amount of oil that can be used per capita on the planet (a reduction in the height of the rectangle).

The implications of the new zero-sum game (if not a shrinking-sum game!) in view of development

Addressing this issue requires moving down to a lower level, looking at how the total exosomatic energy consumption of humankind is divided between different clusters of countries. This analysis is provided in Figure 7.5b. By using the same method of representation used in the previous graph, we can see that at the moment, there are big differences in the EMRs of human activity when moving across types of countries. In fact, the left side of Figure 7.5b shows the profile of energy consumption of different clusters of country, expressing different patterns of exosomatic metabolism (this figure is taken from Ramos-Martin et al, 2007 and it is discussed further in Appendix 2, Figure A2.15). According to this representation, it is easy to see that after assuming a peak oil situation, increases in the overall rate of the consumption of exosomatic energy in certain countries will have to be coupled to equivalent decreases in exosomatic energy consumption in other countries. The alternative is that a major reshuffling of socio-economic organization at the world level will lead to alternative exosomatic metabolic patterns different from the ones expressed so far.

Exosomatic metabolic pattern of modern societies and peak oil: final considerations

- The existing heavy reliance on fossil energy will not disappear in the next few decades. The exosomatic metabolism of humankind will remain stuck on fossil energy for a few decades more (Smil, 2008).
- We should learn how to use the remaining stocks of fossil energy wisely. It is essential to search for feasible and desirable alternatives to fossil energy, but also to explore alternative models of development, making possible the expression of metabolic patterns based on a lower level of energy consumption.
- In the competition for limited resources, the societies that are wiser in their use of fossil energy – for example, that build the infrastructure required to switch in time to alternative exosomatic energy patterns – will be better off at the end of the energetic transition that will take place in the third millennium.
- Nobody knows or can predict the future. Therefore, rather than building mega-models claiming to compute the best thing to do using a single narrative and a single method of quantification (for example, general equilibrium models), it would be wise to develop integrated methods of analysis that are able to characterize the problems in an integrated way, at different levels and in relation to different dimensions of analysis.

- In spite of the heavy doses of uncertainty affecting our ability to predict future events, we can identify some policy strategies as being clearly impractical. In our view, the policy of investing large amounts of financial and technological resources in the development of large-scale agro-biofuels production belongs to this category.

Chapter 8

Agro-biofuel Production is Not Good for Rural Development

Technical Progress in Agriculture and Rural Development

Introduction

Supporters of agro-biofuels call for a large-scale implementation of agro-biofuel production as a tool to stimulate rural development in developed and developing countries. According to this claim, especially in developing countries, agro-biofuels will speed up technical progress in agriculture and boost rural development.

In this chapter, we claim the exact opposite. That is, in the last two decades it has become more and more evident to those who work in the field of agricultural development that the current pattern of technical development in agriculture is locked into a 'Concorde syndrome': innovations aim to do more of the same, in spite of the fact that nobody is happy with the existing performance of agriculture, in either developed or developing countries. The paradigm of industrial agriculture is, at the moment, producing – by generating an excessive stress on the environment – food surplus without a demand in developed countries, and food that is too expensive in developing countries.

To understand the existing problems, it is best to get back to the historical situation that generated the paradigm of industrial agriculture. This paradigm was developed in the context of an agriculture operating within developed countries, and it had two clear objectives:

1. to boost the production of biomass per unit of land well outside the range compatible with natural ecological processes (replacing the natural pattern of ecosystem metabolism with a production of biomass obtained through the unnatural linearization of nutrient flows); and
2. to dramatically reduce the requirement of human labour in agriculture, while reducing as well the local consumption of food by the population of rural communities. The natural pattern of reproduction of rural communities was eliminated, since rural dwellers were supposed to move to the cities, and agricultural production was aimed at feeding urban populations.

As discussed in the previous chapter, because of the industrial revolution, the exosomatic metabolism of modern society changed dramatically. The work performed in a society was no longer aimed mainly at producing the required food, as had been the case in pre-industrial communities. Massive industrialization implied the necessity of performing a variety of new activities, requiring thousands and thousands of new job-roles (Tainter, 1988). With economic development, the bio-economic pressure grew, requiring a dramatic reduction in the number of farmers in modern society. For this reason, capital and fossil energy were poured into the agricultural sector.

If we accept this description of the history of the paradigm of industrial agriculture, is it obvious that a further implementation of this paradigm, with technical progress aimed at doing more of the same, does not seem to be the best policy when the priority is given to rural development.

In the rest of this chapter, we show that the technical progress of agriculture has indeed been driven by these two objectives, and that it can therefore be described as the continuous injection of more technical inputs into the process of agricultural production in order to increase the net supply of food per hectare in response to the growing demographic pressure, and food per hour of labour in the agricultural sector in response to the growing bio-economic pressure (Hayami and Ruttan, 1985; Giampietro, 1997a).

As a matter of fact, technical progress in agriculture was so extraordinary, it became a victim of its own success. As observed in Figure 4.2, right now, in developed countries, 2 per cent of the workforce can produce enough food for the rest of society. Actually, the agricultural sector of developed countries is affected by a systemic excess of supply over demand, in spite of extravagant diets based on an excess of meat and other valuable food products. As a side-effect, rural areas are now depopulated and rural communities no longer need any help from technical progress in further reducing their population. This is already happening spontaneously. Young people, all over the world, are leaving rural areas anyhow. This is to say, the original semantic behind the standard narrative of technical progress in agriculture used in the 1960s and 1970s today no longer makes sense.

However, the industry behind this extraordinary technical progress grew very powerful after decades of endless successes. This industry is resistant to the idea that it is time to direct technical progress towards a different set of goals: the multifunctionality of rural communities based on an integrated set of activities, in which food production should be just one of the relevant tasks to be dealt with (Giampietro, 2003). The industrial sector behind the paradigm of high external input agriculture is clearly suffering from Concorde syndrome (more on this concept in the rest of this chapter).

Because of this situation, the idea of the massive development of agro-biofuels has been perceived by several social actors as a messiah capable of ridding the paradigm of industrial agriculture of its present troubles (more on this point in Chapter 9). As observed in the previous chapter, a fully renewable and zero-emissions agro-biofuel system will never generate surpluses of net ethanol. The systemic presence of surpluses represents the real threat to the paradigm of industrial agriculture, since it could drive away subsidies. In fact, for exactly this reason, discussions about drastic cuts in financial support to food commodities programmes in both Europe and the US are getting more and more intense.

Examples of the effect of technical progress in agriculture

Flevoland is the most recent piece of land to have been claimed in the Dutch people's extraordinary fight against the sea. This fight has meant the costly construction, operation and maintenance of sophisticated dykes, dams, channels and pumping stations (including the famous windmills). Following the gigantic Zuiderzee project, and taking advantage of new technology and the plentiful power available, the Dutch decided to lay claim to the entire surface of the Markermeer, a lake; see Figure 8.1.

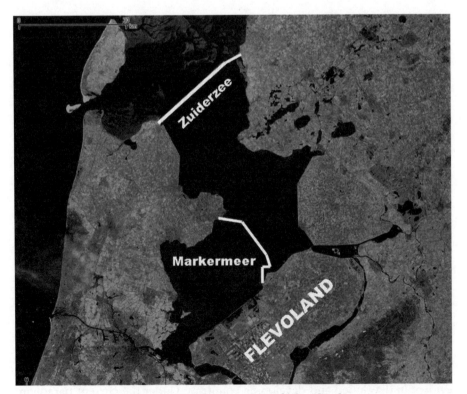

Figure 8.1 *The metaphor of Flevoland*

Satellite picture of Flevoland from the Landsat GeoCover data set at NASA; public domain.

This did not happen though. In fact, after the reclaiming of Flevoland the traditional Dutch narrative about land and agriculture proved to be false. Several Dutch generations grew up with the belief that claiming as much agricultural land as possible from the sea to expand agricultural production was the right thing to do. But the handling of the agricultural sector in the last two decades has made it evident to Dutch policy-makers and public that:

- farming in Flevoland is not perceived as an attractive economic activity at the level of individual households (there are not many Dutch farmers willing to go there);
- the Dutch do not care that nobody wants to farm in Flevoland, because the agricultural sector is no longer perceived as an economic sector driving the growth of the economy; on the contrary, it would represent a drain on the Dutch economy because of the subsidies required by any farmers starting out on marginal land; and
- Dutch society realizes that increasing the amount of farming in Flevoland would represent an additional load on the environment, because of the pollution of the water table.

Thus, in spite of their cultural identity, the Dutch have had to recognize that today, in a post-industrial Europe, increasing land in agricultural production is not necessarily an improvement for farmers or society, let alone the environment. The plans for the future development of Flevoland concern a major integration of the province into the post-industrial activities of The Netherlands, rather than a further specialization in agricultural production (Provincial Plan for Flevoland, 2006). Flevoland reflects the crisis in the agricultural sector across Europe: a crisis that has led to a public acknowledgement of the need to rethink the Common Agricultural Policy (CAP). Plenty of food for thought is presented by the following facts (Giampietro, 2003):

- agriculture is the economic sector with the highest capital investment per job, the lowest return on economic investment, and the largest environmental impact per unit of added value generated; and
- it is more expensive to guarantee a high quality of social services to the rural population than to the urban population.

This last point brings us to another critical aspect of the paradigm of industrial agriculture: its neglected social dimension.

The problematic related to the social dimension is becoming increasingly evident in developed and developing countries. Under the current paradigm, technological progress in agriculture translates into an increased stress on farmers and, in general, on rural communities. In this context, two aspects are particularly relevant. One is the reduced density of farmers in rural areas due to the policy of increasing land per farmer. This is especially true in developed countries. The other is the increased indebtedness of small farmers, due to the continuous need to augment the level of capitalization per farmer in spite of the low return of this capital. This is especially true in developing countries.

In Australia, farmers are so dispersed in the countryside that special TV programmes, for example *Desperately Seeking Sheila* (www.sbs.com.au/sheila/), help male farmers look for wives. In the same way, in some rural areas in Canada schoolchildren have to use the internet to achieve a critical mass of interaction. The situation in Europe is slightly better because of a higher population density. Yet in marginal rural areas, especially in the mountains, the steady rate of emigration of young people causes social problems, including an increase in the average age of farmers and the occasional need to close small villages during the winter because of the lack of a critical mass of inhabitants.

In India, the phenomenon of suicides among farmers unable to pay their debts is massive:

> *According to the National Crime Records Bureau (NCRB), there were 4453 reported farm suicides in Maharashtra state alone in 2006, the latest year for which statistics are available, an increase of 527 compared to 2005. This figure represents over a quarter of the national total of 17,060, and the highest total recorded 'in any year for any state' since the bureau began recording farm suicides. Sharp spikes were also reported in Andhra Pradesh and Chhattisgarh states. Overall, at least 87,567 farmers killed themselves between 2002 and 2006 – or about one farmer suicide every 30 minutes, the bureau says.* (Motlagh, 2008)

Rural areas all over the developing world share a similar situation; the social fabric is disappearing everywhere because of crumbling rural communities. New generations are abandoning the traditional way of farming, since the spread of the paradigm of industrial agriculture is providing neither a viable nor a desirable alternative. As soon as agricultural production is regulated by the market, farmers quickly learn that:

- the cost of inputs grows faster than the price at which they can sell their output;
- the interaction with the rest of the society based on market transactions requires a monetary flow, which often cannot be generated by rural households; and

- specialization in crop production increases the risk of regular failure that would require an economic buffer, which is not available to the majority of small farmers.

In developing countries, the existing stalling in rural development entails a massive abandonment of traditional farming systems that are no longer viable, due to the increase in demographic pressure and the continuous increase in the threshold of densities of monetary flows, which are required to remain economically viable.

For rural communities, this translates into the impossibility of generating both the food and the cash flow (per hectare) that would be required to maintain a decent level of interaction with the rest of society. In fact, more and more rural households, even if living in poor developing countries, depend on the availability of monetary flows for access to critical goods and services required for their daily life. The final result of this increasing pressure is the massive migration of people born in rural areas to the cities or to other countries.

Given this situation, what makes the proponents of agro-biofuels believe that by shifting from monocultures aimed at the supply of food commodities to monocultures aimed at the supply of agro-biofuels, the situation of rural areas will dramatically improve?

In the rest of this chapter, we will make the case that the current impasse in rural development is not due to the specific final use of the products of monocultures (whether the crop is used for food or for fuel). Rather, there are some basic principles associated with the paradigm of industrial agriculture that are at the root of the problem.

The Past Success of the Paradigm of Industrial Agriculture

Definition of the paradigm of industrial agriculture

The paradigm of industrial agriculture can be defined as the existence of an uncontested consensus over the idea that a massive use of technology (capital) and fossil energy in agriculture is justified in order to achieve two key objectives: boosting the productivity of labour in the agricultural sector and boosting the productivity of land in production.

The priority given to these two objectives, under the label of 'technological progress in agriculture', has been driven by two crucial transformations that took place in developed societies in previous decades.

1   A dramatic socio-economic readjustment of the profile of investment of human time, labour and capital over the different economic sectors in

industrial and post-industrial societies. This transformation required the progressive elimination of farmers to free up labour for the workforce in other economic sectors, initially the industrial sector and later the service sector.
2   The demographic explosion that took place, first in the developed world and later everywhere, linked to the globalization phenomenon. This explosion did, and still does, require boosting the yields from the land in production due to the progressive reduction of the available arable land per capita.

In technical/scientific terms, the adoption of the paradigm of industrial agriculture entails adopting a narrative about agriculture that results in a major simplification of the representation of the functional and structural organization of the agricultural process. There are only a few relevant inputs to consider (those that have to be paid); there are only a few relevant outputs to consider (those that can be sold on the market); and there is only one function to be optimized (the profit). In this narrative, agriculture is no longer about reproducing the funds needed for a continuous production. These funds are:

- the community and the people performing agricultural activities;
- the varieties and biota on which the traditional farming system co-evolved; and
- the ecosystems embedding the farming systems.

In the old narrative adopted by the traditional paradigm of agricultural production, these funds were crucial objectives in the definition of development strategies. On the ecological side, the nutrient cycles were guaranteed by the interaction of different species present in traditional pre-industrial agricultural systems. For this reason, the traditional paradigm can be characterized as low external input agriculture (LEIA).

The new paradigm neglects the reproduction of funds, and focuses only on how to optimize the generation of flows! For this reason, the adoption of the paradigm of industrial agriculture implies the abandonment of the traditional, integrated systems of agricultural production, which used to give top priority to the reproduction of the rural community and cycles of nutrients. Industrial agriculture is no longer about reproducing rural communities, but it is about maximizing the net supply of commodities to urban communities. For this reason, it requires a linearization of flows of nutrients obtained with a massive use of commercial fertilizers and irrigation. Such a linearization is possible only because of technical inputs, such as commercial seeds (hybrid or genetically modified seeds with selected properties), pesticides and machinery, which are heavily dependent on fossil energy for their production and/or utilization. As a result of this fact, agricultural produce no longer has any relation to local varieties and local seeds on which the traditional farming system was built. It implies losing the historical identity of the funds for ever. This justifies the definition of the paradigm of industrial agriculture as high external input agriculture (HEIA).

The overcoming of the biophysical constraints associated with natural patterns of nutrient cycling in terrestrial ecosystems certainly represents a major plus for the paradigm of industrial agriculture with regard to the socio-economic dimension of development. HEIA has been able to augment simultaneously the yield per hectare and the productivity of labour to levels that were undreamed of.

But the success of HEIA has been paid for by serious negative effects on both rural development and the environment. The massive use of inputs is based on a progressive depletion of stocks, such as fossil energy, fertile soil and fresh water, and the progressive filling of sinks, such as the accumulation of pesticide residues, the leakage of large quantities of phosphorus and nitrogen into water tables and rivers, and increases in greenhouse gases. It is also directly related to the loss of biodiversity through the destruction of natural habitats and a continuous erosion of the genetic variety of cultivated crop species. Biophysical constraints might have been temporarily overcome by industrial agriculture, but there are clear signs that biophysical constraints on the further expansion of human activities and additional colonization of land do exist (Millennium Ecosystem Assessment, 2005).

However, ironically, the paradigm of industrial agriculture is finally experiencing a crisis, not because of its severe environmental impacts but because of its recent poor performance on the socio-economic side. This general trend seems to have been accentuated by the recent agro-biofuel production policies (see Chapter 1). At the moment, the application of the paradigm of industrial agriculture can produce an excess of food where it is not needed, but cannot produce sufficient affordable food where it is needed. In both cases, this performance is generating an excessive stress on rural and ecological communities. In conclusion, the systemic ignoring of one of the key goals of metabolic systems – the reproduction of the rural community, and the ecological processes that produce the relevant flow – is becoming its Achilles' heel. In fact, in the narrative of the paradigm of industrial agriculture, the flows invested in the reproduction of funds are considered losses (defined as a reduction of output).

How did we arrive at the unconditional acceptance of the paradigm of industrial agriculture?

The paradigm of industrial agriculture is the legacy of one of the most extraordinary successes of technological progress in the history of humankind. This success explains the profound ideological intoxication and uncontested acceptance of the underlying narrative. In the past century, world population has tripled: from 2 billion at the beginning of the 20th century to more than 6 billion at present. It is most impressive that an increase in the productivity of agriculture was able to meet this increased demand for food at a time when land per capita was proportionally shrinking. Moreover, agriculture did not just meet the demand that was due to population growth, but it also succeeded in matching the demand created by more people consuming much more per capita. In fact, at present, grain consumption per capita in developed countries is around 700–1000kg of grain per year, when one includes the consumption of animal production, beer production and other industrial food products. This value is almost four times the annual grain consumption of 250kg per capita that is typical of pre-industrial societies.

But the stunning performance of the paradigm of industrial agriculture is not limited to its ability to produce more food on less land per capita. Equally impressive, but often overlooked, is its ability to produce more food while using much less labour. While in pre-industrial society more than 70–80 per cent of the labour force was typically engaged in the agricultural sector, in the present post-industrial era not a single country with a GDP over US$10,000 per capita allocates more than 5 per cent of the workforce to agriculture (Figure 4.2). The percentage of farmers in the workforce of the richest countries (for example, the US) is consistently lower than 2 per cent. This means, for example, that the entire amount of food consumed in one year by an average citizen of the US is produced using only 17 hours of work in the agricultural sector (Giampietro, 2002a).

Thus, technical progress in food production based on the paradigm of industrial agriculture was very effective in achieving the two main objectives: the use of less arable land because of increases in demographic pressure, and the use of fewer hours of agricultural labour because of increases in bio-economic pressure.

Important drawbacks of the paradigm of industrial agriculture

In relation to the ecological aspects

The massive use of technical inputs aimed at producing more per hour of labour entails:

- the synchronization of farming activities in time and the concentration of farming activities in space; and
- homogeneity in the patterns of application of inputs and harvesting of outputs (to get economies of scale in the development and operation of new technologies).

This translates into a heavy reliance on monocultures, large crop fields (with the loss of the beneficial edge effect, which is important for preserving biodiversity in agro-ecosystems), the erosion of the genetic diversity of crops (due to the use of commercial seeds linked to technological packages), and the spread of the technology of extensive adaptation (not tailored to the location-specific characteristics of different agro-ecosystems) (Giampietro, 1997a; 1997b; 2002b).

A massive use of technical inputs aimed at producing more per hectare of land entails:

- the unnatural concentration of fertilizers and toxic substance in the soil, in the agro-ecosystem and more in general in food and the environment; and
- the local alteration of the water cycle (for example, salinization in crop fields and depletion of underground water reservoirs) and loss of habitat because of the continuous expansion of crop production into marginal areas.

This translates into:

- the depletion of natural resources that are usually considered 'renewable', such as fertile soil and fresh underground water, due to their exploitation rate exceeding the rate of replenishment; and
- a dramatic reduction of biodiversity at the agro-ecosystem level due to the effects of these alterations, such as nitrogen leakages in the water table generating eutrophication in internal water bodies and dead spots in the sea (Giampietro, 1997a; 1997b; 2002b).

In relation to socio-economic aspects

A very specialized system of production requires a lot of technical capital and inputs. As a consequence, it requires a high ratio of cropped land/farmer. This situation (high level of fixed and circulating investment per farmer) entails a series of negative effects on the socio-economic performance of the agricultural sector, when compared with other economic sectors.

1. The technical capital used for cropping is used only a few hundreds hours per year, and therefore tends to have a lower economic return than the capital used in other economic sectors (used for thousands of hours/year).
2. Money enters into the system just once a year, so the system must buffer monetary flows across long periods of time (in the case of a bad performance, for a couple of years in a row).
3. Specialization, in the form of monocultures, implies putting all the eggs in one basket in order to take advantage of economies of scale. This can be risky if insurance against failure and generous economic support from the government are not available.

The Paradigm of Industrial Agriculture Works Against Rural Development

The agricultural technology treadmill

The dramatic reduction in the number of farmers, which was required to make possible the socio-economic transformations leading to modern societies, can be analysed in economic terms. Such an analysis was proposed by Cochrane (1958), who individuated and described the mechanism through which the market can be used in the agricultural sector to get rid of the farmers. Cochrane called this mechanism 'the agricultural technology treadmill'. It works in five steps, iterated in time, which are listed in Box 8.1.

Box 8.1 *The five steps of Cochrane's agricultural technology treadmill*

- Step 1: Many small farms produce the same product. Because none of them can affect the price individually, all produce as much as possible.
- Step 2: A new technology enables innovators to capture a windfall profit – *innovation*.
- Step 3: After some time, others follow – *diffusion of innovations*.
- Step 4: Increasing production and/or efficiency drives down prices. Those who have not yet adopted the new technology must now do so lest they lose income – *price squeeze*.
- Step 5: Those who are too old, sick, poor or indebted to innovate eventually have to leave the scene. Their resources are absorbed by others, who make windfall profits. This is called *scale enlargement*, but it can also be called *farmers' elimination*.

A classic example of the treadmill effect was provided by the bovine growth hormone (BGH/BST) that was used to boost milk production in the 1990s in New York state, a dairy market already suffocated by surplus at that time. The narrative of 'improvement' in the dairy industry involved increasing the efficiency of the physiological process of milk-generation. Scientists delivered the required silver bullet in the form of a hormone capable of boosting milk production in dairy cows. This is a clear example of scientists working successfully on the *how* aspect; the use of BST significantly increased milk production. Unfortunately, belief in the validity of the narrative – *why* this significant boost in the rate of milk production should be considered an improvement – was not shared by most farmers or by the public. The vast majority of small dairy farms in the state were driven out of business by the collapse in the price of milk. Due to an unexpected side-effect – the development of mastitis in super-producing cows – the quality of the milk was also negatively affected in several cases by an unwanted presence of antibiotics.

Indeed, the concept of the *innovation treadmill*, present in agricultural textbooks since 1958, should prompt those proposing additional technical innovations in agriculture to consider the possibility that this innovation could generate the outcome: silver bullet → production up → price down → powerless farmers out. Even less clear is why we should keep boosting the intensity per hour and the density per hectare of the production of food, when experiencing food surpluses and extreme stress in rural areas.

In general terms, we can say the following.

- In developed countries, boosting the production of food commodities above a certain level can be a bad strategy. Because of decreasing marginal returns, it is unavoidable that sooner or later economic losses will rise, and the quality of the output generated by the agricultural sector will fall.
- In developing countries, transforming self-reproducing fund elements (such as traditional farming systems and rural communities) into artificial elements requiring a continuous flow of subsidies (technological units specialized in the generation of food commodities) could be an impossible mission. In any case, it will imply huge economic, environmental and social costs, as well as a huge risk of breakdown in the social fabric.

Can we extend the paradigm of industrial agriculture all over the world?

Differences in the use of technical inputs between developed and developing countries

By assessing the amount of fossil energy backing up an hour of work in agriculture – the level of EMR_{AG} – we can compare the level of capitalization of the agricultural sector. In this way, we find that the difference in capitalization between developed and developing countries is huge. For example, in 1997, EMR_{AG} was 85MJ/h in developed countries versus 1MJ/h in developing countries (summing the fossil energy going into machinery and the fuel for operating it; Giampietro, 2002a). This huge difference, of 85 times, depends on the almost total absence of machinery in developing countries (low socio-economic pressure), and the limited availability of land per agricultural worker, which increases the consumption of irrigation and fertilizers (much higher demographic pressure). As a matter of fact, the difference in hectares of cultivated land per worker in 1997 was considerable: 12ha per worker in developed countries and only 1ha per worker in developing countries. It is remarkable that in spite of the large difference in energy and capital per agricultural worker, the production of food per hectare in developing countries (24.2GJ/ha) is more than double the production achieved in developed countries (10.1GJ/ha) (Giampietro, 2002a). This difference can be explained by the mix of produced crops (mainly cereals in developing countries), multicropping patterns (especially relevant for rice production), and the different yields of the individual crops.

Differences in density of flows of added value generated in different typologies of farming systems

When performing a comparison of the density of flows of added value over different typologies of farming systems, the differences are so big that one has to use a logarithmic scale. Looking at the graph in Figure 8.2, we see three main typologies of agricultural production:

1. production systems operating in developed countries within the industrial paradigm;
2. mixed systems of production operating partially in the market; and
3. subsistence farming systems.

These three typologies are characterized by numerical values separated by orders of magnitude.

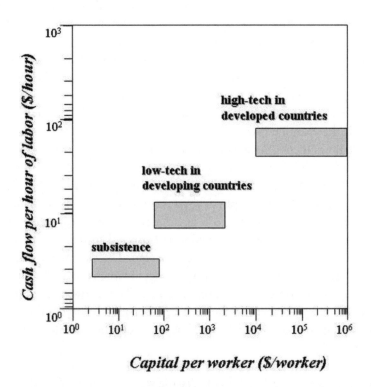

Figure 8.2 *The density in space of monetary flows typical of different systems of agricultural production*

Is it possible to fill the existing gap between the agricultural performance of developed and developing countries?

Looking at existing benchmarks and trends, it seems unlikely that the technical development of agriculture in developing countries will follow the same pattern (a full implementation of the industrial paradigm) experienced by developed countries, for several reasons.

First, because of their high demographic pressure, developing countries are already using, per hectare, more fossil energy per capita than developed countries, for boosting yields. That is, when considering fossil energy relative to fertilizers and irrigation, developing countries used in 1997 more fossil energy per hectare than developed countries: 7.4GJ/ha versus 4.9GJ/ha (Giampietro, 2002a). Therefore, in relation to the use of these technical inputs, developing countries are already in the part of the curve with decreasing marginal returns. An additional intensification in the use of technical inputs will not pay off.

Moreover, in order to be able to increase the bio-economic pressure and follow the pattern of development typical of developed countries, developing countries would have to replace farmers with capital (by boosting their EMR_{AG}). This would require a dramatic increase in the current economic investment in agriculture per worker. The investment would have to be coupled to a massive restructuring of rural infrastructures, since traditional agriculture in developing countries is presently organized in small units of production. In turn, these investments (resulting in an additional elimination of rural people) would result in an additional boost to the process of urbanization, which is already excessive and out of control. But even more perplexing is the idea that it would be convenient to go for such a massive flow of economic investments. As a matter of fact, if one had US$100,000 to invest in a rural area in China, given the limited availability of land and infrastructures, it would be unwise to invest this money in a big tractor.

Also, in relation to the biophysical constraints limiting the expansion of economic performance in agriculture in developing countries, things do not look particularly good. World population is expected to increase by at least another 2 billion people before reaching a plateau. The vast majority of these 2 billion will be born in those areas of the planet in which the problems generated by demographic pressure are already serious. The following relation links the various factors determining the income of farmers. The combination of these factors can be related to the spatial densities of monetary flows indicated in Figure 8.2.

$$\text{Added value/worker} = \text{profit/kg} \times \text{kg/ha} \times \text{ha/h of workload} \times \text{hours of workload/worker}$$

From this relation we can see that:

- the profit generated per kg of product tends to go down in agriculture due to the technological innovation treadmill;
- the value 'kg/ha' is already higher in developing countries, when considering the aggregate values of yields in China, India and other Southeast Asian countries; and
- the number of ha per worker is much lower in developing countries, and will remain so, because of demographic pressure; they do not have enough land to work more hours per year.

This means that we cannot imagine eliminating the differences in performance between developed and developing countries just by increasing the workload. Moreover, due to the physiological rhythms of agricultural production, there are periods in which the available workforce is idle (most of the year) and periods in which it is insufficient (during short periods of peak activities, such as at harvest-time). These periods tend to coincide across contiguous areas, so that it is difficult to cope with this problem with economies of scale. The only solution is to go back to the old multifunctionality of farming systems (Giampietro, 2003).

An additional problem is generated by the rapid demographic changes taking place in developing countries. The particular age structure of China – Figure 4.2 – explains how it is possible for the Chinese economy to deal with a huge demographic pressure in agriculture (using massive quantities of fertilizers and irrigation wherever possible) without using large amounts of machinery. That is, right now in China, the large supply of labour makes it possible to produce both food and labour-intensive industrial products without the use of large machinery (Ramos-Martin et al, 2007). But what will happen when this wave of adults becomes elderly?

Even more complicated is the situation of those countries that experienced in the last decade a demographic boom (lower graph on the right in Figure 4.2). They have a distribution of population over age classes with large waves (or rather tsunamis) of youngsters who will enter in the workforce looking for a job. For these countries, it will be impossible to generate jobs at a high level of capitalization at a pace that will provide a level of employment capable of avoiding social tensions and conflicts. The majority of Islamic countries provide a clear example of this problem. A rural development based on systems of production adopting the paradigm of industrial agriculture will represent the worst case scenario (the highest requirement of capital invested per job generated) for these countries.

So far, agro-biofuels in developing countries have had a negative effect on rural development

In January 2008, the European Commission published its 'renewable energy roadmap' proposing a mandatory target that biofuels must represent 10 per cent of the liquid fuels used in transportation in Europe. In spite of protests from the UN Secretary-General, the President of the World Bank, several national parliaments, hundreds of NGOs and accumulating scientific evidence that this policy is going against the very goals justifying this policy, the European Parliament ratified this decision in December 2008.

> *This target is creating a scramble to supply in the South, posing a serious threat to vulnerable people at risk from land-grabbing, exploitation and deteriorating food security. It is unacceptable that poor people in developing countries bear the costs of emissions reductions in the EU.* (Oxfam, 2007)

This policy would be unacceptable even if it did reduce the emissions of European citizens. However, the policy becomes a real 'crime against humanity' – as suggested by the UN Special Rapporteur for the Right to Food, Jean Ziegler – if the result of this policy is increasing, by several times, the emissions generated by the EU (see Chapter 1).

What have been the consequences so far of this policy? According to reports by Oxfam, 'The destruction of critical ecosystems, such as rainforests, to make way for biofuel plantations has rightly raised serious concerns from an environmental perspective' (www.unep-wcmc.org/climate/mitigation.aspx for a discussion and additional references). But millions of people also face displacement from their land as the scramble to supply intensifies. Those most at risk are some of the poorest and most marginalized in the world. The chair of the UN Permanent Forum of Indigenous Issues recently warned that 60 million indigenous people worldwide face clearance from their land to make way for biofuel plantations (http://mwcnews.net/content/view/14507/235/). Five million of these are in the Indonesian region of West Kalimantan. In Colombia paramilitary groups are forcing people from their land at gunpoint, torturing and murdering those that resist, in order to plant oil palms, often for biofuels ('Massacres and paramilitary land seizures behind the biofuel revolution', *The Guardian*, 5 June 2007), contributing to one of the worst refugees crises in the world (Colombia has the second largest population of internally displaced people in the world after Sudan; see www.unhcr.org/publ/PUBL/4444d3ce20.html). Many of these violent acts occur in the traditional territories of indigenous peoples and Afro-descendent communities, directly affecting the most vulnerable groups in the country. In Tanzania, reports are already emerging that vulnerable groups are being forced aside to make way for biofuel plantations (Oxfam, 2007). Where indigenous people have more protection, the expansion of biofuels based on the ethnic cleansing of traditional communities is leading to institutional confrontations. For example, in Peru the parliament has repealed two land laws, approved by presidential decree, aimed at opening up Amazonian tribal areas for the expansion of agro-biofuel plantations (BBC News, 2008).

> *Once people lose their land, they lose their livelihoods. Many will end up in slums in search of work, others will fall into migratory labour patterns, some will be forced to take jobs – often in precarious conditions – on the very plantations which displaced them.*
> (Oxfam, 2007)

But what is the potential of job creation associated with these policies? An analysis of the density of flows in space, in relation to the ability of reproducing the fund human activity, can be useful to clarify this issue.

When considering the operation of monocultures or plantations, the creation of jobs is very limited. It can be two jobs per 100 hectares of soya, or ten jobs per 100 hectares of sugar cane. In fact, the goal of plantations is not to reproduce human activity (people), but to maximize the net supply of products to be exported. Traditional rural communities can use 100ha of fertile land for reproducing 100–500 people. Moreover, the vast majority of the monetary flow associated with the economic activities of monocultures and plantations certainly does not remain in the rural community. That is, the money generated by stressing the local communities and the local ecosystems (by squeezing the funds) goes abroad or to the urban rulers of the country. In fact, any optimization of monocultures and plantations has the goal of generating profit, by reducing as much as possible either the nutrients or the money flows remaining within the agricultural system generating them.

Can guidelines and protocols guarantee the sustainability of agro-biofuels?

Because of the growing evidence that, so far, the large-scale expansion of agro-biofuels in developing countries has generated negative impacts on both the environment and the social fabric of rural communities, several NGOs have requested a mandatory certification scheme to ensure the correct definition of 'sustainable biofuels'. This certification scheme – endorsed in December 2008 by the European Parliament – should be used to guarantee that biofuel production 'should not cause directly and indirectly social and environment problems'. This certification scheme will have to be based on the implementation of guidelines and protocols, based on the preliminary formulation of long lists of general principles and criteria of sustainability. An example of what is under discussion is provided in Box 8.2. This is a draft document prepared by the Round Table on Sustainable Biofuels (http://EnergyCenter.epfl.ch/Biofuels), which has been developed thanks to contributions from hundreds of stakeholders from around the world who participated in working groups and regional meetings.

Box 8.2 *Proposed certification scheme: major items*

The following points are taken from the Version Zero of the Round Table on Sustainable Biofuels – available at http://cgse.epfl.ch/page79931.html.

1. Legality. Biofuel production shall follow all applicable laws of the country in which it occurs, and shall endeavour to follow all international treaties relevant to biofuels production to which the relevant country is a party.
2. Consultation, planning and monitoring. Biofuels projects shall be designed and operated under appropriate, comprehensive, transparent, consultative and participatory processes that involve all relevant stakeholders.
3. Greenhouse gas emissions. Biofuels shall contribute to climate change mitigation by significantly reducing GHG emissions as compared to fossil fuels.
4. Human and labour rights. Biofuel production shall not violate human rights or labour rights, and shall ensure decent work and the well-being of workers.
5. Rural and social development. Biofuel production shall contribute to the social and economic development of local, rural and indigenous peoples and communities.
6. Food security. Biofuel production shall not impair food security.
7. Conservation. Biofuel production shall avoid negative impacts on biodiversity, ecosystems and areas of high conservation value.
8. Soil. Biofuel production shall promote practices that seek to improve soil health and minimize degradation.
9. Water. Biofuel production shall optimize surface and groundwater resource use, including minimizing the contamination or depletion of these resources, and shall not violate existing formal and customary water rights.
10. Air. Air pollution from biofuel production and processing shall be minimized along the supply chain.
11. Economic efficiency, technology and continuous improvement. Biofuels shall be produced in the most cost-effective way. The use of technology must improve production efficiency and social and environmental performance in all stages of the biofuel value chain.

For each of these main items, there is a list of sub-items. An example of a set of sub-items is given in Box 8.3.

Box 8.3 *Example of a list of sub-items for each main item*

The following list of sub-items is taken from the Version Zero of the Round Table on Sustainable Biofuels, available at http://cgse.epfl.ch/page70380.html.

Sub-items of the list referring to Point 2: Consultation, planning and monitoring

- 2a For new large-scale projects, an environmental and social impact assessment (ESIA), strategy and impact mitigation plan covering the full lifespan of the project shall arise through a consultative process to establish rights and obligations and ensure implementation of a long-term plan that results in sustainability for all partners and interested communities. The ESIA shall cover all of the social, environmental and economic principles outlined in this standard.
- 2b For existing projects, periodic monitoring of environmental and social impacts outlined in this standard is required.
- 2c The scope, length, participation and extent of the consultation and monitoring shall be reasonable and proportionate to the scale, intensity, and stage of the project and the interests at stake.
- 2d Stakeholder engagement shall be active, engaging and participatory, enabling local, indigenous, and tribal peoples and other stakeholders to engage meaningfully.
- 2e Stakeholder consultation shall demonstrate best efforts to reach consensus through free prior and informed consent. The outcome of such consensus-seeking must have an overall benefit to all parties, and shall not violate other principles in this standard.
- 2f Processes linked to this principle shall be open and transparent and all information required for input and decision-making shall be readily available to stakeholders.

Then, for each of the 41 sub-items, there are key guidance statements. An example of key guidance is given in Box 8.4.

A cursory reading makes it extremely clear that there is no chance that:

- any of the agro-biofuels currently produced in either developed or developing countries will ever be able to live up to these standards; and
- it will ever be possible to operationalize through global participatory processes and multilateral agreements this exhaustive and elaborate set of quality checks.

> **Box 8.4** *Example of key guidance for each sub-item*
>
> The following examples of key guidance statements are taken from Version Zero of the Round Table on Sustainable Biofuels, http://cgse.epfl.ch/page 70380.html.
>
> - In relation to the statement 2a:
>   Key guidance: The ESIA shall include the identification of High Conservation Value areas, biodiversity corridors, buffer zones, and ecosystem services; shall evaluate soil health; shall identify potential sources of air, water and soil pollution; shall evaluate potential impacts on water availability; shall cover a baseline social indicator assessment; shall include an economic feasibility study for all key stakeholders; shall identify potential positive and negative social impacts including job creation and potential loss of livelihoods; shall establish any existing water and land rights. Small-scale producers or cooperatives unable to perform ESIAs will need support and/or modified ESIAs. 'Large-scale producers' and 'relevant stakeholders' will be defined in the indicators.
>
> - In relation to the statement 6a:
>   Key guidance: Clear definitions are needed for waste, residues, and degraded or marginal or underutilized land. ESIA should ensure that these lands were not used for livelihoods support, or that benefits of use for biofuels outweigh any loss of livelihoods. All of these definitions are time-dependent; unused land might come into production anyway given climate change as well as population and wealth growth. These criteria and definitions should be periodically re-assessed. The RSB will examine different tools for incentivizing the use of these preferred sources of biofuels.
>
> - In relation to statement 7a:
>   Key guidance: Identification and mapping of HCV areas should be undertaken by governmental, inter-governmental, and conservation organizations, as part of larger processes involving non-biofuel sectors. Where such mapping is occurring, the results shall be respected by producers. Where such maps do not exist, large-scale producers shall use existing recognized toolkits such as the HCV toolkit or the IBAT. Producers or cooperatives unable to perform an environmental impact assessment and/or a land management plan will need support. The use of native crops shall be preferred. Hunting, fishing, ensnaring, poisoning and exploitation of endangered and legally protected species are prohibited on the production site.

Therefore, this preliminary document generates an obvious question: what is the goal of this ongoing discussion over what should be considered a set of sustainability criteria for agro-biofuels?

It is obvious that those concerned about the lack of sustainability of agro-biofuels want to use these guidelines as a tactical tool to stop the agro-biofuel folly. Less evident is the motivation of the promoters of agro-biofuels. One possible explanation is that, on being confronted with a clear case of Concorde syndrome, they have to use wishful thinking to justify, in the short run, the continuation of the operation. In this way they can admit that some problems do exist at the moment, and claim that once the protocols for certifying the sustainability of agro-biofuels production are agreed upon, these problems will disappear. In fact, agreeing on the need to debate and implement a certification scheme can be an effective tactic for buying time. That is, by delaying the formulation and the implementation of these protocols (lobbyists have been very good in this in the past), they will be able to continue with existing processes of agro-biofuel production for another decade, or even more. The implementation of tactics aimed at maintaining the production of agro-biofuels in spite of the growing evidence of their absurdity is discussed in Chapter 9.

The Concorde Syndrome in Agriculture

The Concorde syndrome

The expression 'Concorde syndrome' refers to a form of lock-in that often takes place in the field of technical innovations (Box 8.5). This syndrome occurs when the narrative used to provide the terms of reference for a problem is either wrong, or no longer reflects the current social perception. In the eponymous case, a brilliant technical solution – the supersonic aeroplane Concorde – was generated because of a mistaken belief that people would pay much more to fly from Europe to New York in three and half hours than in seven hours. This awkward situation can easily become a syndrome because, often, brilliant technical solutions require considerable financial investments. For this reason, even when experience shows that the original idea was a bad one, nobody dares halt its implementation. The so-called 'curse of the sunk cost' (further discussed in Chapter 9) leads to a desire to continue with the technical solution despite evidence of its failure in practical applications.

The systemic generation of Concorde syndromes in agriculture

As noted at the beginning of this chapter, the paradigm of industrial agriculture is rooted in the endorsement of a series of assumptions about the role that agriculture should perform in society. The validity of these assumptions is more and more contested by the public.

> **Box 8.5** *The Concorde syndrome*
>
> The building of Concorde, a supersonic passenger airliner, was pursued long after it was known that the plane could never be commercially viable. Concorde's costs were estimated to be competitive with those of the sub-sonic planes that were in service when the go-ahead was given. By the time Concorde came into service, the costs of conventional aircraft had fallen and Concorde was no longer cost-competitive.
>
> Moreover, the Concorde project focused only on one criteria of performance: speed of transportation, considered the only relevant attribute at the onset of the project. As it turned out, other criteria such as convenience, flexibility, frequency and last-minute offers were relevant for customers. These criteria were neglected in the planning of the Concord project, and this gave a competitive edge to the sub-sonic competition.
>
> The Concorde project clearly showed that the most dangerous pitfalls for those working on technological innovations is to assume, first, that the attributes considered relevant by the investors are also considered relevant by the users without double checking; and, second, that today's conditions will last indefinitely. The metaphor is often used to refer to the tendency of policy-makers and organizations to keep pouring resources into projects, after it becomes clear to anybody that wants to see it that there is no hope of succeeding.

The most contested assumptions are as follows.

- Agriculture is just another economic activity whose only goal is to produce commodities and profit.
- Various food products can be effectively considered in generic terms as commodities. In technical jargon, this assumption is based on the claim of 'substantive equivalence', and implies that independently of their taste, texture, smell, cultural traditions and – above all – the place and the way they are produced, food products assigned to the same category all have the same quality: for example, all portions of corn, apples, potatoes and beef are the same (for more, see Giampietro, 1994).
- Existing environmental, social and economic costs associated with the technical progress of agriculture within the industrial paradigm are negligible or can be regulated with precision and rationality. This assumption entails the idea that it is possible to deal with the assessment of negative consequences of technological innovation in terms of risk analysis, and denies the existence of uncertainty and genuine ignorance (Giampietro, 2002b).

These assumptions have led to a series of contested policies in agricultural development, such as the development and use of growth hormones, genetic engineering and crop-based biofuels. These policies were aimed at doing more of the same, which implies continuously increasing the intensity and density of the produced flows. However, this policy is not solving the problem faced by rural areas. In spite of this fact, those developing agricultural policies are systematically denying the relevance of other agricultural attributes, which are increasingly perceived as highly relevant by the public. These neglected attributes of performance are as follows.

- Environmental costs. Soil erosion, the loss of biodiversity, the pollution of the water table, excessive withdrawal from the water table and rivers, and interference at the level of the landscape with bio-geochemical cycles of matter and nutrients. Ecological constraints at the global level have recently been described and documented by the large-scale Millennium Ecosystem Assessment, providing an integrated assessment of the negative impact on the health of ecosystems all over the planet (www.millenniumassessment.org/en/index.aspx).
- Social costs. The loss of social fabric in rural areas, the loss of cultural traditions and the symbolic/cultural dimension of food, negative impacts on rural landscapes.
- Public health costs. Adverse health effects (such as cancer) from the excessive use of pesticides, antibiotics and hormones in the agricultural system, and from the indiscriminate production of food commodities regardless of their quality (obesity, diabetes, cardiovascular diseases).
- Economic costs. Direct costs refer to the requirement of economic resources for subsidies, which keeps increasing in time. Possible indirect costs might include the loss of image of traditional food products, and the loss of traditional rural landscapes (for example, in relation to alternative uses of the land for housing, recreation or tourism).

Conclusions

- It is very unlikely that the paradigm of industrial agriculture will help to solve the problems of rural development in developing countries (except in special niches). Therefore, the call made by many of the proponents of agro-biofuels for a massive implementation in developing countries of this policy for rural development seems to be totally unjustified.
- Even if it were possible to increase the actual productivity of labour in agriculture by 100 per cent, it would not provide a viable and acceptable means of rural development in many areas of the world. In a globalized world, *rural development has to be based on something completely different* (the multifunctional use of the landscape), based on a progressive integration of agricultural activities with non-agricultural activities.

- It is not clear why developing countries should invest an important fraction of their limited financial resources in the type of technical capital required by the paradigm of industrial agriculture. In fact, in developing countries – except in special niches – this capital provides the lowest economic return, the highest requirement of capital per job created and an important environmental impact.
- Due to the need to keep the internal productivity of agro-biofuel production systems as high as possible (discussed in Chapter 7), it is very unlikely that it would be possible to guarantee respect for protocols or guidelines based on global principles, or criteria for sustainable biofuels production which address both the integrity of ecosystems and ethical standards.
- It is very likely that by beginning with a mistaken diagnosis, one will develop the wrong cure. The more expensive the development of the cure, the more likely is Concorde syndrome. It seems that the large investments in, and the policies aimed at, the implementation on a large scale of agro-biofuels production belong to the category of Concorde syndrome, not only because of the fallacy of the analysis on the energetic side, but also because of the fallacy of the analysis on the rural development side.

Chapter 9

Living in Denial

How to Explain the Agro-biofuel Folly

If agro-biofuels are neither feasible nor desirable, why are billions of dollars and euros being poured into agro-biofuel production through investments and subsidies? Have politicians suddenly lost their minds? Have investors suddenly lost their common sense?

As a matter of fact, as is argued in detail further on, neither investors nor politicians have lost their minds. The profits of big companies lobbying for agro-biofuel production are soaring and political decisions on agro-biofuels – as in the case of the European Parliament – are often determined by intense lobbying activities by powerful industrial groups trying to secure their future. Politicians are claiming to protect national interests by preserving the economic assets of their countries. But these two things do not fully explain the agro-biofuel folly. It is also about bad science and misguided public opinion.

In fact, the hundreds of scientific papers published each year on how to improve existing agro-biofuel technologies in order to substitute a relevant fraction of fossil fuels indicate that the academic establishment has ignored available knowledge accumulated over more than a hundred years in the fields of energetics and bio-economics, and that there is a serious failure in the quality control of scientific inputs used for decision-making. To better understand this issue we must ask ourselves the following question: are we dealing with a lapse of memory, or with the selective removal of available information?

Similar situations have occurred throughout history. In the Middle Ages, the European academic establishment experienced a bitter fight over the claim that the Earth was flat, even though the Egyptians calculated the radius of our planet before the birth of Christ, and plenty of other information on the spherical nature of our planet was readily available at that time (including the curved horizon when looking at the sea from a hill). This is to say that it is not uncommon in science that evidence is ignored for non-scientific reasons, whether it regards the Earth's shape or agro-biofuel production. These failures cannot simply be reduced to a mere failure of technical protocols of quality control – e.g. peer review mechanisms – over the scientific inputs used in the process of decision-making. Indeed, we argue that this phenomenon can only be explained by a more systemic problem of cultural and political lock-in that eventually affects what the science does and does not see.

In this chapter we focus on three forms of lock-in that are closely interwoven: ideological, academic and economic. Ideological lock-in refers in general to ideological biases that affect decision-making, including in relation to sustainability issues, and in particular policies such as agro-biofuels. The other two forms, academic and economic lock-ins, are direct consequences of ideological lock-ins and deal with the two most important decision-making processes in our society with regard to sustainability issues. Academic lock-in refers to the interface between the bureaucracy of scientific academia and public administrations where decisions are made about the allocation of funds for research and development. Economic lock-in refers to the actions of powerful lobbies that tend to stabilize the existing economic power structure.

What is common to all these forms of lock-in is the innate reluctance of humankind to change. This profound ideological bias aimed at the preservation of the status quo is manifested at all levels: the individual (for example, the profound fear of death typical of Western civilization), the community (the desire to maintain current cultural identities and lifestyles) and the society. At the level of society, this ideological bias is manifested in the form of processes and filters (bureaucracy) guaranteeing the survival of existing institutions. For this reason, social institutions easily bend under the pressure of lobbies working to preserve existing power structures.

Ideological Lock-in: The Quest for the Holy Grail

The strategy of hiding the identity of the story-teller

The issue of sustainability has gradually gained relevance in the public debate, and has become authentically hysterical in recent years. Seemingly, in the first decade of the third millennium, the larger public finally came to realize – with great surprise and shock – that both fossil energy and environmental services are finite. As a result of this growing hysteria, primed by phrases like 'peak oil' and 'climate change', the hegemonic part of society decided to endorse very dubious policies that sought to preserve the status quo (agro-biofuels is a good example of this), and generated bizarre events such as Al Gore winning both an Oscar and the Nobel Prize in 2007 for his film *An Inconvenient Truth*.

To understand this situation, we have to go back to the roots of modern science: that is, the Enlightenment (in Western culture). Within the narrative adopted by Western science, humankind stands at the centre of the universe. The identity of the story-teller providing the narrative about sustainability coincides with the identity of the hegemonic group ruling the society in which science is developed. As a consequence of this fact, the issue of sustainability has always been framed as the preservation of the pattern of activities associated with the form of civilization defined as relevant by the story-teller. At the moment, the form of civilization endorsed by the hegemonic part of humankind is associated with a metabolic pattern based on fossil energy. Fossil energy is a non-renewable resource, and therefore this pattern should be considered a transient state. Sooner or later, this form of civilization will have to become something else, associated with a different pattern of metabolism. If we agree on this point, then we are forced to note that trying to preserve the status quo is equal to trying to maintain a state of perpetual growth. In this framing of the sustainability issue, the finiteness of natural resources and the fragility of ecological processes – the integrity of natural processes – are perceived as threats to this status quo, against which human ingenuity must take action. This action is interpreted as the continuous search for silver bullets to fix the perceived imperfections of nature. The end result of this is that this framing of the sustainability issue pushes modern societies into a series of impossible missions, such as 'producing' limited resources (for example, providing fresh water through desalinization) or engineering whole ecosystems.

This story-telling has been very successful throughout different knowledge systems, to the point that it was even exported, through Communism, to the ancient culture of China. In this narrative, the definition of sustainability is based on an initial representation of the story-teller – the hegemonic group – interacting with what is perceived to be the external world. Depending on the context, the representation of the story-teller is based on a given selection of attributes (important to the hegemonic group), which are to be preserved for ever. 'Progress' – which many confuse with science and technology – is used to solve problems of sustainability whenever the given set of attributes used to define 'us' (the story-tellers) is at risk, or the given set of goals associated with 'our' cultural identity cannot be achieved.

This means that when dealing with sustainability analysis, the identity of the story-teller is not to be questioned, let alone changed. It never shows up in the models. If CO_2 is accumulating in the atmosphere, then 'progress' must develop tools for sequestrating it. If pandas are becoming extinct, then 'progress' must develop tools for cloning them. If fossil energy is getting scarce and causing GHG accumulation, then 'progress' must find a magic bullet capable of killing two birds with one stone: agro-biofuels.

Within this anthropocentric framing of the sustainability issue, it is the external world (nature and societies that are not associated with the story-teller) that has to be fixed and adjusted to the priorities of the story-teller. Obviously, the priorities considered in the definition of what to sustain are those of the hegemonic group alone. The hidden assumption behind this narrative is that, no matter how serious the sustainability issue, it will be solved by human ingenuity. This naive assumption coincides with the ideological assumptions about the full substitutability of production factors in neoclassical economics. As discussed previously, this idea seems to ignore the facts about the existing power relation between nature and human technology. Our living planet, Gaia, uses for her air-conditioning (the water cycle) about 2000TW (terawatts) of solar energy. Humans, on the contrary, control about 15TW (460EJ/year in the year 2007) of energy for all their activities (BP, 2008). This is to say that humans may have the power to tinker with the electronic controls of the huge train on which they ride – the biosphere – but they have no chance of stopping the train should it run out of control, or getting it started again if it breaks down. Those who believe it possible to engineer the functions of the biosphere according to human wants and by using human technology are uninformed dreamers.

The denial

As discussed earlier, the interest in energy alternatives to oil has been primed in this decade by two major issues: global warming associated with the greenhouse effect, and peak oil. Given these two problems, and ruling out the option that humans will consider alternative patterns of development not based on the maximization of GDP, it is almost unavoidable to conclude that what we need is a primary energy source that does not produce GHGs, and is renewable.

For those not expert in the field of energy analysis (and in the analysis of the metabolism of complex adaptive systems) it is natural to come up with the simple sum $1 + 1 = 2$ and conclude that producing biomass for biofuel kills two birds with one stone. For those supporting this idea, the gospel is always the same:

- producing biomass for biofuel will absorb the CO_2 that will be produced when using that biofuel, therefore this is a zero-emissions procedure; and
- since the production of biomass uses solar energy, the supply of biofuel from biomass is renewable. Hence, the substitution of barrels of oil with barrels of biofuel means that we need no longer question the myth of perpetual economic growth (the maximization of GDP growth and a perpetually expanding human population).

Unfortunately, things are not that simple and the 'magic solution' is magic only in science fiction and in the promises made by politicians. But given that Western civilization is terrified by the idea that it could crumble like the great civilizations of the past, people desperately need to believe in the existence of a silver bullet that can solve sustainability problems. This explains why the myth of biofuels is a fantastic window of opportunity for both academic departments looking for research funds and politicians looking for an easy consensus. This point is well illustrated by fragments from a letter that US Senator Ken Salazar sent to the *Colorado Springs Gazette* (Box 9.1).

> **Box 9.1** *Letter from US Senator Ken Salazar to the Colorado Springs Gazette*
>
> Our national security demands that we meet the challenge of generating 25 per cent of our nation's energy from renewable sources by 2025... According to a recent national survey, 98 per cent of voters feel that meeting 25 per cent of our energy needs from renewables by 2025 is important for the country, and 90 per cent of voters believe this goal is achievable. This kind of bipartisan support is almost unprecedented and signals a willingness to move our country forward toward greater energy independence ... Is it practical? Certainly. (Gazette.com, 'Opinion', 22 June 2006)

In this situation, everyone has to jump on board the biofuel bandwagon to avoid being labelled as being against sustainability. It really does not seem to matter that the presumed economic benefits of biofuels, such as the creation of jobs in rural areas, completely ignore the biophysical foundations of the economic process. A larger requirement of jobs for a given activity not only provides income to families, but also increases the costs of those goods and services requiring too much labour. Suggesting that we move a big chunk of the workforce into agro-biofuel production in a developed country is similar to suggesting a return to harvesting crops manually to increase the number of jobs in agriculture. It is mistaken reasoning, even without considering the weak analyses of agro-biofuel sustainability. But in this situation of denial, scientific knowledge no longer matters. Even those who are sceptical and want to flag up the existence of serious problems with large-scale agro-biofuels must start out by confirming that agro-biofuels are the solution to our sustainability problem in order to gain the legitimacy and attention of the scientific community. This is illustrated by the following passage from a policy statement by the Ecological Society of America. The statement assumes – as did Senator Salazar – that the given target of biofuel production is a feasible one (Box 9.2).

> **Box 9.2** *Passage from the policy statement of the Ecological Society of America*
>
> Supplying the emerging biofuels industry with enough biomass to meet the US biofuel energy target – replacing 30 per cent of the current US petroleum consumption with biofuels by 2030 – will have a major impact on the management and sustainability of many US ecosystems. Biofuels have great potential, but the ecological impacts of their development and use must be examined and addressed if they are to become a sustainable energy source. (ESA, 2008)

The agro-biofuel granfalloon

The term 'granfalloon' was first introduced by Kurt Vonnegut (1963) to indicate 'a proud and meaningless association of human beings'. Wikipedia provides the following definition (among others): 'a group of people who believe that they have a special connection and who believe they are helping to bring about a greater plan, but are actually not'. The idea that the concept of granfalloon is useful to explain the agro-biofuel folly was suggested to us by Vaclav Smil (personal communication).

According to Pratkanis (1995):

> *Granfalloons are powerful propaganda devices because they are easy to create and, once established, the granfalloon defines social reality and maintains social identities. Information is dependent on the granfalloon. Since most granfalloons quickly develop outgroups, criticisms can be attributed to those 'evil ones' outside the group, who are thus stifled. To maintain a desired social identity, such as that of a seeker or a New Age rebel, one must obey the dictates of the granfalloon and its leaders.*

The psychological dynamics behind the formation of granfalloons have important implications for scientific debates. As a matter of fact, when dealing with a scientific controversy, one should stick to the discussion of narratives, models and data, keeping the tone of the discussion focused on these items. Unfortunately this is not the case as soon as we deal with the granfalloon in science. We discussed earlier Professor Dale's (2008) objection to net energy analysis as 'a (mostly) irrelevant, misleading and dangerous metric' in assessing the feasibility and desirability of agro-biofuels. In his paper, Dale accuses David Pimentel of being an 'evil one' outside the granfalloon of agro-biofuels (see Box 9.3).

The point we want to make here is that Dale's personal attack on David Pimentel aims to discredit the information provided by the opponent, but not on the basis of detailed criticism of the semantic behind it. As discussed in Chapter 6, neither Dale (2008) nor Farrell et al (2006), in their papers, provided any semantic discussion of the meaning of the assumptions adopted in their net energy analysis either. Indeed, Pimentel is attacked simply because he does not follow the rules and the way of thinking of the granfalloon. In particular, the author (Dale) is using the judgement of the Berkeley group – a prestigious university – as if this were the ultimate proof that Pimentel must be wrong. In this example we deal with the bizarre situation in which David Pimentel, a world class scientist and one of the pioneers of energy analysis with hundreds of papers published in prestigious scientific journals, is attacked in

> **Box 9.3** *An example of granfalloon dynamics in scientific debate*
>
> *However, Dr. David Pimentel of Cornell University has, for at least 25 years now, grossly misused 'net energy' as a stick with which to beat, first, corn ethanol, and now cellulosic ethanol and biodiesel.*
>
> *Dr Pimentel's net energy work on biofuels is widely accepted and disseminated. His net energy work is also, to put it bluntly: shallow, misleading and borderline dishonest. You will search his writings and his presentations in vain for any discussion of what net energy means or any reference to the net energy literature. He never mentions the energy quality issue. He never compares the net energy of biofuels with their petroleum equivalents. It is easy to show, as I have done, that gasoline's net energy (as defined by Dr Pimentel) is actually lower than ethanol's. I am not the only one to note the blatant and pervasive inadequacies of Dr Pimentel's net energy analysis. The Berkeley group, led by the late Dr Alex Farrell pointed out that Dr Pimentel consistently uses assumptions and data that are 'obsolete', 'inaccurately cited', 'unverifiable', 'invalid', 'unsupported', 'inconsistently used' and 'misreported'. The words in quotations are those of* **the Berkeley group***, not mine.* (Dale, 2008, p496, emphasis added)

relation to his scientific credibility, simply because he is giving advice to his country about what he believes is the right thing to do according to his (pretty good) understanding of the energetics behind agro-biofuels!

Thus, the granfalloon in the biofuel scene implies a tacit agreement to admire the naked emperor's clothes. When this agreement is consolidated inside academic institutions, the academic filter is used against whoever refuses to join the granfalloon. As the majority of research funding comes from the government and private corporations, the work carried out within universities must conform with the belief – biofuel is *the* solution – endorsed by these institutions. When receiving considerable research funds, academics tend to uncritically adopt the story-telling of those who pay (see Box 9.4). Underlining the seriousness of this problem, UNESCO's Chief of Sustainable Water Resources Development, Professor Shahbaz Khan, called for the debate on national water reform in Australia to be 'opened up to include a genuine diversity of opinion', claiming that scientists 'are worried about being crucified' by governments if they express dissenting views. He said that when dealing with certain topics, 'scientists are fearful, to be honest' (Kahn, 2008).

> **Box 9.4** *Press release announcing a large donation by BP to research on biofuels*
>
> BP Selects Strategic Partners for Energy Biosciences Institute: BP today announced that it has selected the University of California Berkeley, the University of Illinois and the Lawrence Berkeley National Lab to join in a $500 million research program that would explore how bioscience can be used to increase energy production and reduce the impact of energy consumption on the environment. The funding will create the Energy Biosciences Institute (EBI), which initially will focus its research on biotechnology to produce biofuels – that is, turning plants and plant materials, including corn, field waste, switchgrass and algae, into transportation fuels. (UC Berkeley News, 2007)

Replicant knowledge

In an earlier paper (Giampietro et al, 2007) written with Timothy Allen, we suggested the use of a new category of knowledge to discuss the use of science for governance: 'replicant knowledge'. This is defined as a form of pseudo-knowledge. It leads to the application of policies indicated by scientific analyses that are lacking in semantic quality in relation to the pre-analytical step of issue definition. That is, these rigorous scientific analyses are based on narratives that may give totally irrelevant results for the social actors affected by the policies. These policies have to be imposed from the outside thanks to the existence of power asymmetry, against the will of stakeholders belonging to a different knowledge system. In this situation, it is obvious that the scientists working on the development of the scientific analysis did not bother to check the quality of the underlying narrative: the semantic behind the grammar used for quantification.

The term 'replicant' comes from the movie *Blade Runner*, in which replicants are pseudo-human forms, with a short shelf-life, used for invading extra-terrestrial colonies. They look and behave like humans, but do not have a history, either as individuals or as a social group. Replicants have no memories, no roots and no future. Their lives are without historical meaning or cultural narrative; they can only live in the present, executing very effectively the received protocols. They cannot evaluate the relevance of the quality of these protocols. Put another way, replicants do not have their own story-telling. They can only adopt the narrative given to them by someone else. In conclusion, we are in the presence of replicant knowledge when knowledge provided by someone else is used – without performing a critical semiotic check on its validity – in a given situation. Replicants cannot take a position; they merely function according to rules and analyses provided by others (Giampietro et al, 2007).

Our point here is that the establishment of a granfalloon can be seen as an autocatalytic loop of belief formation in which public opinion, politicians and the media are reinforcing each other. The final shared belief of the granfalloon is the result of a convergence and final stabilization of a mutual interplay between politicians, media and the public. In our case, the final selected belief is that agro-biofuels are *the* solution to our sustainability predicament. At this point, rigorous scientific analysis that does not bother to check the validity of the narrative should be considered replicant knowledge.

As illustrated by the personal attack on Professor Pimentel, as soon as this mechanism comes into place, it becomes difficult to guarantee the required quality checks of the scientific inputs to policy-making. It becomes hard to maintain control over the pre-analytical assumptions (the semantic behind the narratives) used for quantification (the quality of the grammar). In fact, the granfalloon entails that these assumptions have to be considered valid by default. This systemic bias in the pre-analytical step renders later quality checks on the implementation of the syntax (sampling and data gathering, handling quantitative models) useless. As discussed in Chapter 6, the implementation of a formally correct analysis based on a bad narrative makes the result completely irrelevant (Giampietro et al, 2005).

When dealing with science for governance, it is crucial to establish a process of quality assurance on the generation of scientific analyses. That is, it is necessary to guarantee a semiotic check in terms of a correspondence between the relevance of perceptions within which the scientific analysis has been performed and the pertinence of representation (what models and data are chosen by the analyst to implement the selected narratives?) (Giampietro et al, 2005). As stated by Schumpeter (1954, p42): 'analytical work begins with material provided by our vision of things, and this vision is ideological almost by definition'.

If we admit that the usefulness of any scientific analysis depends on the validity of the perception embodied in the selected narrative, then it is essential to couple wisely the perception and representation of a given issue. This implies that for scientific analyses, worse than being wrong is being irrelevant. The latter is the case with the grammar used, for example, by Dale (discussed in Chapter 6) to compare the energetics of ethanol (from corn) and gasoline (from petroleum). This irrelevant grammar is then used to carry out an elaborate analysis of the output/input energy ratio of ethanol production from corn by assuming credit for co-products. In this example, the problem lies outside the realm of the technical choices performed by the analysts: the implementation of the grammar is rigorous. The results of quantitative models based on irrelevant narratives cannot be tested in syntactic terms for their validity. The problem consists in the systemic lack of meaning of the pre-analytical choice of narrative. When the problem structuring is irrelevant there is insufficiency in the semiotic check.

On the other hand, a scientific analysis may be wrong, but it can still use the right set of categories for a certain narrative. This would be a meaningful grammar that is badly implemented when it comes to deciding how to quantify the chosen set of semantic relations. If this is the case, sooner or later practitioners in the field will learn how to obtain better data and produce better models.

The fallacy of scientific assessments of large-scale agro-biofuel production can largely be ascribed to the choice of irrelevant narratives based on their political convenience. When looking for an alternative energy source to oil, it is completely irrelevant whether the output/input ratio is 1.5/1, 1.2/1, or 1.8/1. Any energy source with an output/input of energy carriers below 5/1 has little chance of becoming a useful energy source for modern society. So it is unclear why scientists should spend their time and precious resources performing elaborate formal treatments of data or scrupulously respecting established protocols to calculate the numbers after the decimal point, when the first digit tells it all.

We use the term 'replicant knowledge' to emphasize the risk of using science to mechanically and indiscriminately apply the quantitative outputs of scientific analyses to the making of policies. When dealing with science for governance, the most important check to be made is not on the choice of the protocol used, but on the validity of the narratives within which the scientific analyses were developed. This point is especially important when there is a clear power asymmetry that could distort negotiations between contrasting legitimate perspectives. This is to say, if powerful lobbies are at work to influence the result of a choice, the development of meaningless formalisms can become a way to muddle the discussion. Such a misuse of quantitative results represents a perfect example of replicant knowledge: a flawless formalization of a semantically empty narrative, which can prove very helpful to the story-teller.

Replicant knowledge at work

We find ourselves in an embarrassing situation in which over the last two decades the European and US academic establishments have provided *unequivocal* scientific evidence to their governments in favour of biofuel production. The huge flow of research funds to agro-biofuels was used to back up policies setting a mandatory target for the internal consumption of liquid fuels in Europe and the US. The massive failure of the academic establishment on both sides of the Atlantic to develop sound analyses and viable solutions in relation to energy and sustainability has become clearly evident even to the general public. This messy situation was allowed to happen because governments wanted to implement a cure without having a clue about the disease, and in this desperate attempt funds and academic glory were allocated only to those promising to find the mythic silver bullet: agro-biofuel as an alternative to oil. Any serious discussion about the actual existence of such a silver bullet has been labelled politically inconvenient; there is no money or credits for those disagreeing with the political agenda. Hall et al (2008, p121) nicely frame it as follows:

> *There are not even targeted programmes in National Science Foundation or the Department of Energy where one might apply if one wishes to undertake good objective, peer reviewed EROI analyses. Consequently much of what is written about energy is woefully misinformed or simply advocacy by various groups that hope to profit from various perceived alternatives.*

Figure 9.1 shows a graph proposed by Tim Allen and inspired by the concept of a Medawar zone (Loehle, 1990) to describe the success of innovative researchers. The shape of the curve shows that truly path-breaking scientists (such as Galileo, Einstein, and Watson and Crick) never had much success when trying to convince the academic establishment about their innovative ideas, and that one should not expect additional acceleration (truly path-breaking results) from senior scientists who are at maximum velocity (those who have academic recognition).

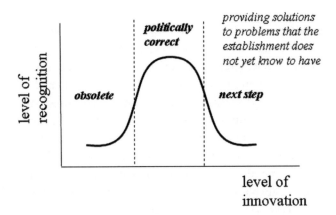

Figure 9.1 *Curve of success for researchers*

Indeed, Figure 9.1 suggests that one should not expect revolutionary innovations leading to a different energy future from famous professors at the peak of their careers or top managers at big corporations. This observation is extremely important, since there is a clear link between funding for research and development and the establishment of economic lock-in. An ideological bias determining a wrong priority in the allocation of R&D funds can move massive economic investments one way rather than another. Dalla Kachan (2008) provides evidence with hard numbers on published patents:

> *Over the last six years a total of 2796 biofuels related patents were published in the US, with the number increasing by over 150 per cent in each of the past two years. In 2007 the number of biofuel patents (1045) was more than the combined total of solar power (555) and wind power (282) patents published in that year.*

To make things worse, investors who do go for innovation are usually not doing so because they believe in the economic validity of the idea (long-term profits), but because of the chance to harvest subsidies (short-term profit) during the consolidation and expansion phase of the new technology. The formation of the granfalloon aggravates this situation since the resulting attitude of the government will guarantee the harvesting of subsidies by private investors and the formation of massive sunk-costs to be paid, later on, either by later investors or by taxpayers (in the case of bail-outs).

Academic Lock-in

A scientific paradigm can be seen as a deeply rooted ideological bias in the academic establishment. Tim Allen (personal communication) defines a scientific paradigm as a tacit agreement not to ask certain questions. Indeed, a scientific paradigm is an uncritical acceptance of the relevance of a given narrative about the world. As discussed under the heading of 'Ideological Lock-in', society has favoured scientific analyses based on those narratives that are compatible with the idea of perpetual growth. It is therefore not surprising that the paradigm of neoclassical economics, which relies by default on the assumption of perpetual growth (e.g. perfect substitutability among production factors), has gained absolute power in the field of sustainability analysis and has provided the basic narrative used in innovation and research.

Unfortunately, the very same ideological lock-in that stabilizes the winning scientific paradigm in the short run causes an *ancien regime* syndrome in the long run. This is to say that scientists rewarded for their commitment and loyalty to the winning paradigm can no longer say that the emperor has no clothes; they can no longer admit bugs in the basic assumptions behind their models.

The battle between cornucopians and prophets of doom

To understand the seemingly universal use of the neoclassical paradigm to frame the issue of sustainability in the scientific discourse and the selective removal of the lessons of energetics and bio-economics from academic memory, one has to go back in time, to the battle between the two opposing views of sustainability that took place in the 1970s. The two sides involved were the cornucopians and the prophets of doom: see Box 9.5.

> **Box 9.5** *Cornucopians versus prophets of doom*
>
> Cornucopians, the stereotype of neoclassical economists, focus exclusively on monetary flows and human welfare without much concern for the environment. They do so because of their unconditional belief in the possibility of perpetual growth. In spite of the laws of thermodynamics, biology and ecology, there are many economists who still believe that if the price is high enough, human ingenuity will produce any required resource – be it energy, water or bees for pollination – according to the needs of the moment. As a matter of fact, Robert Solow (1974, p11), winner of the Nobel Prize in Economics, said: 'If it is very easy to substitute other factors for natural resources, then there is in principle no "problem". The world can, in effect, get along without natural resources.'
>
> The most vocal of the cornucopians has been Julian Simon (not to be confused with Herbert Simon, Nobel Prize winner and father of complexity theory). His abundant scientific production – his most famous book being *The Ultimate Resource* – and the wagers with Paul Ehrlich, a representative of the prophets of doom, earned him the nickname of 'the doomslayer'.
>
> The prophets of doom, on the other hand, focused on the key importance of biophysical flows and ecological welfare. According to their biophysical analyses, they claimed it was necessary to change the pattern of human development to respect the limits imposed by nature. The *ante litteram* pioneer of this group was undoubtedly Thomas Malthus (*An Essay on the Principle of Population*, 1798). Then, in the 1970s, its famous exponents were Paul Ehrlich (author of *The Population Bomb*, 1968) and Garrett Hardin (author of the 1968 *Science* article 'The tragedy of the commons', and the book *Exploring New Ethics for Survival: The Voyage of the Spaceship Beagle*, 1972). Finally, it is important to recall the report of the Club of Rome entitled: *The Limits to Growth* (Meadows et al, 1972).

Energetics and bio-economics were used by the prophets of doom. As a matter of fact, a large part of the theoretical development of these fields in the 1970s had the goal of challenging the assumptions of neoclassical economic theory, which was in full swing at that time. Cornucopians assume by default an endless supply of silver bullets, and as a consequence they postulate that humans cannot run out of energy or other critical resources. Georgescu-Roegen's heavy reliance on thermodynamic principles was aimed at establishing a parallelism between the absurdity of the idea of perpetual motion and that of perpetual growth. In the same way, H. T. Odum applied energy graphs developed in the field of theoretical ecology to the analysis of socio-economic patterns of metabolism to show that the self-organization of human society is perfectly analogous to that of ecological systems, and is therefore subject to the same set of biophysical constraints. That is, both Georgescu-Roegen and H. T. Odum developed their analysis to challenge those, like Solow and Julian Simon, who expressed blind faith in the ability of human ingenuity to 'fix the external world' whenever it is incompatible with human wants.

Another source of confrontation between the two sides was different beliefs about the ability to predict future changes using models based on prices. The view of Georgescu-Roegen on this point is described in Appendix 1. He judged as a systemic error any attempt to predict future events using econometric analysis. This systemic error is generated by neglecting the issue of scale in the making and operation of the model. Econometric analysis deals with a picture based on the production of goods and services within tightly specified institutions, using prices as a numeraire. According to Georgescu-Roegen (see Appendix 1), what is produced by the economic process is the metabolic pattern associated with the production and consumption of goods and services, and not goods and services themselves. If we accept this point then it follows that in the future we will have different products and different services within different economic institutions (different preferences, different processes of production, different natural resources, a different definition of natural resources, and different laws and regulations). Due to the unavoidable occurrence of evolutionary changes – metabolic systems are 'becoming systems' – the actual meaning carried by prices will be completely lost. This is to say that making predictions referring to the year 2050, using general equilibrium theory or any other econometric model, is like trying to study the ecology of elephants by looking at them through a microscope. The quantitative representation used as basis for the analysis adopts a scale which misses the relevant features of the system we want to observe. This is without the problem of predicting right now which currency will be used in the year 2050 at a world level to express the prices: it could be the US dollar, the euro or the Chinese yuan.

Indeed, the systemic failure of econometric models to predict the future is well known. Even Alan Greenspan, after retiring from the powerful position of President of the US Federal Reserve, has openly criticized the predictive power of econometric models (Box 9.6).

> **Box 9.6** *Greenspan criticizing econometric models*
>
> The essential problem is that our models – both risk models and econometric models – as complex as they have become, are still too simple to capture the full array of governing variables that drive global economic reality. A model, of necessity, is an abstraction from the full detail of the real world… Current practice is to introduce notions of 'animal spirits', as John Maynard Keynes put it, through 'add factors'. That is, we arbitrarily change the outcome of our model's equations. Add-factoring, however, is an implicit recognition that models, as we currently employ them, are structurally deficient; it does not sufficiently address the problem of the missing variable. (Greenspan, 2008)

The role of economic models in the process of decision-making

So the real question is: how is it possible that our governments are still justifying policy decisions on the basis of analyses and scenarios generated by economic models? In relation to this question, Funtowicz and Ravetz (1990) pointed at the significance of the evolution of the mechanisms providing legitimization to decision-making.

The organization of the state in Europe before the scientific revolution of the 16th and 17th centuries was based on a hierarchical system of control (power) legitimized directly by God (Figure 9.2a). This 'absolute' source of legitimization was incarnated in the figure of the king, who was in charge of performing semantic checks (quality control) over the validity of the narratives included in the system of knowledge adopted in a given society. In this function the king used a certain number of advisers, who selected, stored and refined better representative tools whenever this activity was in the king's interests. After the Enlightenment, the process of democratization and the reduced influence of religion in determining the legitimization of the organization of the state (hierarchy of power) implied that in many Western states a different mechanism of legitimization was needed (Figure 9.2b). This mechanism was provided by the concept of substantive rationality.

The concept of substantive rationality is based on the assumption that it is possible to define a unique, verifiable and uncontested truth. In fact, it is only under this assumption that it becomes possible to use rationality in order to act for the common good of the community. If this assumption is true, then good governance is about making right choices according to the unequivocal and knowable truth. In this way, it was possible to eliminate God as the ultimate source of legitimacy for the system of control (linked to the organization of the state) and use instead the set of relations indicated in Figure 9.2b. It should be noted that the birth of these Western modern states coincided with an evolu-

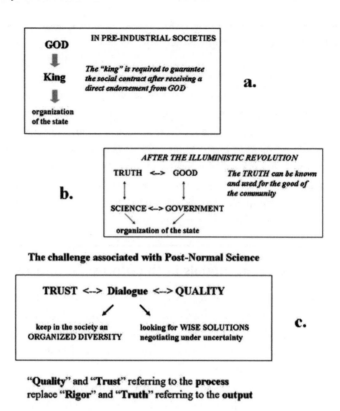

Figure 9.2 *Mechanisms of legitimization of the organization of state, after Funtowicz and Ravetz (1990)*

tionary phase of fast expansion: colonization and the Industrial Revolution. Therefore, these states needed massive investments in efficiency; they needed to amplify as quickly as possible the investments in those patterns that were better at paying back to society. Putting it another way, the fact that Western countries were in an evolutionary phase of rapid expansion implied a hegemonization of the winning narrative. Therefore, the underlying idea that sound governance could be obtained by relying on just one narrative about progress was not particularly disturbing (Giampietro, 2003, p83–84).

It is against this background that the battle between the cornucopians and the prophets of doom should be contextualized. After the Enlightenment and the birth of the modern state, neoclassical economists stuck to a simplified vision of social development that had worked in the past: the richer, the better. In this simplification, the only important goal was to maintain momentum by adopting policies aimed at getting richer (maximizing GDP) at maximum speed. No quality control was applied to the validity of this narrative, since nobody but the powerless losers in the economic battle complained about it.

The prophets of doom, on the other hand, did not deliver any valuable predictions. On the contrary, they rang alarm bells too early. Moreover, they were suggesting policies that were not welcomed by anybody. No wonder that cornucopians won the battle!

The need for a new paradigm: post-normal science

Economists, in their role of modern priests, are no longer effective in supporting the claim that scientific models can provide universal truths. Instead of Latin or other arcane languages, they invented complicated mathematical models to impress ordinary people. However, in spite of these growing levels of complication, the general public is losing confidence in economic analysts. The lack of usefulness and predictive power of their mathematical models is becoming increasingly evident.

In more general terms, we can say that the sustainability predicament is increasingly forcing scientists to fill the growing knowledge deficit. This gap is generated by the need to represent and predict complex large-scale events for which the simplifications typical of reductionism are inadequate. Despite this growing challenge, the academic establishment has ignored the growing knowledge deficit and stuck to business as usual, preserving disciplinary knowledge and the traditional distinctions between specialized departments required to run existing academic institutions.

This modern progressive impasse in the processes providing legitimacy to governmental decision-making seems to justify the call made by Funtowicz and Ravetz (1993) and Ravetz and Funtowicz (1999) for a post-normal science. In this proposed paradigm of science for governance, the mechanism providing legitimization to decision-making should replace the concept of *truth* with that of *quality*. Funtowicz and Ravetz argue that decision-making is no longer about generating optimizing protocols expected to indicate the best course of action. Rather, it is about generating sound and fair procedures for deliberation (weighing pros and cons) about wise strategies to be selected in conditions of uncertainty. This means moving from the concept of substantive rationality to that of procedural rationality (Simon, 1976; 1983). That is, after acknowledging that when dealing with complex large-scale processes it is impossible to represent or define optimal solutions, we can only look for *satisfying* solutions (Simon, 1976; 1983).

The consequences of such a paradigm shift are important for determining alternative ways to organize scientific information in the process of decision-making (Figure 9.2c). Trust, which entails concepts such as reciprocity, loyalty and shared ethical values among the various stakeholders involved in the process of negotiation, becomes the crucial input in the new challenge of governance. It is no longer the final number thrown over the fence that matters in a scientific debate, but the consensus on the usefulness of the narrative and grammars used to calculate the numbers. The proposed paradigm shift is of utmost importance in relation to all the forms of lock-in discussed in this chapter.

Economic Lock-in

The third form of lock-in refers to the action of powerful lobbies that tend to stabilize existing economic power structures. The economic lock-in in the agro-biofuel scene is closely interwoven with the mechanism stabilizing the paradigm of industrial agriculture, which we discussed in detail in Chapter 8. Indeed, the agonizing paradigm of industrial agriculture is being preserved at any cost by existing lobbies for understandable reasons. Direct and indirect food subsidies in the US, the EU and other developed countries together exceed one hundred billion dollars per year. Most of this money goes not to farmers but to large corporations producing external agricultural inputs. A massive switch from the production of food commodities to that of feedstock for agro-biofuels will make the spectre of cuts in food commodity programmes in developed countries disappear.

What is less understandable is the reasoning behind subsidies for agro-biofuels. The governments of the US and the EU justify economic support for the agro-biofuel sector as a start-up investment, assuming that its production will be profitable (or at least viable) in the long term (see the discussion below). The fact that in this early stage, agro-biofuel production is not yet economically profitable is attributed to the low price of oil. They argue that as soon as energy prices increase, agro-biofuel production will become economically competitive.

Mystery of the economic evaluation of agro-biofuels

According to the basic principles of bio-economics discussed in Chapter 4, especially with regard to ethanol from corn, the expectation that agro-biofuel production will become economically profitable when the cost of oil rises is simply misplaced. In fact, even if there is a slight positive return in terms of energy carriers made available in relation to those consumed, there are a lot of additional economic costs, such as those related to importing the required oil and using this oil to produce biofuel feedstock crops. The production of crops in turn requires the use of land and other infrastructures, which have an economic opportunity cost as well. Looking at the overall performance of this process, one must admit that the economic advantage of converting the energy contained in a barrel of oil into more or less the same amount of energy in the form of biofuel is very questionable, to say the least. In the case of an increase in the cost of a barrel of oil, we can anticipate that the real cost of a barrel of ethanol from corn (not considering subsidies) will go up even further in the long run, given that oil is required (directly and indirectly) for its production. That is, the agro-biofuel policy does not provide any protection against the future consequences of peak oil.

As a matter of fact, the increased cost of energy has raised the cost of infrastructures (steel and concrete), and this in turn has already affected the production cost of several alternative energy sources, such as sand tar extraction in Canada (Haggett, 2007).

But if things look bad when the oil price goes up, what happens if the price of oil goes down? If possible, things look even worse. Krauss (2008) reported in the *New York Times* that 'shares of alternative energy companies have fallen even more sharply than the rest of the stock market in recent months. Advocates are concerned that if the prices for oil and gas keep falling, the incentive for utilities and consumers to buy expensive renewable energy will sink.'

Allison and Kirchgaessner (2008) reported a similar story in the *Financial Times* (21 October):

> *Investors, such as Microsoft's Bill Gates, are sitting on billions of dollars in losses after buying into the corn-based ethanol industry that George W Bush embraced as the answer to the US energy woes. Six of the biggest publicly traded US ethanol producers have lost more that $8.7 bn in market value since the peak of the boom in mid-2006 and the beginning of this month, according to an analysis by the* Financial Times. **The boom followed a 2005 law requiring refiners to mix billions of gallons of the biofuel with petrol.** (Emphasis added)

If the actual economic performance of agro-biofuels is that poor, if their production has proved to be an activity merely aimed at harvesting subsidies in the short term, how is it possible that the official economic assessments used by the public administration to justify the relative policies claim the contrary? Indeed, given the economic crisis in the agro-biofuel sector, one cannot but wonder about the quality of the economic evaluations of agro-biofuel scenarios found in official documents that have been used to promote this presumably promising sector.

As a matter of fact, it is worth reading an internal report by the European Commission (EU, 2006) because one gets a clear idea of how these assessments are carried out. The report heavily relies on what we defined earlier as replicant knowledge; experts provide meaningless applications of scientific vocabulary to reach irrelevant conclusions. The assessment of the effects of the large-scale implementation of agro-biofuels on employment nicely illustrates our point (EU, 2006, p24, Section 5.5):

> *In principle, biofuels can have a positive economic impact, locally and regionally, in developed and, probably even more, in developing countries. Employment opportunities are mainly linked to biofuel processing and the provision of agricultural and forestry feedstocks. This is particularly important for rural areas, where other options may be limited. However, since biofuels are still more expensive than fossil fuels, additional cost or public spending might have an impact on the overall economy.*

This passage shows a tendency toward Marie-Antoinette reasoning, but without endorsing it. The elusive construction – 'in principle, biofuels can have a positive impact' – is fascinating, especially since this is an assessment. Of course, biofuels will increase the labour requirement in the energy sector, since agro-biofuels are more labour-intensive than fossil energy fuels. But is this a positive, desirable economic change? At this point economic expertise comes in, and forces the authors to state that these jobs will have to be paid with public money since presently the activity is not economically viable. This is indeed a puzzling example of impact assessment, because:

- in quantitative terms, this assessment does not provide any information about the impact of biofuel on employment; and
- in semantic terms, the assessment should be considered a negative one.

In fact, if the government has public money at its disposal to keep the unemployed busy, why not allocate it to useful social work rather than to damaging the environment by converting oil into ethanol? The authors could not have written this report when they were undergraduate students in economics, and the senior scientists who accepted it would not have done so if it were written in an exam by undergraduate students. However, this assessment is published in an official document that supports the investment of billions of Euros.

This example clearly shows that the function of this report is not to provide a useful assessment. Written in arcane language, its purpose is to make possible the implementation and dissemination of agro-biofuels. Therefore, its semantic content is not really relevant; its function is merely to back up the beliefs supported by the granfalloon. This observation demonstrates the extreme fragility of the interaction between the academic and the administrative bureaucracy when confronting powerful lobbies. This bizarre display and use of scientific expertise in official documents can only be explained by intense pressure from lobbies affecting the outcome of the process.

So what is motivating these powerful lobbies?

WTO agreement on cutting commodity support programmes and powerful lobbies in action

In the last round of World Trade Organization (WTO) trade talks (the Doha round), an agreement was reached (at least in principle) on the need to drastically cut commodity support programmes. The proposed cut includes crop insurance, export subsidies, loan deficiency payments and countercyclical payments, all of which are considered to distort production and trade. If the agreement is ever implemented, it will translate into a subsidy cut of about US$19 billion in the US and US$80 billion in the EU (in 2003 prices). Additional cuts are to take place in other developed countries. The implementation of this agreement could thus make available a large amount of money for improving agricultural practices, such as boosting organic agriculture, protecting local crop varieties, protecting traditional farming practices and protecting the rural landscape. This money could be used to bring about a change in the agonizing paradigm of industrial agriculture.

Obviously, the present beneficiaries of food commodity support programmes – large corporations producing agricultural inputs – have a different perception of the recent WTO agreement. They do not want to phase out of the paradigm of industrial agriculture. A massive switch from an industrial production of food commodities into an equally industrial production of agro-biofuels could avoid the necessity of performing this cut. Indeed, the agro-biofuel option opens up two possibilities: maintaining (if not increasing) the flow of subsidies in spite of WTO agreements; and keeping alive the paradigm of industrial agriculture and, in particular, expanding the use of genetically modified feedstock crops (since the crops are not eaten, they are subject to fewer restrictions). Moreover, as discussed in Chapters 6 and 7, the existing agro-biofuel systems almost entirely consume the energy they produce in their own operation, so that the supply of energy crops for biofuel production will never exceed the demand. For those eager to harvest subsidies for industrial agriculture, subsidized agro-biofuel production comes close to perpetual motion! Evidence that large companies producing agricultural inputs for agro-biofuel production are faring well with the policy supporting agro-biofuels is given in Box 9.7.

Finally, it should be noted that agro-biofuels are supposed not only to save the planet and to help revitalize the paradigm of industrial agriculture, but also to save the car industry (reassuring buyers scared by the issue of peak oil) and companies that invested heavily in genetically modified organisms (opening a whole new market in energy crops). To better understand the nature and mechanism of lobbying, Box 9.8 quotes the description of the process of policy-making in the European capital provided by the watchdog Corporate Europe Observatory.

> **Box 9.7** *Press reports on the benefits of companies producing agricultural inputs*
>
> ### Commodities send Syngenta earn soaring
>
> *Agrochemical company Syngenta AG on Thursday posted a 75 per cent rise in annual net profit on soaring commodity prices and strong demand. The Swiss company reported a profit of US$1.1 billion compared with US$634 million in the same period a year earlier. Nina Baiker, an analyst with Zuercher Kantonalbank, said Syngenta's performance is very good. The company's crop protection division is well positioned to take advantage of farmers' need to boost yield per acre for food and biofuels, she added.*
>
> <div align="right">(<i>International Business Times</i>, 7 February 2008)</div>
>
> ### Grain companies' profits soar as global food crisis mounts
>
> *At a time when parts of the world are facing food riots, Big Agriculture is dealing with a different sort of challenge: huge profits. On Tuesday, grain processing giant Archer-Daniels-Midland Co. said its fiscal third-quarter profits jumped 42 per cent, including a sevenfold increase in net income in its unit that stores, transports and trades grains, such as wheat and corn, as well as soybeans. Monsanto saw its profit in the latest quarter more than double. Rivals DuPont Co. and Syngenta AG recently raised their profit estimates. Deere posted a 55 per cent rise in earnings in the latest quarter. Mosaic's third-quarter net income jumped about twelvefold. ADM's major rivals are notching big profit gains too. Closely held Cargill Inc's profit jumped 86 per cent to $1 billion in the latest quarter. Bunge Ltd's earnings rose about twentyfold to US$289 million. Bunge sells fertilizers in addition to processing and storing grains.*
>
> (Kesmodel et al, 2008)

The affiliation of the board members shows the presence of many representatives of economic sectors that are currently facing an economic crisis, such as the car industry and the biotech industry. No wonder their vision of the future is based on finding a silver bullet capable of guaranteeing the stability of the status quo and therefore of their corporations. The vision of the board members is perfectly understandable. What is more difficult to understand is the choice made by the European Commission when selecting this board. Is it a wise and effective strategy to use top managers from industries in crisis to generate useful visions for an alternative path of development? It seems that the advice provided by this advisory board is basically concerned with the flow of profits going into their industries, rather than with seeking an alternative

Box 9.8 *An insider view of the process of decision-making about biofuels in the EU*

The EU's agrofuel folly: policy capture by corporate interests

The EU's agrofuel folly has been influenced by the strong lobbying activities of interested industries, such as car manufacturers, biotech companies and the oil industry. These industries have forged new alliances with each other, and in recent years have been invited by the European Commission to shape EU policy on agrofuels through several industry-dominated advisory bodies. These include the Advisory Research Council for Biofuels (BIOFRAC), CARS21 and more recently the European Biofuels Technology Platform (EBFTP).

Corporate advice for corporate benefit

The Biofuels Research Advisory Council (BIOFRAC) was created by DG Research in early 2005. A 'group of high level experts representing widely different sectors of the biofuel chain' was invited 'to develop a foresight report – a vision for biofuels up to 2030 and beyond, to ensure a breakthrough of biofuels and increase their deployment in the EU'. In addition to this 'foresight report', the Commission also invited BIOFRAC to prepare the ground for the so-called 'Strategic Research Agenda', and to provide considerable input for the Seventh Framework Research Programme (FP7), the EU's main instrument for funding research in Europe from 2007 to 2013.

The vision: biofuels in the European Union – 2030 and beyond

By 2030, the European Union **covers as much as one quarter (25 per cent) of its road transport fuel needs by clean and CO_2-efficient biofuels.** A substantial part is provided by a competitive European industry. This significantly decreases the EU fossil fuel import dependence. Biofuels are produced using sustainable and innovative technologies; these create opportunities for biomass providers, biofuel producers and the automotive industry.

The panel: affiliation of some of the members of the Biofuels Research Advisory Council (as of the date of publication of the Vision Report)

Chair: Volvo Technology Corporation
Vice-chair: Institut Francais du Petrole and Abengoa Bioenergy
Sample of Members: Total, British Sugar, PSA Peugeot Citroen, EuropaBio, CHOREN, SVEASKOG, Volkswagen AG, European Biodiesel Board, COPA-COGECA, SHELL, CRES, Neste Oil Corporation, IVECO Powertrain, ECN, INRA, VTT Biotechnology, Fraunhofer UMSICHT, Nova Energie, EC-BREC, Lund University

Source: CEO (2007). Emphasis added.

path of development. Should Europeans evaluate their future using as criterion the preservation of leading corporations? Is this what European society needs for a smooth energy transition into the third millennium? Why have these people been allowed to decide the future of European citizens?

This example once again shows the use of replicant knowledge for dealing with very important issues that require semiotic checks. Given the seriousness of the issue, we firmly believe that the European people deserve a better method to discuss society's future than the one adopted so far by the European Commission. In particular, it would be wise for the European Commission – which hands out subsidies – to stop asking the recipients of the subsidies to make decisions about policies that will determine the allocation of subsidies.

Sunk cost leading to the Concorde syndrome

In academic research, the worst case scenario is to discover that basic assumptions (i.e. the narrative) adopted to develop an elaborate assessment were wrong. This leads to the most feared pronouncement in the academic world: 'nice try guys, but let's move on'. All the work done so far goes to waste. It is even worse when the situation has involved investments of considerable sums in technological innovation, as is the case with agro-biofuels. The investor has not only wasted time but has also incurred a sunk cost which, at times, can be pretty high.

By sunk cost, we mean the total amount of losses incurred when a bad investment is abandoned. The standard reaction to large sunk costs is: 'it is impossible to stop now, or we will lose all we have invested so far'. This reaction is certainly understandable, but nevertheless it does not make any sense to keep investing in a bad idea. In fact, continuing to invest in a bad idea will increase the seriousness of the losses incurred. This is straightforward logical reasoning, but it is not easy to write off a considerable investment. For this reason, whenever there is a serious problem of sunk costs there is also a serious risk of the insurgence of the Concorde syndrome (discussed in Chapter 8).

No matter how bad an idea turns out to be, investors will push for its continued implementation for as long as it takes to recover the investment. In this situation, we often witness desperate attempts by investors to shift these sunk costs to others. This solution can only result in even more damage to society in the long run. As explained in Chapter 8, Concorde flew for almost 30 years, in spite of the continuous accumulation of economic losses.

We firmly believe that the corn-ethanol industry in the US is a perfect example of Concorde syndrome: a dead industry walking. For this reason there is a serious risk that the sunk costs will grow and be externalized. This could be in the form of prolonged subsidies and then in a massive bail-out, transferring the overall sunk costs to the public. This is what the governments of the US and EU have done so far with agro-biofuels and probably will be doing in the future.

The lesson to be learned from the predicament associated with sunk costs and the Concorde syndrome is simple: for society, it is more convenient to focus on the early detection of bad ideas, rather than heavily investing in them. Otherwise, sooner or later someone will be forced to pay the bill. Apart from the moral issue, governments in Europe and the US can ill afford to waste large amounts of money on a lost battle. It should be noted that the agro-biofuel industry is not a unique case; other industries, most of them born using heavy subsidies from national governments, belong to the 'dead industries walking' category. This point brings up a last remark about the use of public funds in scientific research that generates profits for the private sector.

Helping a national industry to generate profit should be among the legitimate goals of any government, because the creation of vital industries eventually benefits the economic prosperity of society as a whole. But when dealing with innovations that require large investments, it would be wise (to say the least) to clarify the roles of the various actors involved and the goals of the innovation. In fact, we are increasingly experiencing cases in which public money is invested, through subsidies and direct research funds, in research and development projects that pursue the corporate goals of private industries. In addition to initial investments, the government often stands as guarantor, ready to bail out companies should the original ideas turn out bad. In this way, while aiming to preserve the economic viability of big corporations, national governments help companies that invested in bad ideas to shift the long-term negative consequences (social, economic and ecological) to other actors, usually taxpayers and/or – even worse – poor people in developing countries.

The Lessons to be Learned

Powerful lobbies, sunk costs, Concorde syndrome, obsolete scientific paradigms, replicant knowledge and the successfully established granfalloon add up to a biofuel scene that is hard to change. However, as discussed in Chapter 1, today there are more and more people, some of them highly visible, saying that the emperor has no clothes: producing biofuels from crops was just a blunder. Yet nobody has enough power to stop the runaway train. In the US, it seems unlikely that President Obama will reduce the economic resources committed by the Bush administration to corn-ethanol production; in Europe, the proposal to lift the mandatory threshold for internal biofuel consumption was recently turned down. In Box 9.9, we quote a passage from a report into the capitulation of the European Parliament, as it neatly illustrates the situation.

Any major restructuring of the socio-economic process aimed at changing the narrative employed for the sustainability issue will imply a change in both the power structure and the identity of the incumbents in power. For this

> **Box 9.9** *European Parliament capitulates on biofuel deal*
>
> The policy tennis match between biofuels supporters and opponents in the European Union has all but drawn to a close, with the backers of the controversial fuel source securing almost complete victory.
>
> *Representatives of EU Member States, the European Parliament and the European Commission this week came to a back-room agreement that supports the sourcing of 10 per cent of the EU's road transport fuel from renewable forms of energy by 2020 – the same target figure originally proposed by the EU executive in January of this year. Biofuels, such as palm oil in Indonesia, have been the focus of a battle royale in Brussels over the past year.*
>
> *When the proposal was first unveiled, most policy-makers assumed that biofuels would make up all or most of the 10 per cent figure. But in the wake of reports from the World Bank through to the UN saying that in many cases biofuels produced more greenhouse gases than fossil fuels and threatened global food supplies, EU law-makers were under pressure to slim down or abandon the biofuels element of the 10 per cent renewable transport fuel target.*
>
> *In particular, scientists warned that 'indirect land-use change' – the creation of new farmland on previous grassland or forest to compensate for farmland lost to biofuels – would put the value of even the 'cleanest' biofuels in doubt.*
>
> *While many in the European Parliament had been convinced of the dangers, the Commission and Member States remained adamant that the target go ahead largely unchanged. Under this week's tripartite agreement, consideration of problems caused by indirect land-use change has been completely junked, apart from a caveat that the European Commission will come up with a report by 2010 on how to minimize this process. But since the European Commission has in recent months repeatedly denied that indirect land-use change occurs, it is unlikely to develop strong recommendations to counter the problem.*
>
> Source: Phillips (2008)

reason, those in power tend to control the timing and the topic selection of the sustainability debate to their own advantage.

Finding reasonable alternatives to the current pattern of development would require a wise process of discussion about feasible and desirable scenarios, which should be based on an informed discussion and a participatory process of deliberation about our common future. Before jumping into a frantic search for technological solutions – the mythical silver bullets – that in general have as their only result the giving of public money to large corporations, we need to reflect more, and at times – why not? – use common sense rather than fancy econometric models. In this endeavour, certainly, it would help if we at least tried to better understand the nature of the problems we want to solve.

Chapter 10

Where Do We Go from Here?

Conclusions about Agro-biofuel Policies

Conclusion 1

The idea of producing agro-biofuels on a large scale to replace fossil fuels is a bad one. Excepting ethanol from sugar cane, the implementation of policies based on this idea should be considered a combination of a blunder and a hoax. The special case of ethanol production from sugar cane in Brazil represents a feasible option, but its desirability in relation to economic development and environmental impact is doubtful. We strongly believe that in the short term, developed countries will stop spending public money to support large-scale agro-biofuel production. Therefore, developing countries should realize that in the near future, subsidies might no longer be available to encourage the import of their agro-biofuels. Hence, policies aimed at implementing the cultivation of agro-biofuel in developing countries could be disastrous in the medium to long term.

Conclusion 2

The fact that large-scale agro-biofuel production is a bad idea does not imply that bio-energy – the use of biomass for energetic purposes – is necessarily a bad idea. Biomass has always been used for energetic purposes by humankind, and always will. Bio-energy has great potential, but also serious limitations. When trying to establish a link between bio-energy and rural development we should look for a different paradigm of rural development, based on the concept of multifunctional land use. The integration of a variety of economic activities in rural areas should increase the options for using bio-energy locally. Similarly, the production of bio-energy carriers should be used to diversify the economic activities of rural communities. In this new paradigm of rural development, bio-energy should be directly linked to the preservation of healthy rural communities (the fund element reproducing local human activity) and healthy ecosystems (the fund elements providing the supply of biomass). People and land should be considered as an integrated system of funds to be preserved and enhanced. They should no longer be squeezed to generate energy carriers for urban dwellers and other economic sectors.

Conclusion 3

When discussing the next generation of biofuels, we must avoid repeating the mistakes made with the first generation. Those evaluating its potential should not look only at the characteristics of the production process. As we have attempted to demonstrate in this book, to study the feasibility and desirability of alternative energy sources it is essential to also look at the characteristics of the society that will use them. What has to be verified, using the metaphor of the heart transplant, is the compatibility of the characteristics of the supply of net energy carriers provided by the next generation of biofuels with the type of requirement determined by the desired pattern of societal metabolism.

Conclusion 4

In the discussion of second generation biofuels, we should be aware that the field of alternative energy sources is likely to give rise to many generations of alternatives. As the research and development of alternative energy sources is notoriously expensive, we may anticipate large sunk costs and a high probability of falling victim to the Concorde syndrome.

For example, the nuclear energy industry now claims to be close to the fourth generation of nuclear reactors. While the development of the various generations of new nuclear reactors has witnessed the progressive depletion of uranium reserves, structural solutions to known problems – such as how to handle in a safe and economic way radioactive wastes – have not come any closer. On the contrary, unexpected problems have surfaced, most notably the threat of terrorism and the high costs of decommissioning obsolete plants.

We do not mean to imply that research into alternative energy sources should be stopped. On the contrary, the current lack of a feasible and desirable alternative to fossil energy indicates that this is an objective that deserves top priority. Nobody can predict the future, and therefore nobody can guess what type of technological breakthrough may be achieved by human ingenuity. It is possible that sooner or later one of these 'next generations' will be the right one. But it is also possible that in a long series of 'next generations', none will deliver the required solution. Therefore, in spite of the unavoidable uncertainty associated with any forecasting, we should at least attempt to perform reasonable checks on the feasibility and desirability of proposed alternatives.

Conclusion 5

The conclusion that using bio-ethanol for energetic purposes is a bad idea applies only to the exosomatic energy metabolism of society. It does not necessarily apply to a moderate production and consumption of ethanol within an endosomatic metabolism!

Conclusions about the Energy Analysis of Alternative Energy Sources

Conclusion 1

Thanks to the contribution of many scientists in the fields of energetics and bio-economics, it is possible to perform a robust integrated analysis of the performance of alternative energy sources. In particular, Georgescu-Roegen and H. T. Odum have provided approaches that have proved useful for performing quantitative analyses. We emphasize that any integrated analysis of alternative energy sources has to be based on a combination of biophysical and economic analysis.

Conclusion 2

The quality of alternative energy sources depends on three technical aspects:

1. The output/input ratio of energy carriers used in the energy sector to generate a net supply of energy carriers (often confused with EROI). This information has to be scaled or contextualized in relation to internal and external constraints using the next two factors.
2. The power level (exosomatic metabolic rate per hour of labour) that can be achieved and has to be achieved in this process of the generation of energy carriers. This refers to internal constraints, such as the biophysical feasibility and the economic viability.
3. The availability of an adequate supply of primary energy sources required for this production of energy carriers: that is, external constraints. There is a need for compatibility with available resources and ecological processes, both on the input and on the sink side.

The combined use of these three concepts within a useful grammar – required for tailoring the analysis to the specificity of the case study – makes it possible to check the feasibility and desirability of alternative energy sources for a given metabolic pattern of society.

Conclusion 3

The feasibility and desirability of a given pattern of metabolism in society – the pattern of end-uses across the various compartments making up society – is determined by the quality of its energy sources. This implies that the overall requirement of labour, capital and colonized land for the internal loop of energy-for-energy in the energy sector should be compatible with the stability of the metabolism of the remaining sectors of the society. The overall demand of primary energy sources and environmental services required for the stability of the metabolism of the whole society should not interfere with the stability of the metabolism of the ecosystems embedding the society.

Conclusion 4

The method of accounting for and representing the energy consumption of society found in national and international statistics reflects only the exosomatic metabolic pattern of modern societies based on fossil energy. The template used to represent this pattern is a linear series of conversions that have fossil energy as the primary energy source at the starting point and end-uses at the end point. However, in a future where fossil energy is expected to play a less important role, this grammar might be quite useless. When imagining a combination of alternative primary energy sources – for example, photovoltaic, wind energy, hydroelectric, bio-energy and nuclear – we will have a network of different types of energy carriers defined around different types of impredicative loops. The characteristics of each one of these impredicative loops will depend on the particular primary energy source and the particular technology used for its exploitation. Such a network can only be described as a spider-web. The central point of this web is the accounting of the consumption of energy carriers in relation to their end-use. In such a situation, the use of TOE as the only reference value for quantification will be problematic.

Conclusion 5

The theoretical study of energy analysis needs more attention. As observed by Charlie Hall (Chapter 9), there are no funds for those who want to study energy analysis. Funds are invested in solving the problem – making agro-biofuels! – but not in learning how to study the problem. Politicians believe that throwing money at problems will eventually solve them. However, the special nature of energy proves this belief totally wrong. You cannot beat the laws of thermodynamics, no matter how much money you invest.

Conclusions about Science for Governance

Conclusion 1

For an effective re-discussion of alternative development strategies, it is crucial to admit, first, that there are limits to the expansion of human activity on a finite planet; and second, that no matter how smart the scientists making the predictions, our future is necessarily uncertain. As for the energetic transition of the third millennium, we do not need alternative energy sources to keep alive an obsolete pattern of economic growth. What we need is an alternative pattern of development that will make it possible to use alternative energy sources.

Conclusion 2

If the goal is to go for something completely different in regard to the future pattern of the production and consumption of goods and services, then it is unreasonable to think that the means to effect this revolutionary change will be provided by expert committees made up of representatives of economic sectors that have suffered from Concorde syndrome and are facing large sunk costs. Unfortunately, the decision-making procedure presently adopted in both the US and the EU is easily manipulated by powerful lobbies. This is unreasonable, and can no longer be tolerated by the civil society. It is essential to recognize the agro-biofuel folly for what it is, in order to avoid repeating the same mistakes in a future search for silver bullets. This is especially relevant at the moment; given the current population expansion and limits to natural resources, the search for silver bullets is likely to become increasingly frantic in the near future.

Conclusion 3

A similar reasoning applies to decision-making about priorities over research and development. We need people who are prepared to say 'the emperor has no clothes', people who are capable of thinking outside the box. Decisions cannot and should not be taken by bureaucrats and academicians at the top of their careers. Concorde's usefulness should be decided by the passengers who will use it, not the engineers who built it. When deciding about a desirable common future, decisions that should be based on what civil society wants, not on what the academic establishment thinks should be done.

Conclusion 4

In a context of rapid changes entailing the necessity of large-scale readjustments, it is crucial to develop as soon as possible new processes and mechanisms through which the whole society can be engaged in an informed deliberation about feasible and desirable futures. By 'informed deliberation', we mean an integrated analysis of the feasibility and desirability of the various options, evaluated using a variety of points of view, narratives and story-tellings that make sense to citizens. This is the only approach that can guarantee a robust final decision. This is especially important when considering that no matter how much we study and discuss, a large dose of uncertainty will always affect decisions referring to large-scale changes. When dealing with sustainability, it is crucial to avoid the generation of additional granfalloons at the global level.

Conclusion 5

Finally, the current situation of energy consumption is not as bad as feared by the majority of those belonging to the agro-biofuel granfalloon. For example, in spite of heavy use of fossil energy in agriculture, the fossil energy consumed in food production at the global level is only 5 per cent of total energy consumption (Smil, 2008; Giampietro, 2002). This is to say that modern societies consume so much energy in so many frivolous activities, that there is room – *ample* room – to significantly cut the actual level of energy consumption without going back to the Dark Ages. In the 1960s, many developed countries consumed approximately half the fossil energy they consume now, and there are no reasons to believe that the people living in these developed countries in the 1960s were less happy than the people living in these same countries today.

Actually, it is the ignorance of the cornucopians regarding the biophysical foundations of our economy that generates hysteria about a looming Dark Age. The cornucopians of the 1960s and 1970s have turned into the prophets of doom of the third millennium. So if it is time to change roles, biophysical analysts can and should play the role of the optimists. Accepting that it is impossible to maintain the existing pattern of economic expansion for ever (leading to perpetual growth and a Western lifestyle for everybody on the planet), it is reasonable to assume that it is possible to change to an alternative pattern of development that is feasible and desirable, and does not necessarily entail 'something worse'.

Reasons for Optimism: the Robinson Crusoe Effect

When dealing with human systems, institutional and financial mode-locking are the most persistent and important causes of failure to adapt. But depending on the circumstances, these lock-ins can be broken easily. As a matter of fact, when humans are forced to acknowledge that their behaviour must dramatically change, then their ability to adapt to existing external constraints is simply amazing.

We call this ability to make sudden and dramatic changes 'the Robinson Crusoe effect'. Everybody experiences dramatic changes in life. For example, during a steady-state period, daily life may be experienced as being totally constrained by a particular combination of work and social and family commitments. Then, one can be suddenly hit by a perturbation large enough to generate a collapse in the set of lock-ins determining this steady state: falling in love, getting divorced, losing a job, becoming physically impaired, or – in the case of the eponymous character – becoming shipwrecked on a remote island. When faced with a life-altering experience, the remarkable ability that humans have to adapt becomes evident. After a transitional period, necessary to tune internal characteristics to external boundary conditions, the human system (be it an individual, a small group or a country) will find new steady-state solutions made up of new routines, and new patterns that would have been totally unthinkable before the perturbation took place. A good example is daily life during times of war.

This almost magical ability to adapt to novelties, an incredible flexibility in dealing with disturbance, is probably what was missing in the analysis provided by the prophets of doom in the 1970s. We need to accept without emotional stress that our current civilization will 'collapse' in the near future; the word 'collapse', in this context, means that the existing human civilization will become something else, something we are unable to imagine right now. But the substantial change associated with the term 'collapse' should not necessarily be equated with a major and negative cataclysm for humankind. On the contrary, a dramatic change in the existing situation can also be perceived as an opportunity to make a series of positive changes. Indeed, a necessity for change should be considered as an opportunity to discuss what we would like to do in a different way.

The current predicament of humankind, confronting the issue of energy and sustainability, probably represents a critical situation capable of generating the Robinson Crusoe effect. When discussing how to deal with the issue of sustainability, humans can discuss and reflect on the meaning of their development. Indeed, the virtually unlimited human capability to adapt to changes is not about fixing the planet with silver bullets; it refers to the ability to adjust ourselves to new situations, to question assumptions about technical progress, and to remove the lock-ins hampering our ability to cope with change. In this type of discussion, scientific, political, ethical and socio-economic analyses should be combined in a way that has never been done before in the history of humankind. But this does not mean that this integrated discussion is an impossible task. In the last century, humankind has proved capable of accomplishing a lot of 'unthinkable' things.

Appendix 1

Basic Theoretical Concepts behind the Analysis of Societal Metabolism

Metabolic Systems

Systems able to use energy and material flows to maintain, reproduce and improve their own existing structures and functions. Classic examples are living systems, which must stabilize the expected pattern of metabolism associated with their identity in order to survive and thrive. Human beings, societies and ecosystems are all metabolic systems. For this reason, the availability of energy and resources, together with know-how and technology, are key factors determining the feasibility and desirability of patterns of societal metabolism.

The Metabolism of Human Societies is Analogous to the Metabolism of Ecosystems

'The metabolism of human society' is a notion used to characterize the conversions of energy and material flows occurring within a society that are necessary for its continued existence. It was presented in the mid-19th century by authors such as Marx, Liebig, Boussingault, Moleschott, Arrhenius and Podolinsky (for an overview, see Martinez-Alier, 1987; Fischer-Kowalski, 1997). Parallel to the development of the concept of societal metabolism in social sciences, several ecologists worked on the concept of ecosystem metabolism. In this field, several authors – for example E. P. Odum (1953), Margalef (1968), H. T. Odum (1971; 1983; 1996), Ulanowicz (1986) and Holling (1995) – have pointed out the existence of 'systemic properties' of ecosystems that are useful in studying and formalizing the effects of changes induced in these systems. The concept of the ecological metabolism of ecosystems entails the existence of expected patterns of ecosystem properties over:

- the relative size of functional compartments;
- benchmarks of energy and matter flows per unit of land;
- relative values of turnover times of components; and
- the structure of linkages in the networks and graphs used to represent ecosystems.

Metabolic Systems Belong to the Larger Class of Dissipative Systems

All natural systems of interest for sustainability (for example, complex bio-geochemical cycles, ecological systems and human systems when analysed at different levels of organization and scales above the molecular one) are 'dissipative systems' (Glansdorff and Prigogine, 1971; Nicolis and Prigogine, 1977; Prigogine and Stengers, 1981). That is, they are self-organizing, open systems, away from thermodynamic equilibrium. Because of this they are necessarily 'becoming systems' (Prigogine, 1978), which in turn implies that they:

- are operating in parallel on several hierarchical levels (where patterns of self-organization can be detected only by adopting different space-time windows of observation); and
- will change their identity in time.

Put another way, the very concept of self-organization in dissipative systems (the essence of living and evolving systems) is deeply linked to the idea of:

- *multiple hierarchical levels of organization on different space-time scales*, which entail the need of using multiple identities and non-equivalent descriptive domains for representing the process; and
- *evolution*, which implies that the set of relevant attributes (and then proxy variables) required to describe their behaviour is changing in time.

Metabolic Autopoietic Networks

Maturana and Varela proposed the term *autopoiesis* (Maturana and Varela, 1980; 1998) to indicate the special characteristics of self-organizing systems capable of making themselves. A metabolic autopoietic network can be defined as a network (e.g. a food web) of energy transformations where the components (nodes) of the network are capable of maintaining their own identity of metabolic elements (see next section). Using the terminology proposed by Georgescu-Roegen, metabolic autopoietic networks can be described using a fund-flow model in which the identity of the funds remains the same throughout the duration of the representation, and the identity of the flows is determined by the structure of the graph, the characteristics of the nodes, and the stability of favourable boundary conditions.

The Special Nature of Metabolic Autopoietic Networks

Metabolic autopoietic networks can maintain their own identity simultaneously across different scales and levels. For example, in ecology an ecosystem reproduces itself simultaneously at the level of: the whole ecosystem, the individual species, individual organisms making up a population, individual organs in the individuals, and individual cells. This peculiar ability is typical of living systems and has been described by Rosen using the label of M-R (maintenance and repair) networks (Rosen, 1958a; 1958b). This peculiarity implies that the identity of each of the lower-level elements affects the identity of the whole and vice versa, in a chicken–egg process. In technical jargon, we can say that these networks define themselves via impredicative loops (more on this concept in Rosen, 2000; Giampietro, 2003). H. T. Odum explains this concept by saying that ecological systems can stabilize their own metabolism through informed autocatalytic loops of energy.

The Mechanism Preserving the Identity of M-R Metabolic Networks

M-R systems have a peculiar ability: they produce elements which can be described simultaneously as functional and structural types. In technical jargon, these elements are called 'holons', from the term proposed by Koestler (1968; 1969; 1978). The perception and representation of holons pose a clear challenge to the assumptions typical of reductionist science (Giampietro, 2003; Giampietro et al, 2005). In fact, the concept of holon (an example is given in Figure 5.3) implies that one can have various structural types associated with the same functional type, or a given structural type used to perform different functions. Thanks to these peculiar characteristics, holons can continuously evolve (becoming something else) while remaining the same. That is, they can use a different structural type while keeping the same function (e.g., use a mechanical heart to replace a human heart) or develop a new function while still using an old structural type (e.g., use a functioning heart as a pump for irrigating a bonzai; Giampietro et al, 2006). As a result, it is not easy to have an analytical representation of the behaviour of holons, in the reductionist sense, since they admit a 'many-to-one' mapping. The openness of the relation between semantic (functional relation) and formal realization (structural organization) entails that the structures which realize those functions cannot be predicted from a formal/syntactic study of a system, e.g., being predicted by a mathematical model (on this point, see also Chapter 3 in Polimeni et al, 2008).

An analysis based only on formal and syntactic relations can only handle the representation of machines that do not evolve in time. The concept of holons operating in M-R systems addresses explicitly hidden assumptions of network analyses. That is, when analysing a network structure using a graph of the type proposed by H. T. Odum, the analyst is assuming as *available* and *reliable* a set of information referring to two different hierarchical levels:

- At the *level n−1* about individual components of the network: (1) the identity of the converters at each node is known (can be predicted). Such an identity, in turn, entails the definition of (2) what is the input (or the set of inputs) expected to arrive into that node; (3) what is the output (or the set of outputs) expected to exit from that node; (4) what is the conversion rate between inputs and outputs arriving and exiting each node at a given pace.
- at the *level n* of the entire network made up of nodes: (1) the identity of the graph representing the connections among the different nodes; and (2) the compatibility of the metabolism of the network (seen as a whole) with its associative context. This implies that those nodes of the network that are exchanging inputs and outputs with the outside environment must operate in an expected associative context. That is, the network as a whole must have an environment capable of supplying the required amount of specific input and enough sinking capacity to dispose of its specific outputs (favourable boundary conditions). The environment of a dissipative system must be an 'admissible environment', or the identity of the dissipative system cannot be preserved.
- At the *level n−1*, the *level n* and *level n+1:* the selected identities of converters at each node (level n−1), the selected identity of the graph representing the connections among nodes (level n), and the assumption of admissibility of boundary conditions over the associative context (level n+1) have to remain valid in the future.

After assuming the validity of this information, we can see from the graph in Figure A1.1 that the mutual information expressed by the network entails the definition of an 'essence' (a functional network niche for each element), which is independent from the structural characteristics of the element; e.g. a human heart versus a mechanical heart, as shown in Figure 5.3. Any metabolic element that manages to operate in a given specific position in the network with success can be given that functional identity, which is not dependent on the formal characterization – the blueprint – used to describe its structural organization.

We are getting into these theoretical aspects to show how this approach can check the feasibility and desirability of alternative energy sources. When looking for a substitute for

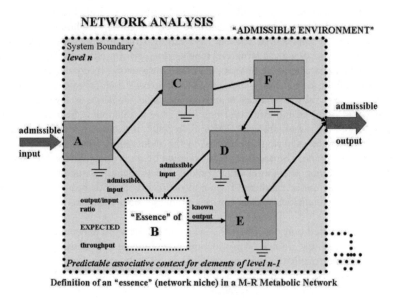

Figure A1.1 *The concept of 'network niche' defined by the mutual information carried out by the M-R network on the identity of an internal element (also known as the 'sudoku effect')*

fossil energy, we can use the metaphor of a heart transplant (illustrated in the text) to evaluate the performance of alternative energy sources.

The Provocative Ideas of Georgescu-Roegen: What Is Produced by the Economy?

When establishing a relationship between the ideas of Georgescu-Roegen and the analysis of metabolic networks proposed by H. T. Odum, we can make a very instructive observation about what is produced by the pattern of ecological metabolism. As observed earlier, ecological metabolism is associated with the ability to reproduce, in an integrated way, individual organisms, species and whole ecosystems (a metabolic network across scales). Therefore, according to the similarity of the typology of graphs illustrated in Figure 2.4 (for ecological systems) and Figure 4.9 (for socio-economic systems), we can say that what is produced by either the whole metabolic network or each one of the elements indicated in the graph (an ecosystem, a plant, a herbivore or a whole functional compartment) is **itself**! This activity of self-reproduction can be represented in the form of a network of interactions, since the ecological networks considered by H. T. Odum are food chains. This means that the various elements interact within the network by eating each other. So when the system represented on the graph can be considered as stable – in a steady state – the overall metabolic pattern indicates that the various elements are produced and consumed over the network at an expected profile of paces. Using economic jargon, we can say that in order to be a legitimate component of the network, an element must match the supply (by making itself) and demand for it (by being eaten by someone else). The integrated matching of a variety of demand and supply definitions translates into the expression of the whole network. Within a stabilized ecological metabolic network, what is transferred from one fund element to another fund element are flows of energy carriers, which are made up of edible biomass. These flows of energy carriers were, formerly, the body of individual members of the equivalence class (funds) making up the element producing the flow. That is, the food (energy input) transferred to an element that is alive (considered as a consumer in the graph) was, before the transfer, a part of another fund element (the supplier of this energy flow). A fund is no longer considered as alive when it becomes a flow of energy input for the next element. Herbivores eat plants, tigers eat herbivores, and when tigers die their bodies are eaten by other living systems in order to close the nutrient cycles.

When applying this narrative to the analysis of the metabolism of an economic process, we have to recall the economic mantra about cycles in the economic process: the cycle is closed in terms of monetary flows, which are produced and consumed by firms and final consumers. Money flows are circulating over the network in the form of wages, investments and the redistribution of added value. In order to get this circular relation of monetary flows, we do not need only goods and services generated by firms (the production side), but also the activity of buyers, investors and institutions controlling the spending of the money (the final consumption side). In human systems, something else is produced and consumed ('eaten', so to speak): *human activity*. A third cycle of production and consumption is linked to the generation of added value. Such a generation entails a coupled process of the *production and consumption of goods and services*: a successful meeting of a given demand and a given supply referring to a particular definition of a specific good and/or a specific service. This is what makes it possible to define a price for it in the first place. If we accept this narrative about monetary flows, and the supply and demand of human activity, then the distribution of monetary flows of added value reflects the need to reproduce the various funds operating in the system. The monetary flows are required to reproduce the households (wages and interest on savings), the exosomatic devices within firms (investments in technical capital), whole economic sectors (investments in financial capital), and the governance of the whole economy (taxes used for running the government). From this perspective, the analogy with the nested self-reproduction of ecological systems is verified. But if we accept the narrative that the performance of an economy should be assessed by its ability to reproduce an integrated set of processes of production and consumption that are required to reproduce the funds, then what are the ecological analogies for the physical flows of products (which sooner or later become obsolete and recycled) and services (which, after being enjoyed, disappear)? When looking at the overall pattern of the metabolism of the socio-economic system, we can see that products and goods are flowing through the funds, where they stay for a while before exiting. The goal of this flow is to help the maintenance and reproduction of these funds. Therefore, in semantic terms, the flow of goods and services should be equated to the flow of nutrients – the biomass of the fund elements – in ecological systems. Therefore, the metabolism of products within socio-economic funds can be associated with the mechanism generating excrements in the ecological process. A refrigerator in a household or a machine in a factory first appear in the system as nutrients in the human body. They remain there for while, performing their useful function. Finally, they are discarded by the fund when they are no longer useful. In this process of utilization, the original input of products and services becomes a flow of wastes.

Georgescu-Roegen's Paradox about Technological Progress

This alternative narrative can explain the famous paradox proposed by Georgescu-Roegen about the biophysical reading of the effect of technological progress: 'technical evolution leads to an increase in the rate at which a society "wastes resources"... the economic process actually is more efficient than automatic shuffling in producing higher entropy, i.e. waste' (Georgescu-Roegen, 1971). In other words, the more developed a society, the higher its rate of generation of wastes per capita. With this paradox, Georgescu-Roegen wants to focus on the need to define in a different way what is achieved by economic development: 'The true product of the economic process is an immaterial flux, the enjoyment of life, whose relation with the entropic transformation of energy-matter – material consumption – is still wrapped in mystery' (Georgescu-Roegen, 1971). That is, the ultimate goal of the economic process is not to produce as many goods and services as possible (increasing the levels of waste generated), but rather the enjoyment of life. When using a fund/flow representation of the pattern of metabolism, we can look for improvements in the enjoyment of life in terms of changes in the expected systemic relations over fund elements. In particular, we can say that economic development tends to increase as much as possible the fraction of fund and flows dedicated to the reproduction of humans and, in particular, to the expression of a diversity of activities: adaptability.

A fund-flow representation of the metabolic pattern of a developed society based on the grammar given in Figure 4.9 is provided in Figure A1.2. The fund is human activity on the left, and the flow is exosomatic energy on the right.

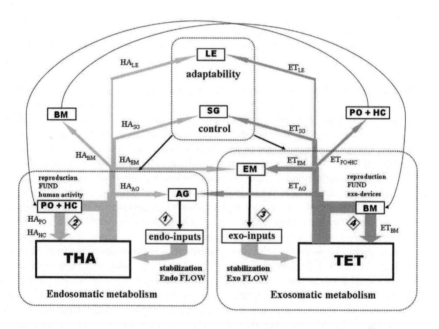

Figure A1.2 *An overview of the exosomatic metabolic pattern: human activity reproduces itself and controls the autocatalytic loop of exosomatic energy, in which exosomatic devices, under human control, also reproduce themselves and provide energy services to humans.*

In Figure A1. 2 we can observe the following.

- The vast majority of human activity goes into the reproduction of the fund human activity. About 65 per cent of human activity is not used for discretionary activities; it must be used in physiological overheads, such as sleeping, personal care, unpaid work for daily chores, and human activity expressed by the dependent population.
- Only a small amount of activity goes into paid work, about 10 per cent of THA. This activity has to be invested in codified roles (job descriptions), required for the functioning of the society. This is especially true for those jobs performed in the productive sector.
- The remaining 25 per cent can be considered as being available for discretionary activities, and goes in leisure time and cultural interactions.

The point made by Georgescu-Roegen is that it is essential to invest as much as possible of the available human activity into those categories (leisure time, education, cultural interactions) that make it possible to maintain a diversity of behaviours. As observed in Chapter 4, this goal also requires moving, within the limited amount of working time, as large a fraction of the workforce as possible into the service sector. This is at the basis of the increase in the level of BEP associated with development. As observed in Chapter 4, in order to obtain this result we need a tremendous boost of technological capital in the productive sector – an accumulation of the fund exosomatic devices in the productive sectors of the economy – which must be capable of delivering all the required products to the rest of society (energy carriers, food, exosomatic devices and consumable products), while absorbing only a limited amount of labour. In conclusion, the paradox indicated by Georgescu-Roegen refers to the consequences associated with a continuous increase of BEP. The boosting of the enjoyment of life in modern societies – a larger fraction of human activity and of exosomatic metabolism invested in leisure, education, healthcare and other social services – must be associated with large investments of exosomatic energy and a large accumulation of exosomatic devices in the society. It requires a tremendous increase in the power level in the PS sector. As a consequence of this fact, due to the well-known trade-off of power against output/input, it entails a major boost in the consumption of resources and the generation of waste.

Georgescu-Roegen's Warning about the Need to Address Properly the Issue of Scale

One cannot describe a complex metabolic network of energy and material transformations without first discussing the implications of scale. Things look different when observed at different scales; quantitative assessments of energy and matter are different when performed at different scales. This implies that the differences between funds, flows and stocks depend on the time duration chosen for the analysis. Moreover, because when dealing with the analysis of the interaction of human societies and ecosystems we are dealing with the activity of two 'becoming systems' (see Glossary), any quantitative representation is the result of a sort of epistemological cheating. We must assume that the set of elements of the economic process, which we are describing, will remain constant within the duration of the analysis (the fund). Because of this choice, other elements will be considered as disappearing over the chosen duration of the analysis (the flows). However, after having made this choice, it is easy to be fooled by the resulting representation. The picture that we obtain by adopting these assumptions is bound to quickly become obsolete. This occurs because in the representation (our written model), the funds remain the same, whereas in the real world, the funds do not remain the same. In the same way, the flows disappearing within the model do not disappear in biophysical processes. They end up somewhere else, generating an effect that is outside the chosen representation.

In conclusion, quantitative models, which are necessarily based on simplified assumptions, cannot be used to make predictions about future scenarios. This is especially true if we want to predict structural changes in the economy; we must assume that the identity of the funds will remain the same in order to be able to study how they will change. For this reason, Georgescu-Roegen was extremely sceptical about the usefulness of econometric models for making predictions about the future:

> *Even more crucial is the absence of any concern for whether the formula thus obtained will also fit other observations. It is this concern that is responsible for the success natural scientists have with their formulae. The fact that econometric models of the most refined and complex kind have generally failed to fit future data – which means that they failed to be predictive – finds a ready, yet self-defeating, excuse: history has changed the parameters. If history is so cunning, why persist in predicting it?* (Georgescu-Roegen, 1976, ppxxi–xxii)

Appendix 2

Examples of Grammars and Applications of Bio-economics

This appendix provides a quick overview of the Multi-Scale Integrated Analysis of Societal and Ecosystem Metabolism (MuSIASEM). A more elaborate presentation of this approach is available at www.societalmetabolism.org/musiasem.html.

Examples of Grammars to be Used to Implement Bio-economics to Perform an Integrated Analysis across Levels and Dimensions

A general overview of the semantic

In Chapter 4, we presented examples of applications of the MuSIASEM method, which is based on the rationale and seminal ideas proposed by Nicholas Georgescu-Roegen and Howard T. Odum.

A multi-level analysis of the metabolic pattern of societies can be obtained by opening the black box of the society and looking at those interactions of its parts which generate the 'emergent property' of the self-reproducing socio-economic system. In order to achieve this goal, we can define different compartments of the society using a multi-level matrix of the fund elements (defined in terms of hours of human activity), and then perform an accounting of the energy and material flows entering and leaving these compartments. Monetary flows can also be included in such an accounting. By using the approach proposed by H. T. Odum based on graphs, we can represent metabolic systems as being composed of fund elements – the nodes of the graph – and flows going through these nodes. Using the graph language proposed by H. T. Odum, the nodes of the graph can be represented either as *producers* (symbolized by a rectangle with a half circle on its side) or *consumers* (symbolized by a hexagon). This representation is consistent with the fund-flow model suggested by Georgescu-Roegen.

In this section we illustrate the basic semantic behind this integrated analysis using the following three figures.

The purpose of these three figures is to illustrate why it is impossible to have a simple representation of the integrated set of relations required for the stabilization of a socio-economic metabolism. In fact, the various fund elements considered in the chosen grammar – e.g. HH, EM, AG, BM, SG – are sometimes producers and sometimes consumers, depending on the flow considered. This situation is illustrated in Figure A2.1: depending on the priority given to the flow to be stabilized or the fund to be reproduced, these different compartments will produce either producers or consumers. This fact requires the use of the five different graphs shown in Figure A2.1.

1   Reproduction of humans (in the middle). This is the most important function of society, assuming that humans are those who decide on the narrative (if they are the

270 The Biofuel Delusion

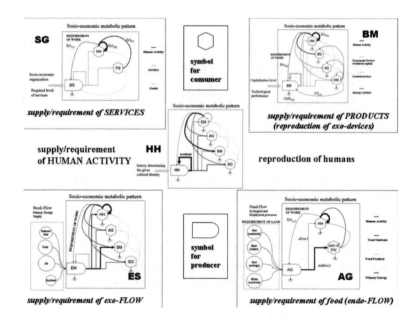

Figure A2.1 *Checking the dynamic budget for each of the five elements producing a critical supply*

rulers!). In this function, the household sector (HH) is the specialized sector that generates the required amount of human activity (FUND). In the short term, the HH produces (using adults) 'working time' for itself (unpaid work), and for the other compartments (paid work). In the long term, the HH produces babies (for replacement) and leisure time (for adaptability). All compartments are consumers of human activity.

2   Production of food (lower-right corner). This is a function required only for the stabilization of the metabolism of humans. The agricultural sector (AG) is in charge of producing the flow of food, which is either consumed directly by HH or processed by the paid work sector (in both PS or SG).

3   Production of exosomatic energy carriers (lower-left corner). This is a function required for stabilizing the metabolism of all sectors of the society. The energy sector (ES) is in charge of producing the flow of exosomatic energy carriers required by society. These flows are then consumed (in different quantities) by the various compartments of the society, including ES itself. All the compartments are consumers of exosomatic energy.

4   Production of products (upper-right corner). This is a function required for stabilizing the metabolism of all the sectors of society. The building and manufacturing sector (BM) is in charge of the production of the flow of exosomatic devices (either infrastructure (B) or products (M)) which are required by society. These products are then consumed (in different quantities) by the various compartments of the society, including BM itself. All the compartments are consumers of products and infrastructures.

5   Production of services (upper-left corner). This is a function required for stabilizing the metabolism of all the sectors of society. The services and government sector (SG) is in charge of producing the flow of services (either in the public or private sector) that are

required by society. These services are then consumed (in different quantities) by the various compartments of the society, including SG itself. All the compartments are consumers of services.

The extreme difficulty of relating the economic prices of goods and services to a biophysical reading of the economic process is explained by the scheme provided in Figure A2.2.

In a market economy, all the compartments of society have a continuous interaction with each other through an exchange of monetary flows. They are getting money when they act as producers of either a fund or a flow associated to a relevant function used by society, and they are paying money when they are consumers of either a fund or a flow associated to a relevant function provided by society. An additional complication is generated by the effect of savings/debts and trade. In general, we can say that prices and the profile of monetary flows reflect the priorities adopted by a given society – preferences included in the economic institutions – when generating the given societal metabolic pattern. But it is never possible to make accurate predictions about this relation across different space-time domains.

As a matter of fact, by combining the two schemes provided in Figures A2.1 and A2.2 it becomes possible to integrate the MuSIASEM analysis with other analytical methods developed in economics, such as input/output tables and social accounting matrices.

To complete the presentation of the rationale behind the MuSIASEM system of accounting, there is a last overview referring to an analysis of the interaction of society – seen as a black box – with its context. When moving from a multi-scale integrated analysis of the metabolic pattern seen from inside the black box (useful for studying the internal constraints) to a multi-scale integrated analysis of the metabolic pattern seen from outside the black box (useful for studying the external constraints), we have to adopt a different representation capable of addressing the interaction of the black box with its context: see Figure A2.3.

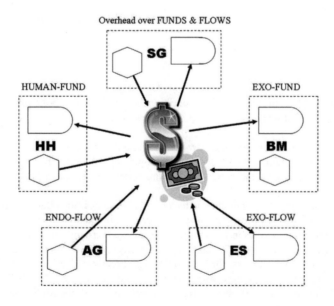

Figure A2.2 *Establishing a bridge between biophysical and economic analysis*

272 *The Biofuel Delusion*

This figure is taken from the technical report of a European project, DECOIN, aimed at the development of new indicators. This specific analysis entails the combination of the MuSIASEM approach with the SUMMA approach (a combination of EMergy Analysis, proposed by H. T. Odum, applied to the assessment of the interaction with ecological processes, and LCA/material and energy accounting for the assessment of the interaction with the other socio-economic systems).

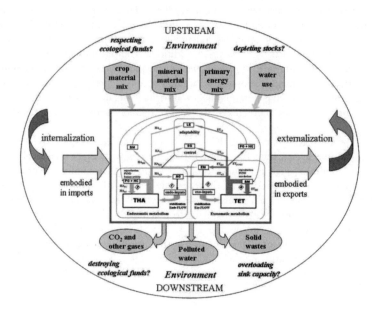

Figure A2.3 *Looking at the interaction of the black box with the context*

The definition of 'favourable boundary conditions' for a given metabolic pattern of matter and energy in a given society depends on two components:

1. The interface with the natural environment (natural resources + ecological processes), which is represented in the vertical flows of inputs into the black box from the context and the outputs into the context. This interaction can be described using EMergy accounting.
2. The interface with the metabolic pattern of other societies (the effect of trade or other transfers of funds and flows due to other forms of interactions), which is represented by two horizontal arrows indicating the flows of products and services. These horizontal flows can be associated with embodied amounts of labour, capital, and material and energy flows. The effect of this interaction can be assessed using material and flow accounting based on life-cycle assessment associated with external trade. This analysis has to be organized over a set of categories compatible with the MuSIASEM representation of metabolism.

From this figure, it becomes obvious that the advantages of externalization – e.g. reducing the consumption of energy and resources by moving the production of goods to developing countries – does not provide a solution that is effective at the global level. When considering the sustainability of the planet, it is not possible to increase the sustainability of societal metabolic patterns through externalization.

A general overview of the grammar used in MuSIASEM

As discussed in the main text, these expected relations between functions and structures have to be translated into a grammar in order to perform a quantitative representation and verify compatibility between the various approaches.

An overview of the division into different compartments of the fund human activity and an example of a metabolized flow (in this case exosomatic energy) is given in Figure A2.4.

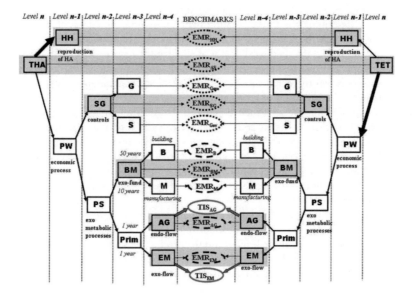

Figure A2.4 *The integrated dendrogram of fund and flow elements within the exosomatic metabolic pattern*

The various compartments into which the fund human activity is split at each level are illustrated on the left-hand side:

- THA at level n becomes $HA_{HH} + HA_{PW}$ at level n−1; $THA = HA_{HH} + HA_{PW}$.
- HA_{PW} at level n-1 becomes $HA_{PS} + HA_{SG}$ at level n−2; $HA_{PW} = HA_{PS} + HA_{SG}$.

The definition of the dendrogram has to guarantee that at each level, the sum of the size of the various fund elements remains the same. That is, the sum of hours making up the various fund elements at each level must be equal to THA.

274   *The Biofuel Delusion*

The various compartments into which the flow exosomatic energy is split at each level are illustrated on the right-hand side.

- TET at level n becomes $ET_{HH} + ET_{PW}$ at level n−1; TET = $ET_{HH} + ET_{PW}$.
- ET_{PW} at level n−1 becomes $ET_{PS} + ET_{SG}$ at level n−2; $ET_{PW} = ET_{PS} + ET_{SG}$.

The definition of the dendrogram has to guarantee that at each level, the sum of the size of the various flow elements remains the same. That is, the sum of MJ consumed by the various fund elements belonging to the same level must be equal to TET.

By implementing this method of accounting, it becomes possible to establish a set of expected relations across the various fund and flow elements sharing the same label. Several examples were discussed in Chapter 4 using four-angle-figures.

- Figure 4.3, describing the forced relation between monetary flows in Spain. The figure considers the consumption of two compartments: the whole country at the level n, and the paid work sector, a compartment defined at the level n−1.
- Figure 4.6a, describing the forced relation between exosomatic energy flows in Spain. The figure considers the consumption of two contiguous compartments: the whole country at the level n, and the paid work sector, defined at the level n−1. This is an example of mosaic effect through the dendrogram, which implies establishing a situation of congruence between the characteristics of funds (the exosomatic metabolic rate of the element) and the relative size of funds and flows.
- Figure 4.7, describing the congruence between the metabolic pace of letters sent and received in a hypothetical society and the mail sector's ability to deliver letters. The figure provides an example of impredicative loop analysis, which implies checking the congruence between the metabolic pace of a given flow metabolized at the level of the whole society – at level n – and the net supply provided by the compartment specialized in the production of that flow – the mail sector – at the level n−3. In this case, the two levels are non-contiguous.
- Figure 4.11, describing the forced relation between exosomatic energy flows in the typical pattern of developed countries, considers the consumption of two compartments: the whole country at the level n, and the productive sector at the level n−2. This figure is another example of mosaic effect, which makes it possible to study the forced congruence, over the dynamic equilibrium, of several relevant characteristics of the society.

In these examples we find two distinct conceptual analytical tools, which are illustrated in Figure A2.5. Both of them are useful in performing an analysis of the constraints affecting the feasibility of the pattern.

Mosaic effect across levels: moving the representation across contiguous levels

Using basic trigonometric rules, we can define a forced relation between the characteristics of the metabolism referring to the fund whole society (size and metabolic rate) at level n, and the characteristic of the fund paid work sector (size and metabolic rate) at the level n−1. This forced congruence is determined by two overheads: reduction over the fund THA (fund/fund – proportional to α), and reduction over the flow TET (flow/flow – proportional to γ) when moving to the lower level.

Using this tool, one can study the forced relation between FUND, FLOW and FLOW/FUND metabolic rate when considering the consumption of two compartments, which are considered within a given definition of the dendrogram. In this way, the forced

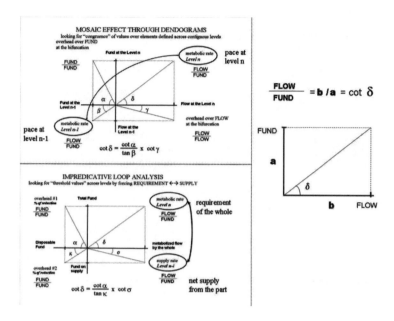

Figure A2.5 *Using trigonometry to study forced relations over funds and flows*

relations of congruence can be used to keep coherence in the integrated representation of the metabolic pattern across levels.

An example of this representation is given in Figure A2.6.

In this example, we can study the split of both the fund human activity and the flow exosomatic energy between the paid work sector (production side) and the household sector (consumption side). In Figure A2.6, we have the simultaneous characterization of the metabolism of commercial energy, characterized in terms of the fund human activity and the flow metabolized energy; and an intensive variable (the pace of metabolism) for the whole society (at the level n), the compartment of production (the paid work sector, at the level n−1 on the right), and the compartment of final consumption (the household sector, at the level n−1 on the left). The example refers to Spain, 1996. By looking at Figure A2.6 it is easy to see that changes taking place in one of the two compartments will be reflected in changes taking place in another compartment. If a society invests more time in producing, it has less time to be invested in consuming (the wisdom of Zipf, discussed several times in the main text).

The example given in Figure A2.7 shows an analysis of the forced relations of congruence on a vertical line within one of the branches of the dendrogram presented in Figure A2.4, i.e. average society (AS) → production (PW) → productive sectors (PS), across different hierarchical levels. We are dealing in this example only with the analysis of the productive side of the multi-level matrix.

By combining the horizontal link illustrated in Figure A2.6 with different vertical links illustrated in Figure A2.7, it becomes possible to have an integrated characterization of the patterns of energy metabolism of the various elements described at different levels, as illustrated in the overview given in Figure A2.1.

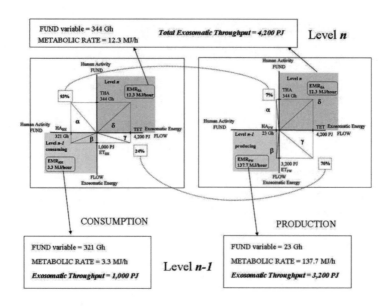

Figure A2.6 *Bridging levels in horizontal (same level)*

Impredicative loop analysis: looking for biophysical constraints on feasibility

After having established a mechanism of accounting that imposes the congruence of flows (money, energy, critical material flows) in relation to the characteristics of the fund elements (size and metabolic rate), it is possible to verify the viability of scenarios by checking the capability of those critical elements which are required to perform critical functions. That is, we can check whether the fund elements that are supposed to guarantee a critical function, when considered as producers, can actually deliver. This can be done by performing an impredicative loop analysis in relation to each of the five functions characterized using the five graphs presented in Figure A2.1. An example of this type of check has been given in Figure 4.7 when studying the factors determining the congruence between the total requirement of letters of a society and the net supply feasible according to the characteristics of the post office. Another example of application of ILA was provided in Chapter 7 – based on the quantification generated by ad hoc grammar – when checking the feasibility of corn-ethanol production within the US. In fact, an ILA is the implementation of a check based on the metaphor of the heart transplant; what is required by the US in terms of qualitative constraints on the supply of liquid fuels (the required net supply per hour and per hectare), and what can be supplied in relation to the definition of these qualitative constraints, by a supply of liquid fuel based on agro-biofuels (the current net supply per hour and per hectare)?

Characterizing the demand side of the metabolic pattern: starting from the household sector

The fund human activity can be characterized in different ways across different hierarchical

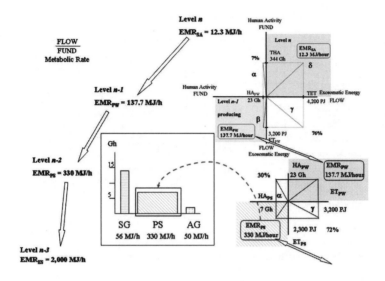

Figure A2.7 *Bridging levels in vertical (on the PW side)*

levels. At the level n−4, human activity refers to individual human beings. At the level n−3, human activity refers to household typologies. At the level n−2, human activity is divided into two categories within the household sector (urban household types and rural household types) and, on the productive side, two categories within the paid work sector (services and government and productive sector). At the level n−1, we saw that human activity is divided into two categories: household sector and paid work sector. In this section, we are dealing with the household sector, so we move within the side of the dendrogram representing the consumption side. In this side it is essential to establish a link across different representations referring to different levels and scales in order to address explicitly the effect of demographic changes on the feasibility of matching simultaneously the five dynamic budgets associated with the expression of the five key functions illustrated in Figure A2.1. But this time, rather than starting from the top (the whole society at the level n), we will start from the bottom of the multi-level matrix.

The level of individuals (level n−4)
The representation of the fund human activity at the hierarchical level of the individuals is illustrated in Figure A2.8 – the analysis refers to a hypothetical society of 100 people.

- What the system is at this level: a definition of the set of metabolic structural types/parts. At this level, we characterize the population as made up of a set of structural types of individuals determining a size (in hours). There are six categories in this example: three for age ($x_1 < 16$, $16 < x_2 < 65$, $x_3 > 65$) and two for gender: (y_1 = males, y_2 = females); this generates a 3×2 matrix, giving six types. The population – 100 people – has to be distributed over these six types.
- What the system does at this level: a definition of the categories of human activity (functional types). There are three semantic categories, which map onto quantitative assessments (hours). Each individual structural type allocates entirely its own endowment

278   *The Biofuel Delusion*

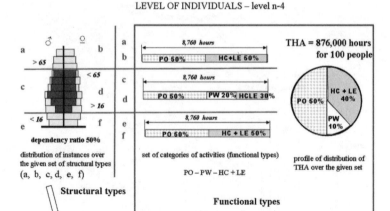

Figure A2.8 *A simple grammar establishing a relationship between demographic structure (structural types) and the profile of human activities (functional types)*

of human activity within a given set of categories (functional types) of human activity. In this example we consider three categories: HA_{PO} = physiological overhead (sleeping, eating and personal care); HA_{PW} = paid work (hours in paid economic activities; in this example this category applies only to adults); and HA_{HC+LE} = household chores + leisure and education (= disposable human activity not invested in paid work).

The very simple grammar in which each structural type maps onto a known pattern of allocation of human activities is shown in the middle of Figure A2.8. In this way, it becomes possible to map the overall profile of the population (over the given set of structural types) onto an overall profile of the distribution of human activity (a distribution of functional types at the level of the whole society) over a given set of semantic categories.

By adopting this representation, 100 people (instances of the chosen types) translate into an overall value for the fund total human activity of 876,000 hours/year – 100 (people) × 8760 (hours in a year) – which is then allocated over the given set of different categories. When applying this method to a modern society, as shown in Figure A2.9, it becomes crystal clear that different types of individuals (e.g. the elderly, children, adults) express different patterns of behaviour, and therefore have different preferences/requirements for goods and services. Put another way, socio-demographic variables matter when studying the patterns of production and consumption in a society.

Bridging the level of individuals (level n−4) with the level of households (level n−3)

At the next level we can characterize human activity within the household in relation to direct exosomatic energy consumption (what is consumed at the level of the household) and the indirect exosomatic energy consumption (what is consumed at the level of the economy

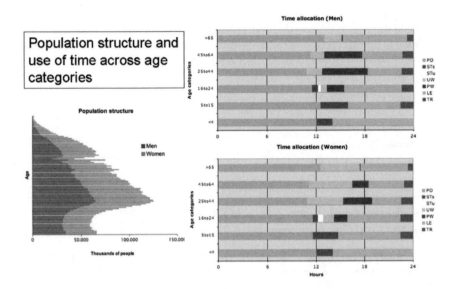

Figure A2.9 *The grammar presented in Figure A2.8 applied to the study of Catalonia*

to produce the goods and services consumed by the household). After choosing a selection of categories (grammar) to be used for such a representation, we can establish a relationship between the structure and functions of different typologies of households: Figure A2.10.

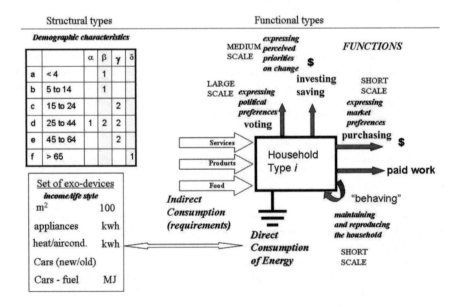

Figure A2.10 *A grammar to map the direct and indirect consumption of household typologies*

280   *The Biofuel Delusion*

At this point, a different profile of population distribution over different typologies of household will change the overall direct and indirect consumption of the household sector in relation to the other sectors of the economy. An example of this relation, performed using only four household types chosen for their relevance to the reality of Catalonia, is given in Figure A2.11.

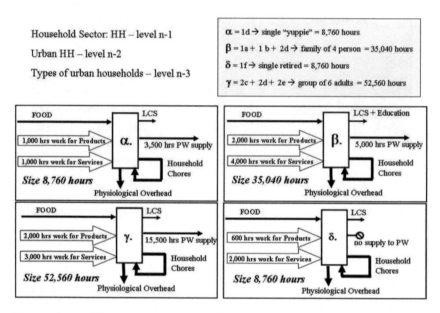

Figure A2.11 *The grammar presented in Figure A2.10 applied to Barcelona*

What is the system at this level; what is the definition of the set of exosomatic metabolic types and parts? At this level we characterize the population as made up of a set of household types mapping into hours of human activity. In this example, referring to people living in a modern society, the chosen categories for defining household types are associated with the size of the fund element, defined in hours of human activity.

- Type α: a single adult, size: 8760 hours/year (1×8760);
- Type β: a couple of adults and two children, size: 35,040 hours/year (4×8760);
- Type γ: a group of six adults, size: 52,560 hours/year (6×8760);
- Type δ: a single retired person, size: 8760 hours/year (1×8760).

Depending on the different size (in terms of hours of human activity) and the profile of distribution of actual instances of households over the chosen set, we can calculate the overall profile of distribution of hours of human activity over these four types.

What does the system do at this level; how do we define the categories of human activity in relation to flows of services, products and added value? The chosen semantic categories map onto quantitative assessments referring to the interaction that households have with the rest of society.

The scheme used in Figure A2.11 provides an example of a possible characterization of these four household types in their interaction with the production side. Within this framework each household type, first, requires a certain amount of services and products, which imply an investment of energy, material and hours in the paid work sector; and, second, supplies hours of paid work to the rest of society. There are household types – such as household type δ – that are net consumers of paid work hours; they require more hours in the paid work sector (embodied in the services and products that they consume) than they deliver. There are other household types – such as household type α (single adults) – that are net providers of paid work hours to the rest of society. As soon as we establish this relation, we can see that a given demographic change – e.g. a larger proportion of elderly people in the population – will translate into a change in the distribution of households over the set of types; e.g. in relation to the representation given in Figure A2.11 it will imply an amplification of type δ. This translates into a sharp increase in the BEP: an increase in the requirement of goods and services and a reduction of the supply of work. For this reason, there is the need to balance this trend with the amplification of 'hypercyclic' household types (those providing much more work than they require). This amplification can be obtained by the introduction of a new typology of household, such as a large household made up of several people, all adults: immigrants! Actually, this emergent type of household – type γ – has been found in our preliminary analysis of the province of Barcelona to be the fastest growing (together with elderly singles) in the last ten years. This is an example in which a phenomenon of emergence – the system becoming something else by adding new typologies – could have been anticipated by looking at the critical situation determined by actual demographic trends. The possibility of anticipating the appearance of new categories that are required for describing emerging typologies – something that is impossible for semantically closed systems of inference such as dynamic systems – is a major plus of the MuSIASEM methodology.

Obviously, it is possible to perform this analysis using a much larger set of household types than the one used in Figure A2.11. In this way, it becomes possible to study how the distribution of human activity and the relative requirement of products, services and metabolic rate are affected by demographic changes and by the movement of HA_{HH} across the various typologies of households (as illustrated in Figure A2.9). That is, a given amount of people – say, 10,000 – will imply a completely different pattern of requirement of services, products and energy if they are living in 1667 instances of the type γ household rather than they will if they are 2500 instances of type β household. Then again, they could all be 10,000 instances of type α (single workers) or type δ (single elderly).

Aggregating the characteristics of household types (level n−3) into an overall characterization of the household sector (level n−1)

In this quick overview, we skip the analysis dealing with the difference in the characteristics of the urban versus rural household types (at the level n−2), since in developed countries both the diversity and the relevance of rural household types in determining the overall pattern of metabolism of the household sector is negligible. The analysis of the difference between rural and urban households, however, would be crucial when dealing with China or India.

Depending on the characteristics of the chosen set of urban household types and the given profile of distribution of the population of households over the chosen set of types there, we can calculate the overall flow of products and services required by the HH sector (final consumption) – at the level n−1 – and the supply of hours of work to the PW sector.

The analysis of the characteristics of the fund human activity, when represented at the hierarchical level of the whole household sector – level n−1 (consumption side) – is illustrated in Figure A2.12. The overall supply of hours of work (in this example, the budget of hours refers to Catalonia) that the household sector can supply to the paid work sector results from the characteristics of the HH sector, when using the chosen representation of what HH is and what HH does. This BEP – the requirement of goods and services per hour of work supply given to the PW sector – must be congruent with the requirement of hours of work of the PW sector determined by the technical coefficients of the PW sector.

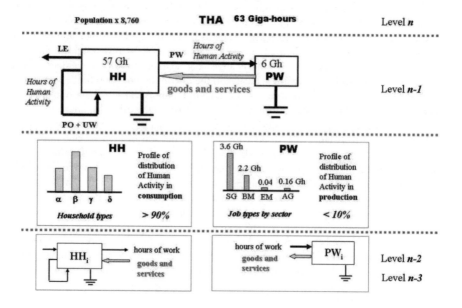

Figure A2.12 *The dynamic budget of hours of paid work (requirement vs supply) between the household sector and the paid work sector*

Checking the feasibility and desirability of scenarios using the sudoku effect

After characterizing the various interacting compartments on the production and consumption sides, we have all the elements we need:

- Feasibility in relation to internal constraints: the congruence between the characteristics of the various compartments' consumers (determining the aggregate requirement) in relation to the characteristics of the compartment that is 'the specialized producer' for each of the functions indicated in Figure A2.1.
- Feasibility in relation to external constraints: the congruence between the overall requirement of environmental services after considering the 'bonus' associated with the net effect of trade, in the interface black box/context; see Figure A2.3.
- The desirability of a given metabolic pattern: the set of performance attributes of the considered pattern of metabolism able to provide the required functions. That is, after having defined a given level of BEP, depending on the choice of dictionaries and grammars used for representing the household typologies and their relative metabolism, we can characterize the level of material standard of living for the various age classes and income classes included in the chosen set of household typologies. In this task, the MuSIASEM semantically open analytical toolkit can be interfaced with conventional economic analysis and with conventional ecological analysis to generate a richer set of

indicators. The resulting integrated set of indicators of performance can cover different dimensions of analysis, and different levels (household, village, province, country, the whole planet).

Examples of Applications of MuSIASEM

Using benchmarks of exosomatic metabolic patterns to compare countries

When using the MuSIASEM approach, it is possible to find useful benchmarks for characterizing typologies of countries (developed, developing, crowded, etc.); see Figure A2.13.

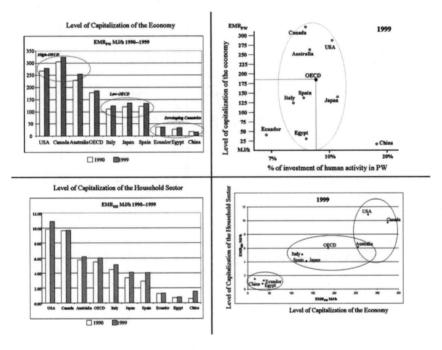

Figure A2.13 *Examples of different benchmark values found in different typologies of exosomatic metabolic patterns*

Looking at the data shown in Figure A2.13, for example, it is possible to note that the OECD country should not be considered as a homogeneous cluster, but rather a combination of two clusters: first, the US, Canada and Australia; and, second, other developed countries such as Japan and European countries, expressing lower metabolic rates at the country level (EMR_{AS}), at the household sector level (EMR_{HH}) and at the paid work sector level (EMR_{PW}). In this graph it is almost impossible to detect differences between developing countries within the same cluster.

After finding benchmarks referring to different countries, it becomes possible to study countries expressing different patterns of energetic metabolism, looking for the factors generating the differences. In Figure A2.14, the four-angle figure template is used to analyse the difference between the performance of OECD countries (considered as a whole) and China. In this way, one can focus on the various factors determining the dynamic budget: demographic characteristics, environmental conditions, quality and quantity of technology, cultural factors. That is, one can open the black box and look for differences between the various compartments that are operating inside the paid work sector.

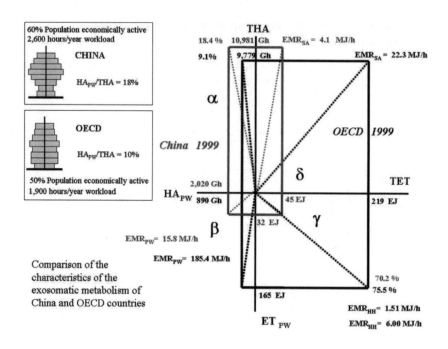

Figure A2.14 *Comparing the exosomatic metabolic pattern of OECD countries and China*

Figures in this section are from Ramos-Martin et al, 2007.

Characterizing the feasibility and desirability of relevant scenarios: linking world characteristics to country typologies characteristics

An analysis of CO_2 emissions at the world level should be based on an analysis of the emissions of different typologies of countries. By 'typology of countries' we mean the individuation of a cluster of countries expressing homogeneous features. Examples of this concept are shown in Figure 7.5a (the entire world) and Figure 7.5b. In this way, it becomes possible to apply the multi-scale approach to establishing a link between the characteristics of the world economy – the global amount of CO_2 emissions – and the characteristics of a set of typologies of countries found in the world economy. For example, the left part of Figure 7.5a shows the total energy consumption of the global economy. This amount of energy consumption can be placed in relation to the aggregate emissions of CO_2 at the world level. To do that, the left-hand graph in Figure 7.5a uses the area of a rectangle as an indicator of total energy consumption; the basis of the rectangle represents the THA (hours of human activity associated with the global population) and the height represents the exosomatic metabolic rate (the pace of energy consumption per hour, on average across the world economy). The same area can be obtained by summing together the areas of smaller rectangles reflecting the characteristics of different typologies of exosomatic country metabolisms: the US, Europe, China, etc. That is, by assuming the world level to be level x, we can express its characteristics using the characteristics of lower level elements – clusters of populations expressing similar patterns of energy metabolism – at the level x−1. This is illustrated in the graph shown in the left part of Figure 7.5b.

By using the MuSIASEM approach, it becomes possible to associate the various emissions reported on the left-hand side of Figure 7.5b (combining population size (a fund variable) and emissions per capita (a flow variable)) to the set of structural constraints determining the dynamic budget of exosomatic energy. This analysis is given in Figure A2.15.

Figures from Ramos-Martin et al, 2007.

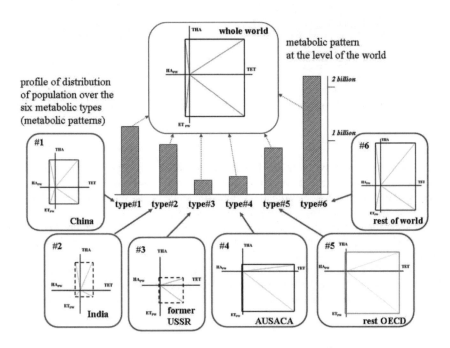

Figure A2.15 *Representing the exosomatic metabolic pattern of the world as determined by a set of metabolic patterns of country typologies*

Studying the trajectories of development of different countries

The case of Spain

In this example, from Ramos-Martin, 2001, one can see the crucial importance of the demographic variables. During its phase of rapid industrialization, Spain remained quite stable in terms of population size. This made it possible to invest a large fraction of the surplus in a

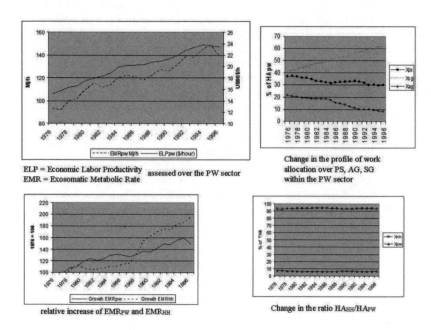

Figure A2.16 *The integrated analysis of Spain: historical series 1978–1998*

quick capitalization of the economy. This phenomenon can be viewed as a 'Franco effect'. During his dictatorship, the sector of final consumption in Spain was compressed, as can be seen by the extremely low exosomatic metabolism value of the household (not shown in Figure A2.16). In the 1970s, Spain had a value of EMR_{HH} – around 1MJ/hour – which was much lower than the relative benchmark in Europe (Ramos-Martin, 2001). For this reason, after reaching a level of capitalization of the economy similar to the rest of Europe, the people of Spain invested heavily in the capitalization of the household sector. In fact, EMR_{HH} kept growing at a rate much higher than the PW sector (this is illustrated in the lower-left graph of Figure A2.16). This explains the peculiar characteristics of Spain, which has a trajectory of development, in energy terms, that is different from the other European countries.

The case of Ecuador

In this example – in Falconi-Benitez, 2001 – it is possible to see that after the baby boom associated with the discovery of oil in the country, linked to a great deal of borrowing, Ecuador experienced a peak in the payment of debt services at the beginning of the 1980s. This prevented more being invested in the economy. Due to the combined effect of a wave of adults entering the workforce and the debt servicing, the economy experienced a significant reduction of capital per worker – EMR_{PW} – in the 1980s, which translated into a reduction in salaries. This generated a compression in final consumption (the reduction in the exosomatic metabolism of the household seen in the graph): EMR_{HH}. It should be noted that this reduction of consumption in the household sector (per hour of human activity) cannot be seen when looking at the aggregate consumption of energy in the final sector, ET_{HH}. Actually, in that period this value increased because of the increase in population. In

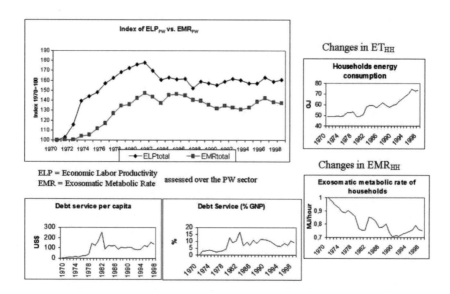

Figure A2.17 *The integrated analysis of Ecuador: historical series 1970–1998*

spite of the increase in ET_{HH} (aggregate consumption of energy of the HH sector), it was the sudden reduction in EMR_{HH} (the level of consumption per hour in the HH sector) that led to social unrest and riots. A sudden reduction in EMR_{HH} is a good predictor of a period of social instability for any country.

References

Chapter 1

Allison, K. and Kirchgaessner, S. (2008) 'Investors suffer as US ethanol boom dries up', *Financial Times*, 21 October, www.ft.com/cms/s/0/8531a8a2-9fa7-11dd-a3fa-000077b07658.html?nclick_check=1, accessed October 2008

Baker, M. M. and Craig, C. (2007) 'Bio-foolery is causing "food shocks"', *Executive Intelligence Review (EIR)*, vol 6, no 4, pp4–13, www.larouchepub.com/eiw/public/2007/2007_1-9/2007_1-9/2007-4/pdf/04-13_704_feat.pdf

BBC News (2008a) 'UN warns on biofuel crop reliance', *BBC News*, 18 July

BBC News (2008b) 'Call for delay to biofuels policy', *BBC News*, 24 March

Brody, S. (1945) *Bioenergetics and Growth*, Reinhold Publications Co, New York, NY, p1023

Burgonio, T. J. (2008) 'Rethink biofuel, says Nobel laureate', *Philippine Daily Inquirer*, 14 January, newsinfo.inquirer.net/inquirerheadlines/nation/view/20080114-112152/Rethink-biofuel-says-Nobel-laureate, accessed January 2009

Cronin, D. (2008), 'Europe: Warnings against biofuels get louder', *IPS News*, 6 May, www.ipsnews.net/news.asp?idnews=42247, accessed January 2009

Danielsen, F., Beukema, H., Burgess, N. D., Parish, F., Bruhl, C. A., Donald, P. F., Murdiyarso, D., Pahlan, B., Reijnders, L., Streubig, M. and Fitzherbert, E. B. (2008) 'Biofuel plantations on forested lands: Double jeopardy for biodiversity and climate', *Conservation Biology*, www.globalbioenergy.org/uploads/media/0811_Danielsen_et_al_-_Biofuel_plantations_on_forested_lands.pdf, accessed December 2008

EUobserver (2008) 'European Parliament capitulates on biofuel deal', *EUobserver*, 5 December, http://euobserver.com/9/27236

Fargione, J., Hill, J., Tilman, D., Polasky, S. and Hawthorne, P. (2008) 'Land clearing and the biofuel carbon debt', *Science*, vol 319, no 5867, pp1235–1238

Funtowicz, S. O. and Ravetz, J. R. (1990a) 'Post normal science: A new science for new times', *Scientific European*, no 266, October, pp20–22

Funtowicz, S. O. and Ravetz, J. R. (1990b) *Uncertainty and Quality in Science for Policy*, Kluwer, Dordrecht, The Netherlands

Giampietro, M., Allen, T. F. H. and Mayumi, K. (2005) 'The epistemological predicament associated with purposive quantitative analysis', *Ecological Complexity*, vol 3, no 4, pp307–327

Giampietro, M., Mayumi, K. and Munda, G. (2006) 'Integrated assessment and energy analysis: Quality assurance in multi-criteria analysis of sustainability', *Energy*, vol 31, no 1, pp59–86

The Guardian (2008a) 'Secret report: Biofuel caused food crisis', 3 July, www.guardian.co.uk/environment/2008/jul/03/biofuels.renewableenergy

The Guardian (2008b) 'MPs call for moratorium on biofuel targets', 21 January, www.guardian.co.uk/environment/2008/jan/21/biofuels.forests

Hecht, L. (2007) 'Smell of gigantic hoax in government ethanol promotion', *Executive Intelligence Review (EIR)*, vol 34, no 4, pp21–26

Koplow, D. and Steenblick, R. (2008) 'Subsidies to ethanol in the United States', in D. Pimentel (ed) *Biofuels, Solar and Wind as Renewable Energy Systems,* Springer, New York, NY, pp79–108

Krugman, P. (2008) 'Grains gone wild', *New York Times*, 7 April
Kutas, G., Lindberg C. and Steenblik, R. (2007) *Biofuels – At What Cost? Government Support for Ethanol and Biodiesel in the European Union*, Global Subsidies Initiative (GSI-IISD), Geneva, www.iisd.org/pdf/2007/ biofuels_subsidies_eu.pdf
Monbiot, G. (2005) 'Worse than fossil energy', *The Guardian*, 6 December
Patung (2007) 'Biofuel production', *Indonesia Matters* (Business & Economy), 22 May, www.indonesiamatters.com/1293/biofuel/
Pingali, P. (2007) 'Westernization of Asian diets and the transformation of food systems: Implications for research and policy', *Food Policy*, vol 32, no 2, pp281–298
Policy Exchange (2008) *The Root of the Matter – Carbon Sequestration in Forests and Peatlands*, Policy Exchange, London, www.policyexchange.org.uk/images/libimages/419.pdf
REAP Canada (2008) 'Analyzing biofuel options: Greenhouse gas mitigation efficiency and costs', Agriculture and Agri-Food Standing Committee Briefing, 26 February, Resource Efficient Agricultural Production (REAP) Canada, www.reap-canada.com/library/Bioenergy/AAFC_Standing_ Committee_Briefing.pdf
REN21 (2007) *Renewables 2007 Global Status Report*, Renewable Energy Policy Network for the 21st Century, Paris, France www.ren21.net/pdf/RE2007_Global_Status_Report.pdf, accessed November 2008
Rosenthal, E. (2008) 'Biofuels deemed a greenhouse threat', *New York Times*, 8 February, www.nytimes.com/2008/02/08/science/earth/08wbiofuels.html
Searchinger, T., Heimlich, R., Houghton, R. A., Dong, F., Elobeid, A., Fabiosa, J., Tokgoz, S., Hayes, D. and Yu, T.-H. (2008) 'Use of US croplands for biofuels increases greenhouse gases through emissions from land-use change', *Science*, vol 319, pp1238–1240
Spiegel Online International (2008) 'World Bank leak: Biofuels may be even worse than first thought', 7 April, www.spiegel.de/international/ world/0,1518,563927,00.html, accessed January 2009
United Press International (2007) 'Chavez warns of biofuel disaster', 9 August, www.upi.com/Energy_Resources/2007/08/09/Analysis_Chavez_warns_of_biofuel_disaster/UPI-50111186685937
World Bank (2008) 'Biofuels: The promise and the risks', *World Development Report, 2008*, econ.worldbank.org/WBSITE/EXTERNAL/EXTDEC/EXTRESEARCH/EXTWDRS/EXTWDR2008/0,,contentMDK:21501336~pagePK:64167689~piPK:64167673~theSitePK:2795143,00.html.

Chapter 2

Boserup, E. (1981) *Population and Technological Change*, University of Chicago Press, Chicago, IL
Cipolla, C. M. (1965) *Guns, Sails, and Empires: Technological Innovation and the Early Phases of European Expansion, 1400–1700*, Minerva Press, New York, NY
Cottrell, W. F. (1955) *Energy and Society: The Relation between Energy, Social Change, and Economic Development*, McGraw-Hill, New York, NY
Daly, H. E. (1994) 'Operationalizing sustainable development by investing in natural capital', in A. M. Jansson, M. Hammer, C. Folke and R. Costanza (eds) *Investing in Natural Capital*, Island Press, Washington, DC, pp22–37
Debeir, J.-C., Deleage, J.-P. and Hemery, D. (1991) *In the Servitude of Power: Energy and Civilization through the Ages*, Zed Books Ltd., Atlantic Highlands, NJ

Georgescu-Roegen, N. (1971) *The Entropy Law and the Economic Process*, Harvard University Press, Cambridge, MA

Georgescu-Roegen, N. (1975) 'Energy and economic myths', *Southern Economic Journal*, vol 41, pp347–381

Giampietro, M. and Pimentel, D. (1990) 'Assessment of the energetics of human labor', *Agriculture, Ecosystems and Environment*, vol 32, pp257–272

Giampietro, M., Bukkens, S. G. F. and Pimentel, D. (1997) 'The link between resources, technology and standard of living: Examples and applications' in L. Freese (ed) *Advances in Human Ecology, Volume 6*, JAI Press, Greenwich, CT, pp129–199

Hoogwijk, M., Faaij, A., van de Broek, R., Berndes, G., Gielen, D. and Turkenburg, W. (2003) 'Exploration of the ranges of the global potential of biomass for energy', *Biomass and Bioenergy*, vol 25, pp119–133

Lotka, A. J. (1956) *Elements of Mathematical Biology*, Dover Publications, New York, NY

Margalef, R. (1968) *Perspectives in Ecological Theory*, University of Chicago Press, Chicago, IL

Mayumi, K. (1991) 'Temporary emancipation from land: From the industrial revolution to the present time', *Ecological Economics*, vol 4, pp35–56

Mendelsshon, K. (1974) *The Riddle of the Pyramids*, Thames and Hudson, London

Odum, E. P. and Odum, H. T. (1971) *Fundamentals of Ecology*, 3rd edition, Saunders, Philadephia, PA

Odum, H. T. (1971) *Environment, Power and Society*, Wiley-Interscience, New York, NY

Odum, H. T. (1996) *Environmental Accounting: Emergy and Environmental Decision Making*, John Wiley, New York, NY

Smil, V. (2003) *Energy at the Crossroads: Global Perspectives and Uncertainties*, MIT Press, Cambridge, MA

Tainter, J. A. (1988) *The Collapse of Complex Societies*, Cambridge University Press, Cambridge

Ulanowicz, R. E. (1986) *Growth and Development: Ecosystem Phenomenology*, Springer-Verlag, New York, NY

Ulanowicz, R. E. (1995) 'Ecosystem integrity: A causal necessity', in L. Westra and J. Lemons (eds) *Perspectives on Ecological Integrity*, Kluwer Academic Publishers, Dordrecht, The Netherlands, pp77–87

UNFPA (United Nations Population Fund) (2008) *UNFPA Annual Report 2007*, UNFPA, New York, NY, www.unfpa.org/upload/lib_pub_file/777_filename_unfpa_ar_2007_web.pdf

Wackernagel, M. and Rees, W. (1996) *Our Ecological Footprint*, New Society Press, Vancouver, BC, Canada

White, L. A. (1943) 'Energy and evolution of culture', *American Anthropologist*, vol 14, pp335–356

White, L. A. (1959) *The Evolution of Culture: The Development of Civilization to the Fall of Rome*, McGraw-Hill, New York, NY

Chapter 3

Cleveland, C. J. (1992) 'Energy quality and energy surplus in the extraction of fossil fuels in the US', *Ecological Economics*, vol 6, pp139–162

Cleveland, C. J., Costanza, R., Hall, C. A. S. and Kaufmann, R. (1984) 'Energy and the US economy: A biophysical perspective', *Science*, vol 225, no 4665, pp890–897

Cleveland, C. J., Kaufmann, R. and Stern, S. I. (2000) 'Aggregation and the role of energy in the economy', *Ecological Economics*, vol 32, pp301–317

Cleveland, C. J., Hall, C. A. S. and Herendeen, R. A. (2006) 'Letters – energy returns on ethanol production', *Science*, vol 312, p1746

COL (Commodity on Line) (2008) *More Drilling Less Oil*, by C. Gilpin, 31 March, www.commodityonline.com/news/topstory/newsdetails.php?id=6839

Conrad, M. (1983) *Adaptability: The Significance of Variability from Molecule to Ecosystem*, Plenum, New York, NY

Cottrell, W. F. (1955) *Energy and Society: The Relation between Energy, Social Change, and Economic Development*, McGraw-Hill, New York, NY

Daly, H. E. (1994) 'Operationalizing sustainable development by investing in natural capital', in A. M. Jansson, M. Hammer, C. Folke and R. Costanza (eds) *Investing in Natural Capital*, Island Press, Washington, DC, pp22–37

Eigen, M. (1971) 'Self organization of matter and the evolution of biological macro molecules', *Naturwissenschaften*, vol 58, no 10, pp465–523

Farrell, A. E., Plevin, R. J., Turner, B. T., Jones, A. D., O'Hare, M. O. and Kammen, D. M. (2006) 'Ethanol can contribute to energy and environmental goals', *Science*, vol 311, pp506–508

Georgescu-Roegen, N. (1971) *The Entropy Law and the Economic Process*, Harvard University Press, Cambridge, MA

Georgescu-Roegen, N. (1975) 'Energy and economic myths', *Southern Economic Journal*, vol 41, pp347–381

Georgescu-Roegen, N. (1979) 'Energy analysis and economic valuation', *Southern Economic Journal*, vol 44, pp1023–1058

Gever, J., Kaufmann, R., Skole, D. and Vörösmarty, C. (1991) *Beyond Oil: The Threat to Food and Fuel in the Coming Decades*, University Press of Colorado, Boulder, CO

Giampietro, M. (2003) *Multi-Scale Integrated Analysis of Agro-ecosystems*, CRC Press, Boca-Raton, FL

Giampietro, M. (2008) 'Studying the "addiction to oil" of developed societies using the approach of Multi-Scale Integrated Analysis of Societal Metabolism (MSIASM)', in F. Barbir and S. Ulgiati (eds) *Sustainable Energy Production and Consumption and Environmental Costing*, Springer, Dordrecht, The Netherlands

Giampietro, M. and Mayumi, K. (2004) 'Complex systems and energy', in C. Cleveland (ed) *Encyclopedia of Energy*, vol 1, Elsevier, San Diego, CA, pp617–631

Giampietro, M., Allen, T. F. H. and Mayumi, K. (2006) 'The epistemological predicament associated with purposive quantitative analysis', *Ecological Complexity*, vol 3, no 4, pp307–327

Hagens, N. J. and Mulder, K. (2008) 'A framework for energy alternatives: Net energy, Liebig's Law and multi-criteria analysis', in D. Pimentel (ed) *Biofuels, Solar and Wind as Renewable Energy Systems*, Springer, Dordrecht, The Netherlands, pp295–319

Hagens, N. J., Costanza, R. and Mulder, K. (2006) 'Letters – energy returns on ethanol production', *Science*, vol 312, p1746

Hall, C. A. S., Cleveland, C. J. and Kaufman, R. (1986) *Energy and Resource Quality*, John Wiley & Sons, New York, NY

Holling, C. S. (1995) 'Biodiversity in the functioning of ecosystems: An ecological synthesis', in C. Perring (eds) *Biodiversity Loss*, Cambridge University Press, Cambridge, pp44–83

Kampis, G. (1991) *Self-Modifying Systems in Biology and Cognitive Science: A New Framework for Dynamics, Information, and Complexity*, Pergamon Press, Oxford

Kauffman, S. A. (1993) *The Origins of Order: Self Organization and Selection in Evolution*, Oxford University Press, New York, NY

Kaufmann, R. K. (2006) 'Letters – energy returns on ethanol production', *Science*, vol 312, p1747

Kleene, S. C. (1952) *Introduction to Metamathematics*, van Nostrand, New York, NY
Leontief, W. (1941) *Structure of the American Economy, 1919–1929*, Harvard University Press, Cambridge, MA
Mayumi, K. (2001) *The Origins of Ecological Economics: The Bioeconomics of Georgescu-Roegen*, Routledge, London
Mayumi, K. and Gowdy, J. (eds) (1999) *Bioeconomics and Sustainability: Essays in Honor of Nicholas Georgescu-Roegen*, Edward Elgar, Cheltenham, UK
Odum, H. T. (1971) *Environment, Power, and Society*, Wiley-Interscience, New York, NY
Odum, H. T. (1996) *Environmental Accounting: Emergy and Environmental Decision Making*, John Wiley, New York, NY
Patzek, T. W. (2006) 'Letters – energy returns on ethanol production', *Science* vol 312, p1747
Pimentel, D. and Pimentel, M. (1979) *Food Energy and Society*, Edward Arnold Ltd, London
Tainter, J. A. (1988) *The Collapse of Complex Societies*, Cambridge University Press, Cambridge
Ulanowicz, R. E. (1986) *Growth and Development: Ecosystem Phenomenology*, Springer-Verlag, New York, NY
White, L. A. (1943) 'Energy and evolution of culture', *American Anthropologist*, vol 14, pp335–356
White, L. A. (1959) *The Evolution of Culture: The Development of Civilization to the Fall of Rome*, McGraw-Hill, New York, NY

Chapter 4

Adriaanse, A., Bringezu, S., Hammond, Y., Moriguchi, Y., Rodenburg, E., Rogich, D. and Schütz, H. (1997) *Resource Flows: The Material Basis of Industrial Economies*, World Resources Institute, Washington, DC
Ayres, R. U. and Simonis, U. E. (1994) *Industrial Metabolism: Restructuring for Sustainable Development*, United Nations University Press, New York, NY
Cleveland, C. J., Costanza, R., Hall, C. A. S. and Kaufmann, R. (1984) 'Energy and the US economy: A biophysical perspective', *Science*, vol 225, no 4665, pp890–897
Cottrell, W. F. (1955) *Energy and Society: The Relation between Energy, Social Change, and Economic Development*, McGraw-Hill, New York, NY
Duchin, F. (1998) *Structural Economics: Measuring Change in Technology, Lifestyles, and the Environment*, Island Press, Washington, DC
Falconi-Benitez, F. (2001) 'Integrated assessment of the recent economic history of Ecuador', *Population and Environment*, vol 22, no 3, pp257–280
Fischer-Kowalski, M. (1997) 'Society's metabolism: On the childhood and adolescence of a rising conceptual star', in M. Redclift and G. Woodgate (eds) *The International Handbook of Environmental Sociology,* Edward Elgar, Cheltenham, UK, pp119–137
Fischer-Kowalski, M. (1998) 'Societal metabolism: The intellectual history of material flow analysis part I, 1860–1970', *Journal of Industrial Ecology,* vol 2, no 1, pp61–78
Georgescu-Roegen, N. (1971) *The Entropy Law and the Economic Process*, Harvard University Press, Cambridge, MA
Giampietro, M. (1997a) 'Socioeconomic pressure, demographic pressure, environmental loading and technological changes in agriculture', *Agriculture, Ecosystems and Environment*, vol 65, pp201–229
Giampietro, M. (1997b) 'Linking technology, natural resources, and the socioeconomic

structure of human society: A theoretical model', in L. Freese (ed) *Advances in Human Ecology*, vol 6, JAI Press, Greenwich, CT, pp75–130

Giampietro, M. (ed) (2000) 'Societal metabolism, part 1: Introduction of the analytical tool in theory, examples, and validation of basic assumptions', *Population and Environment*, vol 22 (special issue), pp97–254

Giampietro, M. (ed) (2001) 'Societal metabolism, part 2: Specific applications to case studies', *Population and Environment*, vol 22 (special issue), pp257–352

Giampietro, M. (2002) 'Energy use in agriculture', in *Encyclopedia of Life Sciences*, John Wiley & Sons, Hoboken, NJ, http://mrw.interscience.wiley.com/emrw/9780470015902/els/article/a0003294/current/pdf

Giampietro, M. (2008) 'Studying the "addiction to oil" of developed societies using the approach of Multi-Scale Integrated Analysis of Societal Metabolism (MSIASM)', in F. Barbir and S. Ulgiati (eds) *Sustainable Energy Production and Consumption and Environmental Costing*, NATO Science for Peace and Security Series: C – Environmental Security, Springer, The Netherlands

Giampietro, M. (2009) 'The future of agriculture: GMOs and the agonizing paradigm of industrial agriculture', in A. Guimaraes Pereira and S. Funtowicz (eds) *Science for Policy: Challenges and Opportunities*, Oxford University Press, New Delhi

Giampietro, M. and Mayumi, K. (2000a) 'Multiple-scale integrated assessment of societal metabolism: Introducing the approach', *Population and Environment*, vol 22, no 2, pp109–153

Giampietro, M. and Mayumi, K. (2000b) 'Multiple-scale integrated assessments of societal metabolism: Integrating biophysical and economic representations across scales', *Population and Environment*, vol 22, no 2, pp155–210

Giampietro, M. and Ramos-Martin, J. (2005) 'Multi-scale integrated analysis of sustainability: A methodological tool to improve the quality of the narratives', *International Journal of Global Environmental Issues*, vol 5, no 3/4, pp119–141

Giampietro, M., Bukkens, S. G. F. and Pimentel, D. (1997) 'The link between resources, technology and standard of living: Examples and applications', in L. Freese (ed) *Advances in Human Ecology*, vol 6, JAI Press, Greenwich, CT, pp129–199

Giampietro, M., Mayumi, K. and Bukkens, S. G. F. (2001) 'Multiple-scale integrated assessment of societal metabolism: An analytical tool to study development and sustainability', *Environment, Development and Sustainability*, vol 3, no 4, pp275–307

Giampietro, M., Mayumi, K. and Ramos-Martin, J. (2006a) 'Can biofuels replace fossil energy fuels? A multi-scale integrated analysis based on the concept of societal and ecosystem metabolism: Part 1', *International Journal of Transdisciplinary Research*, vol 1, no 1, pp51–87, www.ijtr.org/

Giampietro, M., Mayumi, K. and Ramos-Martin, J. (2006b) 'How serious is the addiction to oil of developed society? A multi-scale integrated analysis based on the concept of societal and ecosystem metabolism: Part 2', *International Journal of Transdisciplinary Research*, vol 2, no 1, pp42–92

Hall, C. A. S., Cleveland, C. J. and Kaufman, R. (1986) *Energy and Resource Quality*, John Wiley & Sons, New York, NY

Hopf, F. A. (1988) 'Entropy and evolution: Sorting through the confusion', in B. H. Weber, D. J. Depew and J. D. Smith (eds) *Entropy, Information, and Evolution*, Bradford Book/MIT Press, Cambridge, MA, pp263–274

Lotka, A. J. (1922) 'Contribution to the energetics of evolution', *Proceedings of the National Academy of Sciences*, vol 4, pp409–418

Lotka, A. J. (1956) *Elements of Mathematical Biology*, Dover Publications, New York, NY

Martinez-Alier, J. (1987) *Ecological Economics*, Blackwell, Oxford, UK

Matthews, E., Amann, C., Fischer-Kowalski, M., Bringezu, S., Hüttler, W., Kleijn, R.,

Moriguchi, Y., Ottke, C., Rodenburg, E., Rogich, D., Schandl, H., Schütz, H., van der Voet, E. and Weisz, H. (2000) *The Weight of Nations: Material Outflows from Industrial Economies*, World Resources Institute, Washington, DC

Odum, H. T. and Pinkerton, R. C. (1955) 'Time's speed regulator: The optimum efficiency for maximum power output in physical and biological systems', *American Scientist*, vol 43, pp321–343

Pastore, G., Giampietro, M. and Mayumi, K. (2000) 'Societal metabolism and multiple-scale integrated assessment: Empirical validation and examples of application', *Population and Environment*, vol 22, no 2, pp211–254

Ramos-Martin, J. (2001) 'Historic analysis of energy intensity of Spain: From a "conventional view" to an "integrated assessment"', *Population and Environment*, vol 22, no 3, pp281–313

Ramos-Martin, J. (ed) (2009) 'Anàlisi del metabolisme energètic de l'economia catalana (AMEEC)', *Informes del CADS*, vol 2, no 8, Consell Assessor per al Desenvolupament Sostenible, http://preproduccio.www 15.gencat.net/cads/AppPHP/images/stories/actualitat_cads/ameec_resum_executiu.pdf

Ramos-Martin, J. and Giampietro, M. (2005) 'Multi-scale integrated analysis of societal metabolism: Learning from trajectories of development and building robust scenarios', *International Journal of Global Environmental Issues* vol 5, no 3/4, pp225–263

Ramos-Martin J., Giampietro, M. and Mayumi, K. (2007) 'On China's exosomatic energy metabolism: An application of multi-scale integrated analysis of societal metabolism (MSIASM)', *Ecological Economics*, vol 63, no 1, pp174–191

Tainter, J. A. (1988) *The Collapse of Complex Societies*, Cambridge University Press, Cambridge, MA

Zipf, G. K. (1941) *National Unity and Disunity: The Nation as a Bio-Social Organism*, Principia Press, Bloomington, IN

Chapter 5

Batty, J. C., Hamad, S. N. and Keller, J. (1975) 'Energy inputs to irrigation', *Journal of the Irrigation and Drainage Division, ASCE*, vol 101(IR4), pp293–307

Box, G. E. P. (1979) 'Robustness is the strategy of scientific model building', in R. L. Launer and G. N. Wilkinson (eds) *Robustness in Statistics*, Academic Press, New York, NY, pp201–236

Carnot, S. (1824) *Thoughts on the Motive Power of Fire, and on Machines Suitable for Developing that Power* [original version: *Reflexions sur la Puissance Motrice du Feu sur les Machines Propres a Developper cette Puissance*], Bachelier Librairie, Paris, France

Cottrell, W. F. (1955) *Energy and Society: The Relation between Energy, Social Change, and Economic Development*, McGraw-Hill, New York, NY

Dekkers, W. A., Lange, J. M. and de Wit, C. T. (1978) 'Energy production and use in Dutch agriculture', *Netherlands Journal of Agricultural Sciences*, vol 22, pp107–118

EUROSTAT (1995) *Europe's Environment Statistical Compendium*, Office for Official Publications of the European Communities, Luxembourg

Fluck, R. C. (1981) 'Net energy sequestered in agricultural labor', *Transactions of the American Society of Agricultural Engineers*, vol 24, pp1449–1455

Fluck, R. C. (1992) 'Energy of human labor', in R. C. Fluck (ed) *Energy in Farm Production* (vol 6 of *Energy in World Agriculture*), Elsevier, Amsterdam, pp31–37

Fraser, R. and Kay, J. J. (2002) 'Exergy analysis of eco-systems: Establishing a role for the thermal remote sensing', in D. Quattrochi and J. Luvall (eds) *Thermal Remote Sensing in Land Surface Processes,* Taylor & Francis Publishers, London

Georgescu-Roegen, N. (1979) 'Energy analysis and economic valuation', *Southern Economic Journal*, vol 44, pp1023–1058

Giampietro, M. (2002) 'Energy use in agriculture', in *Encyclopedia of Life Sciences*, John Wiley & Sons, Hoboken, NJ, http://mrw.interscience.wiley.com/emrw/9780470015902/els/article/a0003294/current/pdf

Giampietro, M. (2003) *Multi-Scale Integrated Analysis of Ecosystems*, CRC Press, Boca-Raton, FL

Giampietro, M. (2006) 'Theoretical and practical considerations on the meaning and usefulness of traditional energy analysis', *Journal of Industrial Ecology*, vol 10, no 4, pp173–185

Giampietro, M. and Mayumi, K. (2004) 'Complex systems and energy', in C. Cleveland (ed) *Encyclopedia of Energy*, vol 1, Elsevier, San Diego, CA, pp617–631

Giampietro, M. and Pimentel, D. (1990) 'Assessment of the energetics of human labor', *Agriculture, Ecosystems and Environment*, vol 32, pp257–272

Giampietro, M. and Ulgiati, S. (2005) 'An integrated assessment of large-scale biofuel production', *Critical Reviews in Plant Sciences*, vol 24, pp1–20

Giampietro, M., Mayumi, K. and Munda, G. (2006a) 'Integrated assessment and energy analysis: Quality assurance in multi-criteria analysis of sustainability', *Energy*, vol 31, no 1, pp59–86

Giampietro, M., Allen, T. F. H. and Mayumi, K. (2006b) 'The epistemological predicament associated with purposive quantitative analysis', *Ecological Complexity*, vol 3, no 4, pp307–327

Hudson, J. C. (1975) 'Sugarcane: Its energy relationship with fossil fuel', *Span*, vol 18, pp12–14

Kay, J. (2000) 'Ecosystems as self-organizing holarchic open systems: Narratives and the second law of thermodynamics', in S. E. Jorgensen and F. Muller (eds) *Handbook of Ecosystems Theories and Management*, Lewis Publishers, London, pp135–160

Mayumi, K. and Giampietro, M. (2004) 'Entropy in ecological economics', in J. Proops and P. Safonov (eds) *Modeling in Ecological Economics*, Edward Elgar, Cheltenham, UK, pp80–101

Mayumi, K. and Giampietro, M. (2006) 'The epistemological challenge of self-modifying systems: Governance and sustainability in the post-normal science era', *Ecological Economics*, vol 57, pp382–399

Norman, M. J. T. (1978) 'Energy inputs and outputs of subsistence cropping systems in the tropics', *Agro-ecosystems*, vol 4, pp355–366

Odum, H. T. (1971) *Environment, Power, and Society*, Wiley-Interscience, New York, NY

Odum, H. T. (1996) *Environmental Accounting: Emergy and Environmental Decision Making*, John Wiley, New York, NY

Odum, H. T. and Pinkerton, R. C. (1955) 'Time's speed regulator: The optimum efficiency for maximum power output in physical and biological systems', *American Scientist*, vol 43, pp331–343

Revelle, R. (1976) 'Energy use in rural India', *Science*, vol 192, pp969–975

Schneider, E. D. and Kay, J. J. (1995) 'Order from disorder: The thermodynamics of complexity in biology', in M. P. Murphy and L. A. J. O'Neill (eds) *What Is Life: The Next Fifty Years – Reflections on the Future of Biology*, Cambridge University Press, MA, pp161–172

Stirling, A. (1997) 'Multicriteria mapping: Mitigating the problems of environmental evaluation?', in J. Foster (ed) *Valuating Nature: Economics, Ethics and Environment*, Routledge, London

Williams, D. W., McCarty, T. R., Gunkel, W. W., Price, D. R. and Jewell, W. J. (1975) 'Energy utilization on beef feed lots and dairy farms', in W. J. Jewell (ed) *Energy, Agriculture and Waste Management*, Ann Arbor Science Publishers, Ann Arbor, MI, pp29–47

Zemmelink, G. (1995) 'Allocation and utilization of resources at the farm level', in *A Research Approach to Livestock Production from a Systems Perspective*, Proceedings of the Symposium A Farewell to Prof Dick Zwart, Department of Animal Production Systems, Wageningen Agricultural University, The Netherlands, pp35–48

Chapter 6

Adams, R. N. (1988) *The Eighth Day: Social Evolution as the Self-Organization of Energy*, University of Texas Press, Austin, TX

Allen, T. F. H. and Hoekstra, T. W. (1992) *Toward a Unified Ecology*, Columbia University Press, New York, NY

Allen, T. F. H., Tainter, J. A. and Hoekstra, T. W. (2003) *Supply Side Sustainability*, Columbia University Press, New York, NY

Ayres, R. U. and Warr, B. (2005) 'Accounting for growth: The role of physical work', *Structural Change and Economic Dynamics*, vol 16, no 2, pp181–209

Ayres, R. U., Ayres, L. W. and Warr, B. (2003) 'Exergy, power and work in the US economy, 1900–1998', *Energy*, vol 28, no 3, pp219–273

Bak, P. (1996) *How Nature Works: The Science of Self-Organized Criticality*, Copernicus, New York, NY

Berndes, G., Hoogwijk, M. and van den Broek, R. (2003) 'The contribution of biomass in the future global energy supply: A review of 17 studies', *Biomass and Bioenergy*, vol 25, pp1–28

Cipolla, C. M. (1965) *Guns, Sails, and Empires: Technological Innovation and the Early Phases of European Expansion, 1400–1700*, Minerva Press, New York, NY

Cleveland, C. J. (1999) 'Biophysical economics: From physiocracy to ecological economics and industrial ecology', in J. Gowdy and K. Mayumi (eds) *Bioeconomics and Sustainability: Essays in Honor of Nicholas Gerogescu-Roegen*, Edward Elgar Publishing, Cheltenham, UK, pp125–154

Cleveland, C. J., Costanza, R., Hall, C. A. S. and Kaufmann, R. (1984) 'Energy and the US economy: A biophysical perspective', *Science*, vol 225, no 4665, pp890–897

Cleveland, C. J., Kaufmann, R. and Stern, S. I. (2000) 'Aggregation and the role of energy in the economy', *Ecological Economics*, vol 32, pp301–317

Cleveland, C. J., Hall, C. A. S. and Herendeen, R. A. (2006) 'Letters – energy returns on ethanol production', *Science*, vol 312, p1746

Costanza, R. (1980) 'Embodied energy and economic valuation', *Science*, vol 210, pp1219–1224

Costanza, R. (1981) 'Embodied energy, energy analysis and economics', in H. E. Daly and A. F. Umana (eds) *Energy, Economics and the Environment*, Westview, Boulder, CO, pp119–146

Cottrell, W. F. (1955) *Energy and Society: The Relation between Energy, Social Changes, and Economic Development*, McGraw-Hill, New York, NY

Dale, B. (2008a) 'Commentary', *Biofuels, Bioresources and Biorefining*, vol 2, pp495–496

Dale, B. (2008b) 'Thinking clearly about biofuels: Ending the irrelevant "net energy" debate and developing better performance metrics for alternative fuels', *Biofuels, Bioresources and Biorefining*, vol 1, no 1, pp14–17

Debeir, J.-C., Deleage, J.-P. and Hemery, D. (1991) *In the Servitude of Power: Energy and Civilization through the Ages*, Zed Books Ltd, Atlantic Highlands, NJ

Dogaru, V. (2008) 'Roegenian method and comprehension of the economic processes', in *Proceedings of the 5th International Seminar on Quality Management in Higher Education*, June 2008, Performantica Publishing Houses, Iasi, pp189–194

Farrell, A. E., Plevin, R. J., Turner, B. T., Jones, A. D., O'Hare, M. O. and Kammen, D. M. (2006) 'Ethanol can contribute to energy and environmental goals', *Science*, vol 311, pp506–508

Fischer-Kowalski, M. and Haberl, H. (2007) *Socioecological Transitions and Global Change*, Edward Elgar, London

Georgescu-Roegen, N. (1971) *The Entropy Law and the Economic Process*, Harvard University Press, Cambridge, MA

Georgescu-Roegen, N. (1975) 'Energy and economic myths', *Southern Economic Journal*, vol 41, pp347–381

Georgescu-Roegen, N. (1976) *Energy and the Economic Myth: Institutional and Analytical Economic Essays*, Pergamon Press, New York, NY

Gever, J., Kaufmann, R., Skole, D. and Vörösmarty, C. (1991) *Beyond Oil: The Threat to Food and Fuel in the Coming Decades*, University Press of Colorado, Boulder, CO

Giampietro, M. and Mayumi, K. (2004) 'Complex systems and energy', in C. Cleveland (ed) *Encyclopedia of Energy*, vol 1, Elsevier, San Diego, CA, pp617–631

Giampietro, M. and Ulgiati, S. (2005) 'An integrated assessment of large-scale biofuel production', *Critical Reviews in Plant Sciences*, vol 24, pp1–20

Giampietro, M., Bukkens, S. G. F. and Pimentel, D. (1997) 'The link between resources, technology and standard of living: Examples and applications', in L. Freese (ed) *Advances in Human Ecology*, vol 6, JAI Press, Greenwich, CT, pp129–199

Gilliland, M. W. (ed) (1978) *Energy Analysis: A New Policy Tool*, Westview Press, Boulder, CO

Hagens, N. J., Costanza, R. and Mulder, K. (2006) 'Letters – energy returns on ethanol production', *Science*, vol 312, p1746

Hall, C. A. S. (2000) (ed) *Quantifying Sustainable Development*, Academic Press, London

Hall, C. A. S., Cleveland, C. J. and Kaufmann, R. (1986) *Energy and Resource Quality: The Ecology of the Economic Process*, Wiley-Interscience, New York, NY

Hall, C. A. S., Balogh, S. and Murphy, D. J. R. (2009) 'What is the minimum EROI that a sustainable society must have?', *Energies*, vol 2, pp25–47, www.mdpi.com/journal/energies

Hannon, B. (1981) 'The energy cost of energy', in H. E. Daly and A. F. Umana (eds) *Energy, Economics and the Environment*, Westview, Boulder, CO

Hannon, B. (1982) 'Analysis of the energy costs of economic activities: 1963 to 2000', *Energy Systems and Policy*, vol 6, pp249–278

Herendeen, R. A. (1981) 'Energy intensities in economic and ecological systems', *Journal of Theoretical Biology*, vol 91, pp607–620

Herendeen, R. A. (1998) *Ecological Numeracy: Quantitative Analysis of Environmental Issues*, John Wiley & Sons, New York, NY

Herendeen, R. and Bullard, C. W. (1976) 'US energy balance of trade, 1963–1973', *Energy Systems and Policy*, vol 1, pp383–390

Hoogwijk, M., Faaij, A., van de Broek, R., Berndes, G., Gielen, D. and Turkenburg, W. (2003) 'Exploration of the ranges of the global potential of biomass for energy', *Biomass and Bioenergy*, vol 25, pp119–133

IFIAS (International Federation of Institutes for Advanced Study) (1974) *Energy Analysis*, International Federation of Institutes for Advanced Study, Workshop on Methodology and Conventions, Report no 6, IFIAS, Stockholm, p89

Jevons, W. S. (1865) *The Coal Question* (reprint of the third 1906 edition), Augustus M. Kelley, New York, NY

Kaufmann, R. K. (1992) 'A biophysical analysis of the energy/real GDP ratio: Implications for substitution and technical change', *Ecological Economics*, vol 6, no 1, pp35–56

Kaufmann, R. K. (2006) 'Letters – energy returns on ethanol production', *Science*, vol 312, p1747

Kay, J. (2000) 'Ecosystems as self-organizing holarchic open systems: Narratives and the second law of thermodynamics', in S.E. Jorgensen and F. Muller (eds) *Handbook of Ecosystems Theories and Management*, Lewis Publishers, London, pp135–160

Kay, J. and Schneider, E. (1992) 'Thermodynamics and measures of ecosystem integrity', in D. McKenzie, D. Hyatt and J. McDonald (eds) *Ecological Indicators*, Elsevier, New York, NY, pp159–181

Leach, G. (1976) *Energy and Food Production*, IPC Science and Technology Press Limited, Surrey, UK

Lotka, A. (1922) 'Contribution to the energetics of evolution', *Proceedings of the National Academy of Sciences*, vol 8, pp147–151

Lotka, A. (1956) *Elements of Mathematical Biology*, Dover, New York, NY

Martinez-Alier, J. (1987) *Ecological Economics*, Blackwell, Oxford/New York

Mayumi, K. (2001) *The Origins of Ecological Economics: The Bioeconomics of Georgescu-Roegen*, Routledge, London

Mayumi, K. and Giampietro, M. (2004) 'Entropy in ecological economics', in J. Proops and P. Safonov (eds) *Modeling in Ecological Economics*, Edward Elgar, Cheltenham, UK, pp80–101

Mayumi, K. and Gowdy, J. (eds) (1999) *Bioeconomics and Sustainability: Essays in Honor of Nicholas Georgescu-Roegen*, Edward Elgar, Cheltenham, UK

Odum, H. T. (1971) *Environment, Power and Society*, Wiley-Interscience, New York, NY

Odum, H. T. (1983) *Systems Ecology*, John Wiley, New York, NY

Odum, H. T. (1996) *Environmental Accounting: Emergy and Environmental Decision Making*, John Wiley, New York, NY

Odum, H. T. and Pinkerton, R. C. (1955) 'Time's speed regulator: The optimum efficiency for maximum power output in physical and biological systems', *American Scientist*, vol 43, pp321–343

Ostwald, W. (1907) 'The modern theory of energetics', *The Monist*, vol 17, p511

Ostwald, W. (1911) 'Efficiency', *The Independent*, vol 71, pp867–871

Patzek, T. (2004) 'Thermodynamics of the corn-ethanol biofuel cycle', *Critical Reviews in Plant Sciences*, vol 23, no 6, pp519–567

Patzek, T. W. (2006) 'Letters – energy returns on ethanol production', *Science*, vol 312, p1747

Pimentel, D. and Pimentel, M. (1979) *Food, Energy and Society*, Edward Arnold, London

Pimentel, D., Patzek, T. and Cecil, G. (2007) 'Ethanol production: Energy, economic, and environmental losses', *Reviews of Environmental Contamination and Toxicology*, vol 189, pp25–41

Podolinsky, S. (1883) *Menschliche arbeit und einheit der kraft: Die Neue Zeit*, IHW Dietz, Stuttgart, p413

Polimeni, J., Mayumi, K., Giampietro, M. and Alcott, B. (2008) *Jevons' Paradox: The Myth of Resource Efficiency Improvements*, Earthscan Research Edition, London

Rappaport, R. A. (1968) *Pigs of the Ancestors: Ritual in the Ecology of a New Guinea People*, Yale University Press, New Haven, CT

Rappaport, R. A. (1971) 'The flow of energy in an agricultural society', *Scientific American*, vol 224, pp117–133

Schneider, E. D. and Kay, J. J. (1994) 'Life as a manifestation of the second law of thermodynamics', *Mathematical and Computer Modelling*, vol 19, pp25–48

Shapouri, H., Duffield, J. and Wang, M. (2002) *The Energy Balance of Corn-Ethanol: An Update*, Report 813, USDA Office of Energy Policy and New Uses, Agricultural Economics, Washington, DC
Slesser, M. (1978) *Energy in the Economy*, MacMillan, London
Slesser, M. and King, J. (2003) *Not by Money Alone: Economics as Nature Intended*, Jon Carpenter Publishing, Oxford, UK
Smil, V. (1983) *Biomass Energies*, Plenum Press, New York, NY
Smil, V. (1988) *Energy in China's Modernization*, M. E. Sharpe, Armonk, NY
Smil, V. (1991) *General Energetics*, John Wiley, New York, NY
Smil, V. (2001) *Enriching the Earth*, MIT Press, Cambridge, MA
Smil, V. (2003) *Energy at the Crossroads: Global Perspectives and Uncertainties*, MIT Press, Cambridge, MA
Smil, V. (2008) *Energy in Nature and Society: General Energetics of Complex Systems*, MIT Press, Cambridge, MA
Soddy, F. (1926) *Wealth, Virtual Wealth and Debt*, George Allen & Unwin, London
Steinhart, J. S. and Steinhart, C. E. (1974) 'Energy use in US food system', *Science*, vol 184, pp307–316
Stout, B. A. (1991) *Handbook of Energy for World Agriculture*, Elsevier, New York, NY
Stout, B. A. (1992) (ed) *Energy in World Agriculture* (6 volumes): Singh, R. P. (ed) *Energy in Food Processing*, vol 1; Hesel, Z. R. (ed) *Energy in Plant Nutrition and Pest Control*, vol 2; McFate, K. L. (ed) *Electrical Energy in Agriculture*, vol 3; Parker B. F. (ed) *Solar Energy in Agriculture*, vol 4; Peart, R. M. and Brooks, R. C. (eds) *Analysis of Agricultural Systems*, vol 5; Fluck, R. C. (ed) *Energy in Farm Production*, vol 6, Elsevier, Amsterdam
Tsuchida, A. and Murota, T. (1987) 'Fundamentals in the entropy theory of ecocycle and human economy', in G. Pillet and T. Murota (eds) *Environmental Economics: The Analysis of a Major Interface*, Roland Leimgruber, Geneva, pp11–35
Ulgiati, S. (2001) 'A comprehensive energy and economic assessment of biofuels: When "green" is not enough', *Critical Reviews in Plant Sciences*, vol 20, pp71–106
USDA (US Department of Agriculture) (2006) *Agricultural Statistics*, http://www.usda.gov/nass/pubs/agr05/acro05.htm
Vernadsky, V. (1926) *The Biosphere* (first published in Russian in 1926) (English version: Synergetic Press, Oracle, AZ, 1986)
Watt, K. E. F. (1989) 'Evidence of the role of energy resources in producing long waves in the US economy', *Ecological Economics*, vol 1, pp181–195
Watt, K. E. F. (1991) *Taming the Future*, Contextured Web Press, Davis, CA, p163
White, L. A. (1943) 'Energy and evolution of culture', *American Anthropologist*, vol 14, pp335–356
White, L. A. (1949) *The Science of Culture: A Study of Man and Civilization*, Farrar, Straus and Company, New York, NY
Zipf, G. F. (1941) *National Unity and Disunity: The Nation as a Bio-social Organism*, Principia Press, Bloomington, IN
Zipf, G. F. (1949) *Human Behavior and the Principle of Least Effort*, Addison-Wesley Press, Cambridge, MA

Chapter 7

Boddey, R. M., Soares, L. H. de B., Alves, B. J. R. and Urquiaga, S. (2008) 'Bio-ethanol production in Brazil', in D. Pimentel (ed) *Biofuels, Solar and Wind as Renewable Energy Systems*, Springer, The Netherlands, pp321–356

De Carvalho Macedo, I. (2005) *Sugar Cane's Energy: Twelve Studies on Brazilian Sugar Cane Agribusiness and Its Sustainability*, UNICA (Sugar Cane Agroindustry Union), Berlendis Editores Ltda, São Paulo, Brazil

Farrell, A. E., Plevin, R. J., Turner, B. T., Jones, A. D., O'Hare, M. O. and Kammen, D. M. (2006) 'Ethanol can contribute to energy and environmental goals', *Science*, vol 311, pp506–508

Giampietro, M. and Mayumi, K. (2000a) 'Multiple-scale integrated assessment of societal metabolism: Introducing the approach', *Population and Environment*, vol 22, no 2, pp109–153

Giampietro, M. and Mayumi, K. (2000b) 'Multiple-scales integrated assessments of societal metabolism: Integrating biophysical and economic representations across scales', *Population and Environment*, vol 22, no 2, pp155–210

Giampietro, M. and Ulgiati, S. (2005) 'An integrated assessment of large-scale biofuel production', *Critical Reviews in Plant Sciences*, vol 24, pp1–20

Giampietro, M., Pastore, G. and Ulgiati, S. (1998) 'Italian agriculture and concepts of sustainability', in E. Ortega and P. Safonov (eds) *Introduction to Ecological Planning Using Emergy Analysis with Brazilian Case Studies*, LEIA-FEA Unicamp, Campinas, Brazil

Patzek, T. W. and Pimentel, D. (2005) 'Thermodynamics of energy production from biomass', *Critical Reviews in Plant Sciences*, vol 24, no 5–6, pp327–364

Pimentel, D., Patzek, T. and Cecil, G. (2007) 'Ethanol production: Energy, economic, and environmental losses', *Reviews of Environmental Contamination and Toxicology*, vol 189, pp25–41

Ramos-Martin, J., Giampietro, M. and Mayumi, K. (2007) 'On China's exosomatic energy metabolism: An application of multi-scale integrated analysis of societal metabolism (MSIASM)', *Ecological Economics*, vol 63, no 1, pp174–191

Smil, V. (2003) *Energy at the Crossroads: Global Perspectives and Uncertainties*, MIT Press, Cambridge, MA

Smil, V. (2008) *Energy in Nature and Society: General Energetics of Complex Systems*, MIT Press, Cambridge, MA

USDA (2007) 'NP 307 Action Plan 1999-2007', United States Department of Agriculture web page, www.ars.usda.gov/research/programs/programs.htm?np_code=307&docid=281&page=1

USDE (2008) 'About the program', United States Department of Energy Biomass Program web page, www1.eere.energy.gov/biomass/economic_growth.html

World Bank (2008) 'Biofuels: The promise and the risks', in *World Development Report, 2008*, http://siteresources.worldbank.org/INTWDR2008/Resources/2795087-1191440805557/4249101-1191956789635/Biofuels-(large).gif

Chapter 8

BBC News (2008) 'Peru throws out Amazon land laws', http://news.bbc.co.uk/2/hi/americas/7578040.stm

Cochrane, W. (1958) *Farm Prices: Myth and Reality*, University of Minnesota Press, Minneapolis, MN

Giampietro, M. (1994) 'Sustainability and technological development in agriculture: A critical appraisal of genetic engineering', *BioScience*, vol 44, no 10, pp677–689

Giampietro, M. (1997a) 'Socioeconomic pressure, demographic pressure, environmental loading and technological changes in agriculture', *Agriculture, Ecosystems and Environment*, vol 65, no 3, pp201–229

Giampietro, M. (1997b) 'Socioeconomic constraints to farming with biodiversity', *Agriculture, Ecosystems and Environment*, vol 62, pp145–167

Giampietro, M. (2002a) 'Energy use in agriculture', in *Encyclopedia of Life Sciences*, John Wiley & Sons, Hoboken, NJ, http://mrw.interscience.wiley.com/emrw/9780470015902/els/article/a0003294/current/pdf

Giampietro, M. (2002b) 'The precautionary principle and ecological hazards of genetically modified organisms', Ambio, vol 31, no 6, pp466–470

Giampietro, M. (2003) *Multi-Scale Integrated Analysis of Ecosystems*, CRC Press, Boca-Raton, FL

The Guardian (2007) 'Massacres and paramilitary land seizures behind the biofuel revolution', *The Guardian*, 5 June

Hayami, Y. and Ruttan, V. (1985) *Agricultural Development: An International Perspective*, 2nd edition, Johns Hopkins University Press, Baltimore, MD

Millennium Ecosystem Assessment (2005) *Synthesis Report: Ecosystems and Human Well-Being*, www.maweb.org/documents/document. 356.aspx.pdf

Motlagh, J. (2008) 'Suicides plague Indian farmers struggling to adapt to changing economics', *World Politics Review*, 11 March, www.worldpoliticsreview.com/article.aspx?id=1756

Oxfam (2007) *Biofuelling Poverty: EU Plans Could Be Disastrous for Poor People, Warns Oxfam*, 29 October 2007, www.oxfam.org/es/node/217

Provincial Plan for Flevoland (2006) *Summary of the Provincial Plan for Flevoland 2006–2015*, http://provincie.flevoland.nl/omgevingsplan/documenten/Samenv-OGP%20Eng.pdf

Ramos-Martin, J., Giampietro, M. and Mayumi, K. (2007) 'On China's exosomatic energy metabolism: An application of multi-scale integrated analysis of societal metabolism (MSIASM)', *Ecological Economics*, vol 63, no 1, pp174–191

Tainter, J. A. (1988) *The Collapse of Complex Societies*, Cambridge University Press, Cambridge, MA

Chapter 9

Allison, K. and Kirchgaessner, S. (2008) 'Investors suffer as US ethanol boom dries up', FT.com (*Financial Times* online), www.ft.com/cms/s/0/8531a8a2-9fa7-11dd-a3fa-000077b07658.html?nclick_check=1

BP (2008) *Statistical Review of World Energy 2007*, www.bp.com/productlanding.do?categoryId=6929&contentId=7044622

CEO (Corporate Europe Observatory) (2007) *Briefing Paper*, CEO, June 2007, Brussels, www.corporateeurope.org/agrofuelfolly.html

Dale, B. (2008) 'Thinking clearly about biofuels: Ending the irrelevant "net energy" debate and developing better performance metrics for alternative fuels', *Biofuels, Bioresources and Biorefining*, vol 1, no 1, pp14–17

Ehrlich, P. R. (1968) *The Population Bomb*, Ballantine Books, New York

ESA (Ecological Society of America) (2008) *Policy Statements: Biofuel Sustainability*, http://www.esa.org/pao/policyStatements/Statements/biofuel.php

EU (European Union) (2006) *Annex to the Communication from the Commission An EU Strategy for Biofuels Impact Assessment*, http://ec.europa.eu/agriculture/biomass/biofuel/sec2006_142_en.pdf

Farrell, A. E., Plevin, R. J., Turner, B. T., Jones, A. D., O'Hare, M. O. and Kammen, D. M. (2006) 'Ethanol can contribute to energy and environmental goals', *Science*, vol 311, pp506–508

Funtowicz, S. O. and Ravetz, J. R. (1990) *Uncertainty and Quality in Science for Policy*, Kluwer, Dordrecht, The Netherlands

Funtowicz, S. O. and Ravetz, J. R. (1993) 'Science for the post-normal age', *Futures*, vol 25, pp735–755

Gazette.com (2006) 'Letters, 22 June 2006', www.gazette.com/opinion/energy_12959___article.html/marriage_mountain.html

Giampietro, M. (2003) *Multi-Scale Integrated Analysis of Ecosystems*, CRC Press, Boca-Raton, FL

Giampietro, M., Allen, T. F. H. and Mayumi, K. (2005) 'The epistemological predicament associated with purposive quantitative analysis', *Ecological Complexity*, vol 3, pp307–327

Giampietro, M., Allen, T. F. H. and Mayumi, K. (2007) 'Science for governance: The implications of the complexity revolution', in A. Guimaraes-Pereira, S. Guedes-Vaz and S. Tognetti (eds) *Interfaces between Science and Society*, Greenleaf Publishing, Austin, TX, pp82–99

Greenspan, A. (2008) 'We will never have a perfect model of risk', *Financial Times*, 16 March 2008, www.ft.com/cms/s/0/edbdbcf6-f360-11dc-b6bc0000779fd2ac.html?nclick_check=1

Haggett, S. (2007) 'Oil sands forecast cut as costs increase 50%', Reuters New Agency, www.thestar.com/article/276971

Hall, C. A. S., Powers, R. and Schoenberg, W. (2008) 'Peak oil, EROI, investments and the economy in an uncertain future', in D. Pimentel (ed) *Biofuels, Solar and Wind as Renewable Energy Systems*, Springer, Dordrecht, The Netherlands, pp109–132

Hardin, G. (1968) 'The tragedy of the commons', *Science*, vol 162, pp1243–1248

Hardin, G. (1972) *Exploring New Ethics for Survival: The Voyage of the Spaceship Beagle*, Viking, New York

International Business Times (2008) 'Commodities send Syngenta earns soaring', www.ibtimes.com/articles/20080207/commodities-send-syngenta-earns-soaring_1.htm

Kachan, D. (2008) *Biofuels Patents are Booming*, Cleantech Group, http://cleantech.com/news/2329/biofuel-patents-are-booming

Kahn, S. (2008) CSIRO public lecture of Professor Shahbaz Khan, Canberra, 21 July 2008, www.canberratimes.com.au/news/local/news/general/population-bomb-ticks-louder-than-climate/1173782.aspx#

Kesmodel, D., Etter, L. and Patrick, A. O. (2008) 'Grain companies' profits soar as global food crisis mounts', *The Wall Street Journal*, 20 April, pA1, http://online.wsj.com/article/SB120949327146453423.html?mod=todays_us_nonsub_page_one

Krauss, C. (2008) 'Alternative energy suddenly faces headwinds', *New York Times*, www.nytimes.com/2008/10/21/business/21energy.html?partner=rssnyt&emc=rss

Loehle, C. (1990) 'A guide to increased creativity in research: Inspiration or perspiration?', *BioScience*, vol 40, pp123–129

Malthus, T. R. (1798) *An Essay on the Principle of Population*, J. Johnson, London

Meadows, D. H., Meadows, D. L., Jogen, R. and Behrens III, W. W. (1972) *The Limits to Growth*, Universe Books, New York, NY

Phillips, L. (2008) 'European Parliament capitulates on biofuels deal', *The EUobserver*, 5 December, http://euobserver.com/9/27236

Pratkanis, A. R. (1995) 'How to sell a pseudoscience', *Skeptical Enquirer*, vol 19, no 4, pp19–25

Ravetz, J. R. and Funtowicz, S. O. (1999) (eds) *Futures* (Special issue dedicated to Post-Normal Science), vol 31

Schumpeter, J. A. (1954) *History of Economic Analysis*, George Allen & Unwin Ltd, London

Simon, H. A. (1976) 'From substantive to procedural rationality', in J. S. Latsis (ed) *Methods and Appraisal in Economics*, Cambridge University Press, Cambridge, MA
Simon, H. A. (1983) *Reason in Human Affairs*, Stanford University Press, Stanford, CA
Solow, R. (1974) 'The economics of resources or the resources of economics', *American Economic Review*, vol 64, pp1–14
UC Berkeley News (2007) Press release, 1 February, *UC Berkeley News* website, www.berkeley.edu/news/media/releases/2007/02/01_ebi.shtml
Vonnegut, K. (1963) *Cat's Cradle*, Holt, Rinehart and Winston, Austin, TX

Chapter 10

Giampietro, M. (2002) 'Energy use in agriculture', in *Encyclopedia of Life Sciences*, John Wiley & Sons, Hoboken, NJ, http://mrw.interscience.wiley.com/emrw/9780470015902/els/article/a0003294/current/pdf
Smil, V. (2008) *Energy in Nature and Society: General Energetics of Complex Systems*, MIT Press, Cambridge, MA

Appendix 1

Fischer-Kowalski, M. (1997) 'Society's metabolism: On the childhood and adolescence of a rising conceptual star', in M. Redclift and G. Woodgate (eds) *The International Handbook of Environmental Sociology*, Edward Elgar, Cheltenham, UK, pp119–137
Georgescu-Roegen, N. (1971) *The Entropy Law and the Economic Process*. Harvard University Press, Cambridge, MA
Georgescu-Roegen, N. (1976) *Energy and the Economic Myth: Institutional and Analytical Economic Essays*, Pergamon Press, New York, NY
Giampietro, M. (2003) *Multi-Scale Integrated Analysis of Ecosystems*, CRC Press, Boca-Raton, FL
Giampietro, M., Allen, T. F. H. and Mayumi, K. (2005) 'The epistemological predicament associated with purposive quantitative analysis', *Ecological Complexity*, vol 3, pp307–327
Giampietro, M., Mayumi, K. and Ramos-Martin, J. (2006) 'Can biofuels replace fossil energy fuels? A multi-scale integrated analysis based on the concept of societal and ecosystem metabolism: Part 1', *International Journal of Transdisciplinary Research*, vol 1, no 1, pp51–87, www.ijtr.org/
Glansdorff, P. and Prigogine, I. (1971) *Thermodynamics Theory of Structure, Stability and Fluctuations*, John Wiley & Sons, New York, NY
Holling, C. S. (1995) 'Biodiversity in the functioning of ecosystems: An ecological synthesis', in C. Perring, K. G. Maler, C. Folke, C. S. Holling and B. O. Jansson (eds) *Biodiversity Loss: Economic and Ecological Issues*, Cambridge University Press, Cambridge, MA, pp44–83
Koestler, A. (1968) *The Ghost in the Machine*, MacMillan, New York, NY
Koestler, A. (1969) 'Beyond atomism and holism – the concept of the holon', in A. Koestler and J. R. Smythies (eds) *Beyond Reductionism*, Hutchinson, London, pp192–232
Koestler, A. (1978) *Janus: A Summing Up*, Hutchinson, London
Margalef, R. (1968) *Perspectives in Ecological Theory*, University of Chicago Press, Chicago, IL

Martinez-Alier, J. (1987) *Ecological Economics: Energy, Environment and Society*, Blackwell, Oxford, UK
Maturana, H. R. and Varela, F. J. (1980) *Autopoiesis and Cognition: The Realization of the Living*, D. Reidel Publishing, Dordrecht, The Netherlands
Maturana, H. R. and Varela, F. J. (1998) *The Tree of Knowledge: The Biological Roots of Human Understanding*, Shambhala Publications, Boston, MA
Nicolis, G. and Prigogine, I. (1977) *Self-Organization in Nonequilibrium Systems*, Wiley-Interscience, New York, NY
Odum, E. P. (1953) *Fundamentals of Ecology*, W. B. Saunders, Philadelphia, PA
Odum, H. T. (1971) *Environment, Power, and Society*, Wiley-Interscience, New York, NY
Odum, H. T. (1983) *Systems Ecology*, John Wiley, New York, NY
Odum, H. T. (1996) *Environmental Accounting: Emergy and Environmental Decision Making*, John Wiley, New York, NY
Polimeni, J., Mayumi, K., Giampietro, M. and Alcott, B. (2008) *Jevons' Paradox: The Myth of Resource Efficiency Improvements*, Earthscan Research Edition, London
Prigogine, I. (1978) *From Being to Becoming*, W. H. Freeman and Company, San Francisco, CA
Prigogine, I. and Stengers, I. (1981) *Order out of Chaos*, Bantam Books, New York, NY
Rosen, R. (1958a) 'A relational theory of biological systems', *Bulletin of Mathematical Biophysics*, vol 20, pp245–260
Rosen, R. (1958b) 'The representation of biological systems from the standpoint of the theory of categories', *Bulletin of Mathematical Biophysics*, vol 20, pp317–341
Rosen, R. (2000) *Essays on Life Itself*, Columbia University Press, New York, NY
Ulanowicz, R. E. (1986) *Growth and Development: Ecosystem Phenomenology*, Springer-Verlag, New York, NY

Appendix 2

Falconi-Benitez, F. (2001) 'Integrated assessment of the recent economic history of Ecuador', *Population and Environment*, vol 22, no 3, pp257–280
Ramos-Martin, J. (2001) 'Historic analysis of energy intensity of Spain: From a "conventional view" to an "integrated assessment"', *Population and Environment*, vol 22, no 3, pp281–313
Ramos-Martin, J., Giampietro, M. and Mayumi, K. (2007) 'On China's exosomatic energy metabolism: An application of multi-scale integrated analysis of societal metabolism (MSIASM)', *Ecological Economics*, vol 63, no 1, pp174–191

Glossary

Agro-biofuels Biofuels generated by agricultural crops. The most common agro-biofuels are ethanol (alcohol generated by the fermentation of crops such as corn or sugar cane) and biodiesel (diesel fuel derived from crops such as soybean, rapeseed, sunflower and palm oil).

Autopoietic systems A class of systems capable of producing themselves, conceptualized by Maturana and Varela. *Autopoiesis* literally means 'self-production' (from the Greek *auto* for 'self' and *poiesis* for 'production'). The term expresses a fundamental complementarity between structural types and functional types found in biological and social systems. An autopoietic system must continuously re-define in time the set of formalizations adopted when storing experience, making anticipatory models and developing mechanisms of controls, aimed at preserving its own identity (see Appendix 1).

Becoming systems A class of systems introduced by Prigogine to flag a peculiar characteristic of dissipative systems. In order to guarantee their long-term survival they have to continuously change their current identity.

Bio-economics The application of basic principles of energetics to the study of the feasibility and desirability of metabolic patterns associated with socio-economic systems. The concept of bio-economics was proposed by Lotka and was operationalized by Georgescu-Roegen with the fund-flow model. Bio-economics makes it possible to integrate the representation of monetary flows with matter and energy flows. In this way, it provides a bridge between economic, ecological, social and demographic analysis.

Bio-economic pressure (BEP) The term 'bio-economic pressure' indicates the existence of a link between changes in the economic performance of an economy and changes in the metabolic pattern of society. An increase in the monetary flows within the economy has the goal of increasing the enjoyment of life of the members of a society. It also promotes an increase in the intensity of the throughputs of matter and energy per hour of labour in the productive sector of the economy. The productive sector includes the primary sectors (agriculture, forestry, fisheries, energy and mining) plus the secondary sector (building and manufacturing). In plain terms, this means that a richer society will demand more services and will work less because of ageing and increased education and leisure time. Facing these changes, the productive sector must be able to supply a large amount of products, energy and materials to society using only a very limited amount of work hours.

Biofuels Liquid fuels generated from biomass (non-fossil biomass).

Colonized land This refers to all those land-use categories that make it possible to predict a certain density of metabolized flows (energy, water, monetary flows) per hectare. The term 'colonized' indicates that human activity associated with control and management entails the expression of an expected pattern over typologies of land uses, e.g. corn fields in Iowa, compact apartments in Paris, car factories in China.

Concorde syndrome This refers to a form of lock-in that often takes place in the field of technical innovations (see Box 8.5). It occurs when a technical solution has been generated by a mistaken narrative. This awkward situation can easily become a syndrome because, often, brilliant technical solutions require considerable financial investments. For this reason, even when experience shows that the original idea was a bad one, nobody dares to halt the implementation of the technical solution. See also 'sunk cost'.

Desirability of a metabolic pattern A feasible pattern of metabolism that is advantageous, beneficial and wise according to a given set of performance criteria.

Dissipative systems All natural systems that are relevant to sustainability (e.g. complex biogeochemical cycles on this planet, ecological systems and human systems when analysed at different levels of organization and scales above the molecular one) are 'dissipative systems'. That is, they are self-organizing, open systems, operating away from thermodynamic equilibrium (see Appendix 1).

Endosomatic metabolism Physiological conversions of different types of energy inputs – i.e. food items – into end-uses that take place inside the human body.

Energetics A systemic analysis of the set of energy transformations associated with the maintenance and reproduction of metabolic systems, such as the biosphere, ecosystems and human societies. There are general principles of energetics that dictate the features of metabolic patterns, and hence they can be used to study the characteristics of socio-economic systems that are based on the expression of these metabolic patterns.

Energetic transition This encompasses the time that elapses between an introduction of a new primary energy source and its claim to a substantial share (20 per cent to 30 per cent) of the overall market, or even its becoming the single largest contributor or an absolute leader (with more than 50 per cent) in national or global energy supplies (after a definition by Vaclav Smil).

Energy carriers The various forms of energy inputs required by the various sectors of a society to perform their functions. Energy carriers are produced by the energy sector using primary energy sources. Examples of energy carriers include liquid fuel in a furnace, gasoline in a pump, electricity in a factory, and hydrogen in the tank of a car.

Energy end-uses This expression refers to the useful tasks/work performed by the various sectors of society when converting energy carriers into applied power. Examples of end-uses include moving goods, melting iron, building roads and air-conditioning rooms.

Energy return on investment (EROI) This is one of the most important concepts in the study of the quality of alternative energy sources, but also one of the most controversial. Economic return on investment – which inspired the concept of EROI – is an accounting valuation method used to assess the convenience of a financial investment. In financial analysis, this concept addresses at least two different relevant issues: the payback time of the investment, and the size of the required investment (input), which the investor must be able to handle. However, very often, in many applications to energy analysis, the EROI is considered as merely an output/input ratio determining a net surplus of energy, without consideration being given to the time dimension, or the power level at which the flows are invested and supplied.

As explained in the text (Chapter 5), the EROI is a semantic concept useful for studying the quality of primary energy sources. According to Cottrell, 'societies adopted a new energy technology only if it delivered a greater energy surplus, and hence a greater potential to produce goods and services'. However, to implement this semantic definition one has to use a set of indicators and not just a simple ratio. In particular, two factors should be considered when using EROI as an accounting evaluation method:

1. The energetic burden associated with the conversion of primary energy sources (PES) into a net supply of energy carriers (EC). This burden depends on the overall loss associated with two conversions: PES → gross EC, and gross EC → net EC. The second conversion refers to the internal consumption of energy carriers in the production of energy carriers within the energy sector.
2. The power level at which energy carriers are invested in the exploitation of PES. This is a crucial factor determining the final intensity of the net supply of energy carriers to society.

Energy sector The specialized sector of society whose goal it is to deliver the required mix of energy carriers to society using primary energy sources. The mix of energy carriers supplied by the energy sector has to match in quantity and quality the demand of the various sectors of the society.

Exosomatic metabolism Technical conversions of different types of energy inputs (energy carriers) into end-uses that take place outside the human body, but under direct human control.

External constraints to metabolism Refers to the availability of the required energy input resulting from the interaction of the metabolic system with its context. For example, the availability of gas stations along the road from New York to San Francisco will determine the feasibility of a trip involving a truck carrying 20 tonnes of material from one city to the other in less than a week. An external constraint is present when the system has plenty of technical capital but does not have enough primary energy inputs. This is a situation in which the amount of fish catch is determined by the availability of fish in the face of an excess of fishing boats.

Feasibility of a metabolic pattern A metabolic pattern that is possible under a given definition of constraints.

Fossil fuels Liquid fuels generated from fossil energy. Fossil energy is organic material generated in prehistoric times and stored below the surface of the Earth.

Fund/flow model for metabolic systems Georgescu-Roegen proposed a fund-flow model useful for representing, in biophysical terms, the metabolism of socio-economic systems.

- Fund elements are those that remain the same during the analytical representation (they reflect the choice made by the analyst when deciding *what the system is* and *what the system is made of*).
- Flow elements are those that are either produced or consumed during the analytical representation (they reflect the choice made by the analyst when deciding *what the system does* and *how it interacts with its context*). Flow elements can be described in terms of relevant monetary, energy and material flows.

In this model, fund elements are metabolic converters; they must be able to maintain and reproduce themselves in order to keep their original identity. Thus, fund elements entail:

- an overhead on the flow they process, for their maintenance and reproduction;

- a definition of what should be considered as an admissible input (their identity entails that they can only metabolize a specified type of inputs); and
- a set of biophysical constraints on the relative conversion pace of metabolized flows (their identity can be associated with an expected power level).

For this reason, the fund-flow model is particularly suited to studying the metabolic pattern of socio-economic systems.

Fund-flow energy supply (renewable energy sources) A flow of energy originating from a fund does not imply a change in time of the characteristics of the system. For example, we can milk a healthy cow every day, and if we do not overdo it, the cow will remain healthy. A self-reproducing dairy farm – producing milk with sufficient calves guaranteeing the replacement of cows and enough pasture for feeding them – would represent a fund providing a stable supply of milk. As long as the fund is able to repair and reproduce itself, the resulting flow can be considered stable. Hence, this milk supply can be called a renewable resource.

Grammar A set of expected relations between a given set of semantic categories and a given set of formal categories. The definition of a grammar requires a pre-analytical choice of a lexicon, and production rules (how to calculate names from the values of tokens). The grammar has to be implemented through a set of decisions about the choice and use of data (the values assigned to tokens before operating the inferential system associated with production rules; see Chapter 5).

Granfalloon The term 'granfalloon' was first introduced by Kurt Vonnegut (see Chapter 9) to indicate 'a proud and meaningless association of human beings'. Wikipedia provides the following definition (among others): 'a group of people who believe that they have a special connection and who believe they are helping to bring about a greater plan, but are actually not'.

Impredicative loop The formal definition of impredicativity is 'when a property is possessed by an object whose definition depends on that property'. The term 'impredicative loop' indicates a situation – very common in autopoietic/self-organizing systems – in which the identity of the whole (e.g. society) affects the identity of the parts (e.g. energy sector) and vice versa, in a pattern of causality recalling the chicken–egg puzzle (see Box 3.1).

Impredicative loop analysis (ILA) Using the fund-flow model, it becomes possible to define a forced set of relations between the relative sizes of funds and flows (those generating the overall demand and those providing the supply), which can be described over a dynamic equilibrium across hierarchical levels (parts and whole). An ILA is a quantitative analysis based on the forced congruence over numbers, but it is not deterministic; the same overall dynamic budget can be obtained by different combinations of fund/flow characteristics. Therefore, ILA makes it possible to explore the option space of different feasible solutions capable of obtaining congruence over the overall metabolic pattern.

Internal constraints to metabolism Refers to the internal capacity of a given converter to convert an energy input into useful work at a given rate, e.g. a truck required to transport a given load uphill, or a person capable of running over a given distance in a given time, or the ability to pull a given weight. An internal constraint is present when the system does not have enough technical capital to take advantage of available energy inputs. This is a situation in which plenty of fish are available, but the shortage of fishing boats determines the overall catch.

Metabolic system Any system that is able to use energy, materials and other natural resources to maintain, reproduce and improve its own existing structures and functions (see Appendix 1).

Paradigm of industrial agriculture This can be defined as the existence of an uncontested consensus in society over the idea that a massive use of technology (capital) and fossil energy in agriculture is justified in order to achieve two key objectives: boosting the productivity of labour in the agricultural sector, and boosting the productivity of land in production.

Post-normal science An expression proposed by Silvio Funtowicz and Jerome Ravetz to indicate a critical situation in the production and use of science for governance. In contrast with 'normal science' – as defined by Khun – a post-normal science situation indicates that 'facts are uncertain, values in dispute, stakes high and decisions urgent'. This implies changing the focus of the discussion from truth to quality by enlarging the variety of methods, criteria and actors involved in the assessment of the validity and relevance of the scientific output. In relation to sustainability science, this occurs when available narratives for explaining the present sustainability predicament are no longer valid, and validated narratives for making useful predictions about possible futures are not available.

Primary energy sources This expression refers to the energy forms required by the energy sector to generate the supply of energy carriers used by human society. According to the laws of thermodynamics, primary energy sources cannot be produced. They must be available to society in order to make possible the production of energy carriers. Examples of primary energy sources include below-ground fossil energy reserves (coal, gas, oil), blowing wind, falling water, the sun and biomass.

Societal metabolism A notion used to characterize the set of conversions of energy and material flows occurring within a society that are necessary for its continued existence. This concept implies that we can expect given patterns of energy metabolism to be associated with the different structures and functions of a socio-economic system (see Appendix 1).

Stock-flow energy supply (non-renewable energy sources) A flow of energy originating from a stock entails a change in the characteristics of the system. If we start with a stock of 1000 units and we consume for one year a flow of 100 units per year, the stock from which we obtained the input will have changed its identity. After one year, the original stock of 1000 units will have become a stock of 900 units. For this reason we can call the supply of a given input coming from a stock-flow a non-renewable resource.

Sunk cost The sunk cost refers to the total amount of losses incurred in abandoning a bad investment. The sunk cost takes place when an investment is abandoned before reaching its payback point. The 'curse of the sunk cost' refers to the unavoidable inertia of economic investments, which have to be paid back before moving to a new type of investment: 'we cannot stop now, because everything we have invested so far will be lost'.

Index

academic lock-in 15, 228, 238–244
accounting *see* energy analysis
acid rain 162–163
agricultural technology treadmill 213–214
agriculture 80, 137–144
    flow densities comparison 215
    and labour productivity 70–71, 148, 231
    high external input agriculture (HEIA) 210
    low external input (LEIA) 209
    nutrient issues *see* fertilizers
    surpluses/subsidies in 205, 206, 244
    varieties/seeds in 209–210
    see also industrial agriculture paradigm
Allen, Timothy 234, 237, 238
Allison, K. 1, 245
alternative energy sources 12–13, 14, 40–43, 172, 176
    analysis of *see* bio-economics; EROI
    quality issues 41, 255
anthropology 21, 150, 157
antibiotics 213, 225
appropriateness/compatibility issues 40–43
Argentina 3
Australia 207, 233
autocatalytic loops 50–51, 56–57, 64, 81, 89, 132, 149, 151, 235
autopoietic systems 49, 154

Bailey, Robert 4–5
Baker, M. M. 7
Ban Ki-moon 4
Bangladesh 5
benchmarks 86–87, 94, 95–96, 110, 119–120, 122–123, 153, 167, 174, 189
BEP (bio-economic pressure) 14, 80–89, 98–103, 123, 190, 191–193
    defined 80–81
    and development 98–100, 103
    and EROI 118–120, 133–135
    and exosomatic flows 82–85
    and industrial agriculture 204
    in livestock production 142, 143
    and overheads *see* overheads
Berkeley group 232–233
Berndes, G. 161
bio-economics 12, 13–14, 40, 69–103, 105, 244
    benchmarking in 86–87, 94, 95–96
    capital/capitalization 85–89
    and capitalism 88–89
    contextualization in 139–140, 142
    dynamic budgets in *see* dynamic budgets
    and EROI *see* EROI
    exosomatic metabolism patterns in 92–98, 100–103
    grammar in *see* grammar
    hierarchies in 72, 74–77, 82, 87, 89, 96, 100–102, 135–137
    impredicative loop analysis in 89–92
    and lock-ins 244–246
    multi-level accounting in 72–77
    MuSIASEM approach *see* MuSIASEM
    origins/development of 147–157
    and stock flow exploitation 88–89
    see also energetics; fund-flow model; metabolism, societal
biodiesel 2, 128, 173
biodiversity 5, 39, 128, 163, 210, 212, 222, 225
BIOFRAC (Biofuels Research Advisory Council) 249
Biofuels, Bioresources and Biorefining 165
biomass exploitation 18–23, 26, 253
biomass feedstock 137–144
biophysical analysis/constraints 18–20, 21–26, 255

biosphere 149
Boddey, R. M. 181–182, 189
Boserup, E. 20
bovine growth hormone 213
Box, G. E. P. 106
BP 107, 234
Brazil, biofuel production in *see* sugar cane ethanol
Britain 5–6, 147
building sector 80–81, 97, 119, 129
Burgonio, T. J. 6
Bush, George W. 1, 245, 251
by-product feeds 159, 163

Canada 6, 96, 245
CAP (Common Agricultural Policy) 206
capital/capitalization 85–88
capitalism 9, 10, 88–89
carbon dioxide *see* CO_2 emissions
Carnot, S. 144–145
cattle feed *see* livestock production
CEO (Corporate Europe Observatory) 7, 247
certification schemes 219–223
ceteris paribus hypothesis 131
Chavez, Hugo 9–10
China 2, 10, 72, 73, 81, 189–190, 229
    agriculture in 216, 217
    oil consumption of 82
Chu, Steven 7
civil society 6–7, 256
Cleveland, C. J. 85–86, 152
climate change 9, 157, 228, 230
    see also CO_2 emissions
closed/open systems 24–26
CNOOC (China National Offshore Oil Corp) 2
co-production credits 159
CO_2 emissions 5–6, 157, 160, 162
    see also climate change
coal 21, 28, 37, 43, 134, 147–148, 159
Cochrane, W. 213
Colombia 218
commodity support programmes 247
Concorde syndrome 16, 203, 204, 223–225, 226, 250–251, 254, 256
conservation 161, 220, 222
conversion losses 107, 109, 115, 124–128, 158–159, 178
    types of 124–126
corn ethanol 2, 4, 5, 9, 43, 128, 178–181, 188, 189, 234, 244
    benchmarks for 179–181
    corn production for 164
    labour input for 180–181
    promotion/subsidies for 1, 3, 7, 244, 245, 250–251
    quantitative analysis study 14–15, 157–164, 167–169
corn stoves 40, 41
cornucopians 239–240, 242–243
Corporate Europe Observatory (CEO) 7, 247
corporations 244–252
    see also lobby groups
Cottrell, William Frederick 85, 152
Craig, C. 7
critical organization 151
Cronin, D. 4
crop residues 234
    see also livestock production
cultural evolution 150–151

Dale, B. 165, 232–233, 235
Daly, Herman 28, 47

Danielsen, F. 2
DDG (dried distiller grains) 159, 163–164
De Carvalho Macedo, I. 181, 189
de Schutter, Oliver 4
debt 150, 198, 207
decumulation 31
deforestation 5, 9, 163
demographic pressure *see* population growth
demographic structure/change 72, 73, 74, 77, 81, 101, 217
developed countries 3, 11
dictionaries *see* grammar
displacement method 163–164
dissipative networks 22–23, 51, 54, 56–57, 62, 151
Dogaru, V. 155
Doha round 247
dynamic budgets 54–56, 71–77, 82–85, 96, 117, 118, 120, 122, 175, 192
    and EROI 131, 132, 133
    and impredicative loops 89–92, 151

EBI (Energy Biosciences Institute) 234
ecological theory 18, 29, 152–153
ecology
    industrial 69
    systems 152–153
    theoretical 21–22, 89
econometric models 240–241
economic analysis 12, 14, 255
    fund-flow/stock-flow supply 18
    models in 241–243
    neoclassical 239–240, 242
    Roegenian method 153–155
    three factors in 59
    see also bio-economics; socio-economic systems
economic growth 9, 13, 86–87, 191, 230
    and BEP 80–81, 98–100, 103
    and energy consumption 147–148
    perpetual 197–201, 238–240, 241–243, 258
economic lock-in 16, 228, 244–251
economic models 240–243
ecosystems 22–26, 39, 218, 253
    as dissipative/metabolic networks 22–23, 25–26
    hierarchy in 23
    see also rainforests
Egypt 5, 35
Ehrlich, Paul 239
EI (energy input) 32–33, 41–43, 106, 135–137
ELP (economic labour productivity) 75–77, 85, 86, 134–135
EMergy analysis 114
empires 32, 34–35, 89
employment 72, 80–81, 190, 231, 245–246
end-uses 44–46, 60, 96, 136–137, 255–256
    see also metabolism, societal
endosomatic metabolism 13, 27–28, 31, 35, 36, 50–51, 63, 70, 79–80, 142, 143, 150, 195, 255
    dynamic budget of 32–33
energetics 8, 11–14, 16, 39–67
    basic concepts 13, 39–40, 43–54
    compatibility issues 40–43
    constraints in 54–55, 56
    energy sources/carriers in 43–44
    EROI *see* EROI
    metabolism in *see* metabolism
    misunderstandings in 44
    origins/development of 147–157
    semantics in *see* semantics
    sources/carriers/end-uses in 44–45
    see also bio-economics; energy analysis
energy analysis 11, 14, 40, 45, 48, 50–52, 72–77, 135–145
    applications of 156
    by-product credit in 163–164
    decline of 105, 156
    displacement method 163
    equivalence issue 111–112
    gross/net energy in 127, 165–166, 167, 173, 186–187

heart transplant metaphor 115–117, 120, 129, 254
integrated 169–174, 255
models in 106
net *see* net energy analysis
recent debate on 157–166
scale in 61, 113–115, 117–118, 144, 163–164
semantics/grammars in *see* grammar
shortcomings in 158–166
statistics used for 107, 108, 175, 176
see also bio-economics; energetics; EROI
energy carriers 13, 14, 44–46, 107, 109, 115–116, 135–137, 138, 255
    gross/net supply of 53, 60, 167, 186–187, 191
    GSEC/NSEC 124–126
    and losses *see* conversion losses
    output/input ratio *see* output/input ratio
    see also energy flows
energy conversions/converters 44–50, 152
    semantics/grammar in 48–50
energy density 36, 166, 177–178
energy flows 18, 21, 22–35, 69, 77–80, 148
    of ecosystems 18–26
    and energy security 46–47
    fund/stock 28–31
    linear representation of 109, 110
    maximum, law of 149
    of modern societies 27–28
    in multi-purpose grammar 167–169
    and power levels 50–52, 84, 89, 133, 135–137
    of pre-industrial societies 18–21
    sources/carriers 44
    TET *see* TET
    see also metabolism
energy graphs 49, 170–174
energy intensities 95–96
    threshold values 166, 176
energy sector *see* primary sector
energy security 2, 9, 15, 46–47
energy sources *see* primary energy
energy transformation 44–48
energy transitions 152
energy-for-energy 131, 168, 178, 186, 256
Enlightenment 241–242
entropy 114, 153
environmental audit committee (UK) 6
epistemological issues 40, 50–51, 105, 111, 144, 153–155
    replicant knowledge 234–238, 245
equivalence criterion 45, 50
EROI (energy return on investment) 13, 14, 40, 54–67, 105, 117–123, 131–135, 148, 176
    and BEP 118–120, 133–135
    and constraints 118–120, 122–123
    and dynamic budgets 131, 132
    of energy forms 60, 62, 129–131
    ignored in literature 59
    loss categories 60
    systemic problem in 61–66
    time element in 59, 61, 122
ESIA (environmental and social impact assessment) 221, 222
Europe
    biofuels production in 2, 4
    biofuels promotion in 1
    economic models in 241–243
    scientific establishment in 236
    THA in 81
European Union (EU) 2, 3, 5–7, 218, 227, 245–246
    biofuel subsidies in 244, 247, 248, 250–252
    biofuels critics in 6–7
    Directive EC2003/30 6
    and lobby groups 248–250
    sustainable biofuels certification 219–223
evolutionary theory 149, 150–151, 240
exergy 114
exosomatic devices 88, 131, 150
exosomatic metabolism 13, 27–28, 31–32, 36, 37, 50–51, 77–79,

84–85, 120–122, 186
    and capitalization 85–88
    and constraints 118
    of energy sector (EMRES) 120–122, 129–131, 134, 166, 175
    in hierarchical levels 100–102
    multi-purpose grammar for 170–174
    of society (EMRAS) 84–85, 118–120, 254
    TET *see* TET
    typology of 92–98, 128, 177

Fargione, J. 5
farmers 204, 206–207, 211, 213
Farrell, A. E. 59, 157, 158–164, 166, 169, 232
fermentation/distillation 49, 180, 181, 183
fertilizers/pesticides 35, 48, 112, 160, 162, 181, 187, 188, 212, 248
fish stocks/fishing industry 28, 47, 80
Flevoland 205–208
flows 28–31, 154
    see also energy flows
food crisis 3–5, 9, 248, 252
food energy 33, 51
food production 9, 38, 156
    and commodity support programmes 247
    fossil fuels in 258
    and industrial agriculture 203, 204, 210–211
    in pre-industrial societies 18, 20, 33, 64–65
food security/insecurity 3, 5, 142
forest ecosystem 24–25
forestry/logging 2, 80, 246
fossil fuels
    early use of 21–22
    flows 36
    in food production 258
    and power levels/economic growth 134–135, 150
    quality of 66–67, 127–128
    substitution of 129–131
    used to make biofuel 125, 128, 161, 166, 188
    see also coal; natural gas; oil/petroleum
Friends of the Earth 7
fuel prices 5, 244, 245
fund elements 29–30, 40, 42, 52, 62, 65, 74, 85, 96, 115, 118–120, 132, 137, 149, 153
fund-flow model 18, 28–31, 70, 72–80, 94–95, 117, 132, 133
    constraints on 29, 36, 88–89
    defined/elements of 28–30
    and multi-level accounting 72–77, 170
    reduction in 84
Funtowicz, Silvio 13, 241, 243
future of biofuel 253–259

Gates, Bill 245
general equilibrium theory 240
genetic modification 247
Georgescu-Roegen, N. 26, 27, 30–31, 40, 52, 58, 75, 80, 87–88, 127, 154–155, 240
Germany 134, 148
GHG (greenhouse gas) 2, 5, 15, 31, 161, 162–163, 220, 230, 252
Giampietro, M. 24, 105, 168, 233
globalization 5, 197, 198, 199, 209, 225
Gore, Al 228
Gossen, Hermann Heinrich 147
governance 13, 241–242, 243–244, 256–258
grain production 3, 248
grammar 48–50, 97–98, 105–111, 131–132, 176, 255
    critical appraisal of 158–159
    limitations of 108–109, 158–159, 160–161, 166–167
    mono-purpose 106–111
    multi-purpose/meta- 110–111, 166, 167–174
    in scientific debate 232, 235–236
    see also semantics
granfalloons 15, 232–233, 235, 238, 246, 258
greenhouse gases *see* GHG
Greenspan, Alan 240–241
GSEC (gross supply of energy carriers) 124–126, 129, 167

GSI (Global Subsidies Initiatives) 2
Guardian, The 3–4

Hall, C. A. S. 85, 237, 256
Hardin, Garrett 239
HCV (High Conservation Value) areas 222
heart transplant metaphor 115–117, 120, 129, 254
Hecht, L. 7
HEIA (high external input agriculture) 210
Hopf, F. A. 89
household sector 44, 73, 83–84, 87, 96–97, 129
household-level analysis 34, 55–56, 58, 70, 70–71, 94
human density 20–21
human fat 40–42, 58
human labour energetics 113, 114
human rights 220
hydroelectric power 27, 28, 46, 256
hypercycle 54–57

ideological bias 228, 238
ideological lock-in 15, 228–238
IFIAS (International Federation of Institutes for Advanced Study) 156, 160
impact assessments 221, 222, 245–246
impredicative loops 47, 89–92, 118, 127–128, 132, 151, 256
India 189–190, 207
indigenous people 218–219
Indonesia 2, 4, 218, 252
industrial agriculture paradigm 15, 191, 203–226, 244, 247
    acceptance of 210–211
    and BEP 204
    certification schemes 219–223
    and Concorde syndrome *see* Concorde syndrome
    crisis in 210
    defined 208–210
    in developed/developing countries 214–217
    ecological drawbacks of 211–212, 218
    effect of 205–208
    and fossil energy 209
    global applicability of 214–217
    labour issues of 203, 204, 217
    and land seizures/violence 218–219
    nutrients/seeds used in 209–210, 212
    performance-related costs in 225
    and rural development 213–223
    socio-economic drawbacks of 212, 218–219
    sustainability of 219–223
    treadmill effect of 213–214
    two objectives of 203
    water pollution in 206, 210, 212, 222, 225
    water use in 209, 214, 216, 217
Industrial Revolution 13, 27, 35, 36–37, 43, 46, 70, 152
industrial sector 44, 96
industrialization 21, 27, 204
informed deliberation 256–257
infrastructure 181, 190, 244–245
innovation 46, 47, 70, 88, 150, 151, 237–238
    treadmill 213–214
    see also industrial agriculture
investment in biofuels 1–2, 227, 238, 244–252
    payback time for 59
    and performance 47–48
irrigation 181, 182, 190, 209, 214, 216, 217
Italy 72, 73, 81, 96, 176, 186–187

Japan 96
Jevons paradox 148
Jevons, William Stanley 147–148

Kachan, Dalla 238
Kahn, S. 233
Kauffman, S. A. 49
Kesmodel, D. 248
Keynes, John Maynard 241
Kirchgaessner, S. 1, 245

Kleene, S. C. 47
knowledge gap 243–244
Koplow, D. 7
Krauss, C. 245
Krugman, Paul 3, 9–10
Kutas, G. 2

labour 13, 35, 80–81, 126, 128, 143, 180–181, 255
   demand 167, 169
   EMR of 122, 126, 128, 131, 133, 175–176, 178–179, 182–184, 186, 189, 191–192
   energetic equivalent of 114
   food energy for 33
   Marxist view of 148
   in multi-purpose grammar 167–169
   productivity (ELP) 70–77, 79–80, 85–86, 134–135, 142, 148, 152, 186
   requirement 166, 170
   slave 32, 34
land 18–20, 22, 67, 161, 162, 186–187, 188
   emancipation from 37
   and energy density 177–178
   as fund element 29, 31, 36, 38, 89, 109, 158, 164, 166, 169
   Ricardian 29
   seizures 218–219
   typologies 26
land-use change 5–6, 9, 13, 15, 37–38, 161, 252
Lawrence Berkeley National Laboratory 234
LEIA (low external input agriculture) 209
Leontief, W. 40
Lindberg, C. 2
liposuction 40–42, 58
livestock production 137–144, 213
   BEP in 142, 143
   hierarchies in 143
   socio-economic context of 142–143
lobby groups 4, 7, 227, 228, 236, 237, 244–251
lock-ins 15–16, 227–251
   academic 15, 228, 238–244
   breaking 258–259
   cornucpians/prophets of doom in 239–240, 242–243
   economic 16, 228, 244–251
   economic models in 240–243
   and granfalloons 232–233, 235, 238, 246, 258
   ideological 15, 228–238
   and post-normal science 243–244
   and replicant knowledge 234–238, 245
   and research funding 236–238
   see also Concorde syndrome
Lotka, Alfred 27, 149–150, 151

Malaysia 2
Mali 141, 143–144
Malthus, Thomas 239
Malthusian trap 35
manufacturing sector 80–81, 97, 119, 129
Margalef, Ramon 153
market economy 10
Martinez-Alier, Joan 147, 148
Marxism 148, 229
maximum power principle 89
meat consumption 3, 204
Medawar zone 237
Menger, Carl 147
metabolism, Earth's 149, 230
metabolism, energy 12, 13–14, 17–18, 39, 153
   of ecosystems 22–26
   energy forms in 44–45, 129–131
   evolutionary pattern of 147–148
   funds/flows in 28–31
   hypercycles in 54–57
   of pre-industrial societies 17–23, 35–36
   typologies 114
   and urbanization 18, 35
   see also endosomatic metabolism; EROI; exosomatic metabolism
metabolism, societal 69–103, 254, 255–256
   and fund-flow model 70, 74–80, 84
   and labour productivity 70–74
   multi-level accounting in 72–77
   see also MuSIASEM
Michell, Halmuth 6
milk production 213
Millenium Ecosystem Assessment 225
mining industry 80, 191
Monbiot, George 7
monetary flows 74–77, 150, 239
monitoring 7, 221, 247
monocultures 5, 15, 191, 208, 212, 218–219
multicropping patterns 215
MuSIASEM (multi-scale integrated analysis of societal and ecosystem metabolism) 69–74, 106, 117–118, 172, 176
   and labour productivity 70–74

natural gas 21, 28, 37, 43, 159
net energy analysis 13, 55, 57–58, 61, 148, 152, 165, 232–233
   see also EROI
Netherlands 205–208
network analysis 57
New York Times 3
New Zealand 40
NGOs 6–7, 218
nitrogen leakage 160, 162, 188
NSEC (net supply of energy carriers) 124–126, 129
nuclear energy 174, 254, 256

Obama, Barack 7, 251
Odum, E. P. 23, 24, 153, 155
Odum, H. T. 23, 24, 26, 89, 96, 114, 144, 149, 152–153, 155, 170–174, 240
OECD countries 2, 73
oil/petroleum 2, 13–15, 47–48, 155–164, 159–160
   crisis 155–156
   peak 15, 157
   quantitative study of 14–15, 157–164
   used in biofuel production 8
organizations, hierarchies in 14
Ostwald, Wilhelm 148
output/input ratio 15, 43, 45–46, 48, 51–52, 53–54, 59, 105, 111–113, 123–124, 175, 255
   in animal feed 137–144
   equivalence issue 111–112
   and grammar 235–236
   joint production dilemma 111, 112–113
   losses in *see* conversion losses
   and power levels 117–118
   trade-off in *see* trade-offs
   truncation problem 111, 113
   see also EI; EROI
overheads 81, 82, 83–84, 88, 90, 91, 101, 124, 126, 195
Oxfam 4–5, 218

palm-oil 2, 4, 135, 173, 252
paradigm, scientific 238–239
   post-normal 243–244
   see also industrial agriculture paradigm
Pastore, G. 100
patents, biofuel-related 238
Patung 2
Patzek, T. W. 181
peak oil 15, 157, 193–202, 228, 245, 247
   and exosomatic metabolic patterns 193–197
   fund/flow supply alternatives 195–197, 199
   and perpetual growth 197–201
   and population growth 197, 199–201
   and stock depletion 194
perpetual growth 197–201, 238–240, 241–243, 258
Peru 219
PES (primary energy sources) *see* primary energy
pesticides *see* fertilizers/pesticides

petroleum *see* oil/petroleum
Philippines 6
phosphorus leakage 160, 162, 188
photosynthesis 23, 24, 44, 160–161, 162, 190
photovoltaics 107, 127, 172, 174, 256
    see also solar energy
Pimentel, David 24, 180, 181, 232–233, 235
Pinkerton, R. C. 89, 144, 149
Podolinsky, Sergei 148
policy strategies 3, 9, 10–11, 253
    and science *see* scientific establishment
Polimeni, J. 148
political aspects 1, 9, 227, 229, 231, 235, 236–237
pollution 160, 162–163, 220
population density 20–21, 178, 207
population growth 3, 19, 37, 204, 208, 209, 211, 214, 216, 217
    and fossil fuels 21
    and peak oil 197, 199–201
post-normal science 13, 243–244
power 51–53, 61, 96, 123, 241
power levels 50–52, 84, 89, 129–131, 133, 134–137, 175, 255
    and evolutionary theory 149, 151
Pratkanis, A. R. 232
pre-analytical stage 48–50, 106–111, 158, 162, 235
pre-industrial societies 17, 18–22, 192
    biophysical constraints in 18–20, 21–22
    as closed system 26
    economic models in 241–243
    EI of 32–33
    elite groups in 20–21, 23, 33, 62
    energy metabolism of 17, 20–21, 27, 32–36
    and EROI 61–66
    food production/energy in 18, 20, 33–34, 195–197
    human density in 20–21, 178
    limits/collapse of 32, 34, 36, 61–62
    rural-urban input of 19
    slaves in 32, 34
    taxation in 34
primary energy sources (PES) 13, 14, 32, 39, 41, 44–48, 57–58
    consumption, assessment 107
    and energy losses *see* conversion losses
    misperception of 160
    quality of 123–131, 152, 169–174
    quantitative analysis study 14–15, 157–160
primary sector 34, 44, 60, 61
    energy per hour of labour 78, 81–82
    power levels in 129–131
prophets of doom 239–240, 243
public opinion 227, 231, 235

quantitative analysis 14–15, 157–164

rainforests 2, 4, 5, 218
Ramos-Martin, J. 87, 201
rapeseed 4, 128
ratchet effect 34
Ravetz, Jerome 13, 241, 243
'real' wealth 150
reciprocal constraints 91–92
Rees, W. 26
Renewable Fuels Standard (US) 7
replicant knowledge 234–238, 245
research funds 236–238, 251, 256
restriation 151
Ricardian land 29
rice straw 139
Robinson Crusoe effect 258–259
Roman Empire 34, 89
Rome, Club of 239
Rosenthal, E. 6
Round Table on Sustainable Biofuels 220–223
rural development 2, 15, 191, 220, 225, 253
    and industrial agriculture *see* industrial agriculture
rural landscape 225, 247
rural population 204, 206–207, 225

rural societies *see* pre-industrial societies
rural-urban migration 204, 207, 208
rural-urban ratios 18–20, 26

Sachs, Jeffrey 4
Salazar, Ken 230–231
salmon oil 40, 53
Schumpeter, J. A. 235
Science 5, 147, 157
scientific establishment 13, 15, 227–228, 231–233
    and innovation 237–238
    and replicant knowledge 234–238, 245
    see also granfalloons
Searchinger, T. 5
self-sufficient production 185–186
semantics 12, 14, 47, 105–106, 131–132, 154
    'dictionaries' 96–97, 176
    in scientific debate 232, 235
    see also grammar
service/government sector 44, 77, 78, 81, 129
Shapouri, H. 164
Simon, Julian 239, 240
Smil, Vaclav 177, 232
SO_2 162–163
socio-economic systems 10, 13, 16, 28–32, 36, 39–40, 46, 51, 141–143, 149–150, 156
    benchmarks for 86, 95, 153
    complexity in 53–54, 65, 105
    constraints in 56, 139, 187
    integrated analysis of 69, 72, 140–141
    post-industrial 154
    variables in 74
Soddy, Friederick 150, 198
soil degradation/protection 163, 187, 188, 220, 225
solar energy 48, 49, 107, 230
    see also photovoltaics
Solow, Robert 239, 240
South Korea 189–190
Spain 74–77, 82–85, 86, 87, 96
spatial density 23–26
standard of living 17, 18
Starkey, Bob 1
statistics, international 107, 108, 175, 176
Steenblik, R. 2, 7
Stirling, A. 113
stock market 245, 248
stock-flows 18, 28–35, 88–89, 160
    in multi-purpose grammar 170–174
    utilization of 32–35
subsidies for biofuels 1–2, 3, 5, 7, 238, 245, 247
    and food crisis 9
    future of 253
substantive rationality 241–242, 243
sugar cane ethanol 2, 9, 123, 178, 188–191, 253
    benchmarks for 181–185, 189
    labour input for 182, 184, 185, 189
    quantitative analysis study 14–15, 173
sunflowers 128, 173, 187
sustainability 11–12, 16, 39, 46, 61–62, 153, 185–191, 219–223, 252, 258
    analysis 106, 153, 156, 238
    and lock-ins 228–231, 238–240, 243
    syntax/semantics in 12
switchgrass ethanol 5, 173, 234
Syngenta AG 248
syntax 12, 131

Tainter, J. A. 32, 34, 61–62
Tanzania 218–219
targets 3, 5–6
technology innovation *see* innovation
TET (total exosomatic throughput) 77–78, 80–81, 82, 83, 84, 94–95, 118–122, 129–130, 166, 192–193
THA (total human activity) 74, 81, 82, 84, 94, 118–120
Thermodynamics, First/Second Laws of 43–44, 123

threshold value 91
TOE (tonnes of oil equivalent) 45, 48, 51, 96, 107, 109, 127, 128–129, 175, 256
trade-offs 31, 139, 144, 173–174
transport sector 8, 218, 252
transportation costs 164, 181, 186, 188

Ukraine 3
Ulanowicz, R. E. 56, 57
Ulgiati, S. 105, 168
unemployment 70, 72, 73
UNICA study 181–185, 190
United Nations (UN) 4, 218, 252
United States 47, 86, 96, 110, 151, 164, 176, 236
  biofuel critics in 7
  biofuels production in *see* corn ethanol
  biofuels promotion in 1, 3, 7, 244, 245, 250–251
  biomass feedstock in 138, 141, 142, 144
urbanization 18–21

van der Sluijs, Jeroen 11
Vernadsky, Vladimir 149
Vivien, Franck-Dominique 80
Vonnegut, Kurt 232

Wackernagel, M. 26
Walras, Leon 147
water
  irrigation 181, 182, 190, 209, 214, 216, 217
  pollution 187, 188, 206, 210, 212, 222, 225
  rights/reform 222, 233
  salinization/desalinization 212, 229
  shortage 39, 210, 212, 220
  usage/demand 22, 109, 158, 160, 164, 166, 169, 187, 188
water power *see* hydroelectric power
Watson, Robert 5
White, Leslie 150–151
wind power 27, 151, 172, 174, 256
World Bank 1, 5, 218, 252
  food crisis report (2008) 3–4
WTO (World Trade Organization) 247–250

Zemmelink, G. 138, 140
zero-emission myth 18, 115, 164, 166, 185–191, 205, 230
Ziegler, Jean 4, 218
Zipf, George Kingsley 87–88, 151
Zoellick, Robert 3